THE POSTMODERN TURN

CRITICAL PERSPECTIVES

A Guilford Series

DOUGLAS KELLNER, Editor
University of Texas at Austin

A THEORY OF HUMAN NEED
Len Doyal and Ian Gough

POSTMODERN THEORY: CRITICAL INTERROGATIONS
Steven Best and Douglas Kellner

PSYCHOANALYTIC POLITICS, SECOND EDITION: JACQUES LACAN AND FREUD'S FRENCH REVOLUTION
Sherry Turkle

POSTNATIONAL IDENTITY: CRITICAL THEORY AND EXISTENTIAL PHILOSOPHY IN HABERMAS, KIERKEGAARD, AND HAVEL
Martin J. Matuštík

THEORY AS RESISTANCE: POLITICS AND CULTURE AFTER (POST)STRUCTURALISM
Mas'ud Zavarzadeh and Donald Morton

POSTMODERNISM AND SOCIAL INQUIRY
David R. Dickens and Andrea Fontana, Editors

MARXISM IN THE POSTMODERN AGE: CONFRONTING THE NEW WORLD ORDER
Antonio Callari, Stephen Cullenberg, and Carole Biewener, Editors

AFTER MARXISM
Ronald Aronson

THE POLITICS OF HISTORICAL VISION: MARX, FOUCAULT, HABERMAS
Steven Best

ROADS TO DOMINION: RIGHT-WING MOVEMENTS AND POLITICAL POWER IN THE UNITED STATES
Sara Diamond

LEWIS MUMFORD AND THE ECOLOGICAL REGION: THE POLITICS OF PLANNING
Mark Luccarelli

SIGN WARS: THE CLUTTERED LANDSCAPE OF ADVERTISING
Robert Goldman and Stephen Papson

REVOLUTION OF CONSCIENCE: MARTIN LUTHER KING, JR., AND THE PHILOSOPHY OF NONVIOLENCE
Greg Moses

POSTMODERN WAR
Chris Hables Gray

THE POSTMODERN TURN
Steven Best and Douglas Kellner

THE POSTMODERN TURN

Steven Best
Douglas Kellner

THE GUILFORD PRESS
New York London

Published by The Guilford Press
A Division of Guilford Publications, Inc.
72 Spring Street, New York, NY 10012

Printed in the United States of America

This book is printed on acid-free paper.

Last digit is print number: 9 8 7 6

Library of Congress Cataloging-in-Publication Data

Best, Steven.
The postmodern turn / Steven Best, Douglas Kellner.
p. cm.—(Critical perspectives)
Includes bibliographical references and index.
ISBN 1-57230-220-8. — ISBN 1-57230-221-6 (pbk.)
1. Postmodernism—Social aspects. 2. Sociology—Philosophy.
I. Kellner, Douglas, 1943– . II. Title. III. Series: Critical perspectives (New York, N.Y.)
HM73.B468 1997
301′.01—dc21 97-15234
CIP

In memory of Bill and H. A. Best.—S. B.

In memory of C. A. and Toni Kellner.—D. K.

Contents

	Preface and Acknowledgments	viii
ONE	The Time of the Posts	3
TWO	Paths to the Postmodern: Kierkegaard, Marx, and Nietzsche	38
THREE	From the Society of the Spectacle to the Realm of Simulation: Debord, Baudrillard, and Postmodernity	79
FOUR	Postmodernism in the Arts: Pastiche, Implosion, and the Popular	124
FIVE	Entropy, Chaos, and Organism in Postmodern Science	195
SIX	Between the Modern and the Postmodern: Paradigm Shifts in Theory and Politics	253
	Bibliography	283
	Index	300

Preface and Acknowledgments

The past several decades have witnessed a postmodern turn in theory, the arts, and the sciences, one that is part of a major paradigm shift and, some would argue, an epochal transformation from a modern to a postmodern world. The dramatic changes and turmoil in every dimension of life have led many to claim that we have left behind the modern era and are entering a new postmodern epoch.[1] These arguments provoked an explosion of postmodern discourses over the past two decades (surveyed in Best and Kellner, 1991), producing theory wars between advocates of modern and postmodern theory. Key postmodern theorists argue that contemporary societies, with their new technologies, novel forms of culture and experience, and striking economic, social, and political transformations, constitute a decisive rupture with previous ways of life, bringing to an end the modern era. In the sphere of culture, there has been a repudiation of modernism and an explosion of postmodernism in the arts, which has permeated every aesthetic domain from architecture to film to new multimedia artifacts. In addition, various forms of postmodern theory have circulated through every domain of academic discourse and have challenged and transformed intellectual practice in a plethora of fields, including science.

We are calling this dramatic transformation in social life, the arts, science, philosophy, and theory "the postmodern turn" and argue that we have now entered a new and largely uncharted territory between the modern and the postmodern. The postmodern turn involves a shift from modern to postmodern theory in a great variety of fields and the move toward a new paradigm through which the world is viewed and interpreted. The postmodern turn includes as well the emergence of postmodern politics, new forms of postmodern identities, and novel configurations of culture and technology. Most dramatically, many theorists claim that we have left modernity behind, that we have entered a new historical space with new challenges, dangers, and possibilities, while for others, we are experiencing the end of history itself.

The postmodern turn is exciting and exhilarating in that it involves an encounter with experiences, ideas, and ways of life that contest accepted modes of thought and behavior and provide new ways of seeing, writing, and living. The postmodern turn leaves behind the safe and secure moorings of the habitual and established, and requires embarking on a voyage into novel realms of thought and experience. It involves engaging emergent forms of culture and everyday life, as well as confronting the advent of an expanding global economy and new social and political order. Indeed, the postmodern turn is global, encompassing by now almost the entire world, percolating from academic and avant-garde cultural circles to media culture and everyday life so as to become a defining, albeit highly contested, aspect of the present era.

Although many at first dismissed the postmodern turn as a fad and have been predicting its demise for years, postmodern discourses continue to proliferate and attract interest, winning fervent advocates and passionate opponents. The term "postmodern" is thus increasingly taken as a synonym for the contemporary social moment and as a marker to describe its novelties and its differences from modern culture and society. Yet there is no agreement on what constitutes the postmodern, on whether we are indeed in a new postmodern era, or on what theories best illuminate the dynamics and experiences of the contemporary moment. Accordingly, in this and forthcoming studies, we will map some of the many twisting and winding pathways into the postmodern, exploring some of the theoretical discourses that have undertaken the postmodern turn in theory, the arts, and the sciences.

We will therefore map and analyze some defining moments in the postmodern turn in an attempt to illuminate our current situation. Our goal is to interrogate major transformations in theory, culture, and society to provide insights into the passage from the modern to the postmodern. We combine social theory with cultural criticism to contribute to writing the "history of the present" (Michel Foucault) and to developing "a theory of contemporary society" (Max Horkheimer). We believe that mapping the transition from the modern to the postmodern requires sociological and historical perspectives that relate the current moment both to the past out of which it arose and to the future it anticipates.

Our entry into the conceptual field of the postmodern in Chapter 1 provides a provisional mapping of the contours of the postmodern turn and explains some of the principal concepts, issues, and problems that we will engage. In Chapter 2, we explore some important sources of postmodern theory in 19th-century thinkers such as Søren Kierkegaard, Karl Marx, and Friedrich Nietzsche, and we show how these theorists anticipated contemporary forms of the postmodern turn, demonstrating our claim that postmodern discourses do not emerge in vacuo but, rather, have a complex history of anticipations in modern theories and developments. In Chapter 3, we explore one significant path to the postmodern, through Guy Debord and the Situationist International to Jean Baudrillard and French postmodern theory. Debord and the Situationists updated the Marxian critique of capitalism within the context of consumer and media society, providing a transitional link from the modern to the postmodern and influencing the work of Baudrillard, who developed one of the first and most compelling descriptions of a new postmodern era.

After these analyses of some trajectories of postmodern theory, Chapter 4 maps the origins of postmodern culture in literature, the visual arts, and architecture and charts the movement from modernism to postmodernism in the arts. Our studies suggest that in a postmodern society of image and spectacle, culture is playing an increasingly important role and thus that it is imperative to develop critical theories of culture that interrogate the meanings, effects, and consequences of this transition in culture and technology.

Chapter 5 analyzes developments in postmodern science that constitute a major break with modern science. In particular, we claim that the transition from modern to postmodern science is a key route into the postmodern turn and examine how the concepts of entropy and chaos figure importantly in contemporary scientific discourse, along with new understandings of organism, ecology, and the cosmos as a whole that put in question the misguided modern project to dominate nature. These shifts are parallel to developments in postmodern social theory, philosophy, and the arts, and their resemblances lead us in Chapter 6 to chart the contours of a new postmodern paradigm. In this concluding chapter, we argue that we are entering a new conceptual and social field, which the discourse of the postmodern is attempting to articulate.

By now, there are many genealogies, many narratives, and many ways of presenting the postmodern turn, each with its own precursors, privileged disciplinary focus, path of development, and point of view. No genealogy of the postmodern is neutral and unmotivated, and responses range from the celebratory discourses of those like Ihab Hassan who affirm the postmodern turn, to the critical ones of Jürgen Habermas and others who deplore it. The postmodern turn, as we shall see, moves through many different fields and crosses a varied terrain of theory, the arts, the humanities, science, politics, and diverse areas of social reality. We will trace the pathways through some of these thickets, mapping territories traversed by the discourse of the postmodern and illuminating various ways that the postmodern has captured the contemporary imagination.

Though we show how the postmodern turn in the arts and sciences parallels in certain respects the transition from modern to postmodern society and from modern to postmodern theory, we will also show differences among postmodern discourse in the fields of theory, the arts, and science, as well as variations within these domains themselves. There is thus a specificity to each path, different accounts that can be given of the postmodern turn in each field, and intense struggles over how to portray the postmodern in particular domains, as well as more generally. Yet despite these differences and contestations, we wish to show that there is a shared discourse of the postmodern, common perspectives, and defining features that coalesce into an emergent postmodern paradigm.

How we trace the genealogy of the postmodern helps determine how we see contemporary society and whether we have a positive or negative, simplistic or complex, vision of the vicissitudes of contemporary history, the problems of the present age, and prospects for the future. Thus, it is important to have many genealogies and numerous perspectives in order to acquire a dynamic and complex account of the postmodern that grasps both continuities and discontinuities, as well as progressive and regressive lines of development.

Our goal throughout is to delineate the postmodern turn in a variety of fields and to show how the disparate trajectories of the postmodern, despite their differences, are coalescing into a *new paradigm* that we see as emergent, not yet dominant, and that therefore is hotly contested. As Thomas Kuhn (1970) defined it, a "paradigm" is a "constellation" of values, beliefs, and methodological assumptions, whether tacit or explicit, inscribed in a larger worldview. Kuhn observed that throughout the history of science there have been *paradigm shifts,* conceptual revolutions that threw the dominant approach into crisis and eventual dissolution, a discontinuous change provoked by altogether new assumptions, theories, and research programs. In science, Kuhn argued, a given paradigm survives until another one, seemingly having a greater explanatory power, supersedes it. At any one time, in other words, certain assumptions and methods prevail in any given discipline, until they are challenged and overthrown by a new approach that *emerges* through posing a decisive challenge to the status quo; if successful, this new approach becomes *dominant,* the next paradigm, itself ready to be deposed by another powerful challenger as the constellation of ideas continues to change and mutate.

Kuhn limited his focus to scientific paradigms, but obviously there can be a paradigm for any theoretical or artistic field as well as for culture in general, such as Foucault (1972) attempted to identify for different stages in the development of modern knowledge through his concept of *episteme.* As we conceptualize it, the "postmodern paradigm" signifies *both* specific shifts within virtually every contemporary theoretical discipline and artistic field *and* the coalescing of these changes into a larger worldview that influences culture and society in general, as well as the values and practices of everyday life. We wish to contextualize paradigm shifts not only in the history of ideas, as Kuhn and others have done, but also as effects of *developing social and institutional factors,* as driven by changes in industry, technology, economics, politics, and often science itself. Thus, we will analyze paradigm shifts both as internal responses within a given conceptual domain; as changes in the tacit, underlying "rules of knowledge," such as the early Foucault tried to identify as informing conceptual ruptures within the human sciences; *and* as part of broader mutations in society and history that influence changes in culture and thought. Philosophy, art, literature, and science have their own histories, problems, debates, and conceptual dynamics, but they are also deeply conditioned, whether acknowledged or not, by larger social and political conditions.

We will thus examine changes both "inside" and "outside" given fields, avoiding both a sociological determinism that reduces ideas to forces such as economics or technology, and a conceptual idealism that fails to contextualize intellectual alterations as part of broader social patterns and movements. Thus, in our view, postmodern paradigm shifts arise in different fields as critical responses to ideas and methods perceived to be staid, dogmatic, erroneous, or oppressive, as well as in response to developments in society, technology, economics, culture, and politics. In Chapter 1 we begin delineating features of the now widely contested modern paradigm and some conceptual distinctions to elucidate some defining features of the postmodern paradigm. In succeeding chapters, we will depict emergent postmodern paradigms in the fields of theory, art, the sciences, and politics. We sort out the claims for and against postmodern theory and advance our own position that we are currently in an era between the modern and the postmodern and that therefore both modern and postmodern perspectives are relevant, requiring new syntheses and a transdisciplinary approach to capture the complexity and turbulence of current events and developments.

We endeavor to follow up our previous book *Postmodern Theory: Critical Interrogations* (Best and Kellner, 1991), which interrogated the discourses of the postmodern in its now classical theorists from Foucault through Jameson, with studies that supplement and go beyond that earlier work. After its positive reception, we began by collecting some essays already published separately and jointly on postmodern culture and theory. We then reworked all of these texts, while adding new ideas and undertaking further collaborative studies, so that the text now appears, as in our first book, as a joint authorship, forming an articulated whole that delineates our shared perspectives on contemporary culture, society, and politics. We work in a transdisciplinary space and develop critical reflections on a wide range of topics in social theory, philosophy, media culture, painting, architecture, literature, science, technology, and politics. The text reflects our position that social reality can be analyzed most adequately through multiple methodological and theoretical perspectives. Building on recent work in Kellner's *Media Culture* (1995) and Best's *The Politics of Historical Vision* (1995), we seek to present new insights into both postmodern theory and contemporary society and culture, which we argue is a borderland between the modern and something new for which the term "postmodern" has been coined. Providing conceptual content and articulation to this vastly overused and abused concept is one of the main goals of the following studies.[2]

* * *

For their critical readings of these essays we would like to thank Carl Boggs, David Hall, Rhonda Hammer, Darrell Hamamoto, Sandra Harding, Ali Hosseini, Fred Kronz, Valerie Scatamburlo, Steven Seidman, and Peter Wissoker. We would like to thank Jon Epstein and Pat Lichty for developing illustrations for the book and are grateful to Pat Lichty for cover design and the finalization of the illustrations. For extremely competent and constructive help with the editorial process we would like to thank Judith Grauman of The Guilford Press. Once again, we want to thank Keith Hay-Roe for helping us with copious computer and word-processing problems and for always exhibiting patience and good humor in the face of our quandaries. For Internet navigation aid, we are grateful to Jay Ashcraft and Robert Prescott for helping us to move to new frontiers and regions of cyberworld. We dedicate this book to the memory of Bill and H. A. Best (S. B.), and of C. A. and Toni Kellner (D. K.), who died as we were concluding this text.

NOTES

1. The books on postmodernity and the concept of the postmodern now fill a small library and are still growing. They include Baudrillard (1983a, 1983b, 1993 [1976]); Lyotard (1982 [1979]); Kroker and Cook (1986); Harvey (1989); Turner (1990); Lash (1990); Best and Kellner (1991); Jameson (1991); Bauman (1992); Smart (1992, 1993); Lyon (1994); Bertens (1995); and Ritzer (1997).

2. Toward the completion of writing this book, we rediscovered Ihab Hassan's book *The Postmodern Turn: Essays in Postmodern Theory and Culture* (1987). Although our books share the same title, and Hassan's subtitle implies the engagement of a postmodern paradigm shift, to which he contributes a certain amount of anecdotal evidence, our orientations are completely different. Hassan's approach is symptomatic precisely of what we reject—an ironic, detached, playful, and wilfully cryptic and allusive use of postmodern discourse. A paradigmatic example of a certain postmodern style, Hassan's book is a pastiche of fragmentary essays, none of which clarify or define the main tendencies of "postmodern theory and culture." Indeed, Hassan is a master of "definition" by lists (both names and concepts), quotes, examples, and arcane terminology. As he proudly admits: "I have not defined Modernism: I can define Postmodernism less" (1987: 40). He invokes Thomas Kuhn's name and the notion of paradigm shifts at various places, but doesn't engage the concept of paradigm except to reject it (120), while still declaring that we live in a "postmodern moment of unmakings" (121) that transcends the postmodern turn in any one discipline. Our approach, by contrast, is to recognize the complexity, plurality, and unfinished nature of

postmodern discourse while nevertheless attempting to clarify and map numerous genealogies, approaches, and styles of the postmodern, examining its various uses and abuses, regressive and progressive aspects. While postmodern discourse is open, evolving, and unstable, it is not totally indeterminate or closed to definitions and analysis. And while Hassan seeks to flee from "rightist or leftist cant" for "a tough and limber pragmatism" and a "posthumanism" (xvii), in fact, pragmatism and humanism are conceptually linked for William James and John Dewey. Moreover, while Hassan fails to contextualize the postmodern turn, we argue that transformations of capitalism are a major source of the postmodern paradigm shift and are the source of global social and environmental crisis today. Furthermore, we seek to revive the egalitarian, democratic, and humanist norms that constituted the best of modern and progressive traditions, which Hassan and other postmodernists are willing to leave behind. Crucially, whereas Hassan's frame of reference is primarily literary theory, with some allusions to philosophy and science, our privileged perspective is that of critical social theory.

THE POSTMODERN TURN

Post No Signs

CHAPTER ONE

The Time of the Posts

> Maybe the most certain of all philosophical problems is the problem of the present time, of what we are, in this very moment.
>
> —MICHEL FOUCAULT

According to many, we live in the time of the "posts"—postindustrialism, postFordism, postMarxism, posthumanism, posthistory, and postmodernism. The term "post" signifies a historical sequencing in which a previous state of affairs is superseded and thus functions in the first instance as a periodizing term. The discourse of the post is sometimes connected with an apocalyptic sense of rupture, of the passing of the old and the advent of the new. Indeed, numerous books have been written describing the end of the industrial era and the coming of a new postindustrial era in which knowledge and communications become the new organizing principles of society (Bell, 1976; Frankel, 1987; Poster, 1990). Others argue that the Fordist model of mass production and consumption has been superseded by a new form of global and more flexible production, with parts of products manufactured throughout the world and assembled in sites with cheap labor, leading to the breakup of older industrial units and communities, accompanied by a decline of unions and massive regional unemployment (Bluestone and Harrison, 1982; Harvey, 1989). With the collapse of Soviet communism, some commentators have declared the end of Marxism (Fukuyama 1992), and theorists like Lyotard (1984) pronounce the end of the grand narratives of modern theory and the need for new postmodern theories and politics.

In addition, some scientists and historians of science are arguing that there can be no more great scientific breakthroughs, that we are at "the end of science" (Horgan, 1996). Leading postmodern theorists declare that we are at the end of modernity itself, that the modern era is over, and that we have entered a new postmodern world (Baudrillard, 1983a, 1983b, 1993).[1] There have even been arguments that we are at "the end of history" (Baudrillard, 1986; Fukuyama, 1992), that no more fundamental political or historical changes will occur, leading some postmodernists to claim that we are now in a new stage of posthistory in which the theories, politics, and worldviews of the past are obsolete. Such discourses put in question all previous theories and call for new theories and politics to deal with the striking novelties of the present.

Yet no one has yet explained satisfactorily *why* all of these discourses of the "post" have been proliferating and attracting attention and even devotion in recent years. We maintain that there is no single and simple explanation for the emergence of postmodern discourse and the belief that we are living in the "time of the posts." Rather, clarification of the postmodern turn requires discussion of the sociohistorical environment in which the discourse of the "post" first emerged. Accordingly, our account of the emergence of the postmodern turn in the contemporary era will begin with some shared social and political experiences that would lead some former 1960s radicals to develop discourses of the postmodern in the 1970s and 1980s. We then discuss experiences of the later "Generation X," which undertook a postmodern turn in the 1980s and 1990s, and discuss a variety of reasons why the discourse of the postmodern has mobilized such passionate support and evoked such frenzied attack in our current situation. Subsequent chapters will present more detailed historical and analytical accounts of the postmodern turn in theory, the arts, the sciences, and politics, producing the matrix for an emergent postmodern paradigm shift that we believe is one of the salient features of the present age.

THE EMERGENCE OF THE "POST"

> Suddenly, there is a curve in the road, a turning point. Somewhere, the real scene has been lost, the scene where you had rules for the game and some solid stakes that everybody could rely on.
>
> —JEAN BAUDRILLARD

During the 1960s, a group of radical intellectuals and activists who became the first major postmodern theorists experienced what they believed to be a decisive break with modern society and culture. These theorists believed that significant changes were occurring in history with the advent of new social movements opposing the Vietnam War, imperialism, racism, sexism, and capitalist societies in their entirety, demanding revolution and an entirely new social order. At the same time, an oppositional counterculture emerged that called for a society which renounced the materialist ethos and success-oriented norms of capitalism. Third World revolutionary movements generated hopes that emancipatory alternatives were grounded in the very dynamics of history, leading to more egalitarian, just, and democratic societies. Many believed that a decisive break with the past had taken place, that a revolution in morals, politics, and perception was underway, and that a new era of history was dawning.

Most of the major postmodern theorists—Foucault, Lyotard, Baudrillard, Deleuze, Guattari, Jameson, Laclau, Mouffe, Harvey, and others—participated in, and were deeply influenced by, the tumult of the 1960s, and these experiences of rupture helped produce a readiness, an openness, to the discourse of historical breaks and discontinuities. One cannot exaggerate the role of May '68 in France in producing a sense of rupture with the past, a sense that an irreversible turning point had occurred, that a new world was being born. In a series of spectacular uprisings, students and workers occupied the universities, factories, and other institutions of Paris, fought police in the streets, and forced French president Charles de Gaulle to flee France to come up with a solution to the crisis.[2]

The equivalent of this experience in the United States and elsewhere in the Western world were the student uprisings and major protests against the Vietnam War, including spectacular events like the student takeover of Columbia University in spring 1968. These events also led many to believe that revolution was underway, that a rupture with the previous society and culture was in the making, that a new day was dawning. In the United States and elsewhere, the counterculture also believed that it was creating an entirely new society and culture, based on a new set of values, sensibilities, consciousness, culture, and institutions, which produced a rupture with mainstream, or "establishment," society.[3]

The subsequent theories of the postmodern turn, however, first developed in the 1970s, argued that the break was caused by developments in the economy, technology, culture, and society, rather than by mass struggle and revolutionary upheaval as advocated in the 1960s (Baudrillard, 1993 [1976]; Lyotard, 1984 [1979]). Moreover, their discourses, especially those of Baudrillard, had pessimistic rather than optimistic overtones (see Kellner, 1989b). This pessimism has roots in an experience of defeat. Most of the French theorists who were the first to produce theoretical and political analyses of the postmodern in the 1970s were indeed deeply disappointed by the betrayal of the 1968 events in France. Just as students and workers stood on the verge of overthrowing the Gaullist state, the French Communist Party supported de Gaulle's call for new elections and a return to normalcy, a conformist policy that helped return the president to power and restore the status quo, thus dashing the hopes of revolution.

Deeply angered by communist accommodation to the existing system, numerous French radicals came to associate Marxism with communist bureaucracy and sought new theories and politics. Many were put off by the deterministic "scientificity" of the "structuralist Marxism" developed by Louis Althusser and his followers,[4] and so a generation of new French theorists attacked both Marxism and structuralism to develop "poststructuralist" theories that replaced core tenets of modern theory with strong emphases on difference and multiplicity themes, later advocated by postmodern theorists. In place of Marxism, which appeared totalizing and reductive, the poststructuralists turned to Nietzsche to regenerate their radical aspirations. They subsequently rejected modern visions of revolution and emancipation as dangerous and totalizing and turned to individualist programs of liberated subjectivity and nomadic desire.

Thus, the first versions of French postmodernism were marked by the failures of Marxism and contain a conflicted matrix of clearly anti-Marxian features with a transcoding of other Marxian ideas into inventive and hybrid theoretical discourses. As noted, the pessimism and nihilism of some early postmodern theorists like Baudrillard is in part a product of the failure of communism and 1960s radicalism, while the critique of grand narratives of history and totalizing theories is clearly an attack on Marxism. Yet the emphasis on rupture and discontinuity is congruent with Hegelian–Marxian modes of thought, and the concept of postmodernity arguably transcodes the Marxian theory of revolution, while its critical and radical spirit incorporates Marxian motifs of critique, oppositionalism, and social transformation.[5]

Moreover, new theoretical impulses appeared in France during the 1960s, and this theoretical revolution erupted concurrently with the political upheaval and helped foster a postmodern turn throughout the world. Thinkers like Foucault, Derrida, Lyotard, and Deleuze were turning to Nietzsche and Heidegger and appropriating their critical discourses against modern theory and modernity itself. Postmodern assaults on Enlightenment rationality and universalism, as well as postmodern emphases on relativism, perspectivalism, difference, and particularity, stem as much from philosophical critiques of Western thought that begin with Nietzsche and continue through Dewey, Wittgenstein, Heidegger, and feminism, as from particular political experiences. Thus, there was a turn away from modern discourses of truth, certainty, universality, essence, and system and a rejection of grand historical narratives of liberation and revolution. A vogue of deconstruction carried out a demolition of major classical and modern philosophical systems and challenged the very premises of modern thought.

William Barrett's *Irrational Man* (1958) provides symptomatic articulation of the ferment in philosophy, the arts, science, and society that shaped the context in which postmodern theory emerged. Highly popular in the 1960s, written at the very cusp of postmodern breakthroughs in architecture, literature, painting, and theory, Barrett's book breathlessly describes the emergence of "a new philosophy" bound up with a paradigm shift in science, literature, and the arts. He is speaking not of poststructuralism but, rather, of good old existentialism, which did not register in U.S. life until after World War II and did so mainly in French, especially Sartrean, form. On Barrett's reading, existentialism emerged in the context of 19th-century Europe, but it carries a larger message for the United States and for modern individuals in general. The ultimate meaning of existentialism initially exploded—quite literally—on the global scene on July 16, 1945, with the first successful testing of the atomic bomb: "The bomb reveals the dreadful and total contingency of human existence. Existentialism is the philosophy of the atomic age" (Barrett, 1958: 65).

According to Barrett, existentialism arose as the product of European modernity in crisis, as the main values of civilization and of the Enlightenment were being called into question by both philosophical and historical events. By the 1880s, for example, Nietzsche (1968a: 3ff.) had diagnosed nihilism as the sickness of the times ("Nihilism stands at our door"), anticipating the grotesque spectacle of world wars and genocide played out within the next 60 years. Kierkegaard had already explored the feelings of dread, anxiety, and "sickness" unto death that shook his soul, and Heidegger would soon emphasize finitude and contingency as the essence of *Dasein*, as he and Karl Jaspers were analyzing the depersonalizing forces creating a standardized mass society that required a struggle for an authentic and genuine life. For Barrett, the beginning of World War I marks an important break in history. Coming at the end of a period of relative peace and prosperity in Europe, "August 1914 shattered the foundations of that world. It revealed that the apparent stability, security, and material progress of society had rested, like everything else, upon the void. European man came face to face with himself as a stranger" (1958: 34).

Hence, Barrett argues, existentialism is "a product of bourgeois society in a state of dissolution" (1958: 34). The existential themes of anxiety, dread, contingency, conformity, death, and apocalypse could hardly have been the product of U.S. thinkers, since the new-born country was intoxicated with unlimited growth, expansion, and Manifest Destiny. Indeed, U.S. academics in the first half of the 20th century were steeped in the meliorism of pragmatism or the technical sterility of positivism and analytic philosophy. Barrett affirms the importance of liberalism, democracy, reason, and Enlightenment values, but he also sees rationalism—the Cartesian position that reason is the "essential substance" of human life, that it can have absolute knowledge of reality, and that its purpose is to dispel ambiguity and unclarity, replacing them with the transparency of logic and truth—as a grossly inadequate position that fails to acknowledge the dark side of reason, its repressed and passionate dimension, as well as its limitations and sociohistorical rootedness.

As Barrett argues, existentialism is linked to a paradigm shift that he sees as unfolding in philosophy, art, literature, and science, all of which evince a sense of crisis, breakdown, loss of absolutes and foundations, and dissatisfaction with rationalism. Relativity theory, quantum mechanics, complementarity in physics, the incompleteness principle in mathematics, and new forms of indeterminacy in the sciences undermined belief in absolute foundations of knowledge and a well-ordered universe. The flattening of pictorial space and climaxes in cubism, for Barrett, is "paralleled" in literature by the flattening of time in the works of James Joyce, T. S. Eliot, Ezra Pound, and William Faulkner (Barrett, 1958: 49–56), as a sense of nothingness and the breakdown of the modern image of man haunts not only the work of Jean-Paul Sartre but also the literature of Ernest Hemingway and Samuel Beckett (62–63) and the sculpture of Alberto Giacometti.

Thus, for Barrett, science, philosophy, and the arts share the same themes of the contingency of knowledge and limitations of reason (1958: 36–41), undermining modern dreams of power, certainty, and progress. While the new paradigm Barrett describes is modern, not postmodern, the main themes he emphasizes are embodied, in intensified form, in postmodern thought. Indeed, as we argue below, existentialism is a direct forerunner of postmodern thought, and both belong to a larger antirationalist and antimodern tradition. Postmodern thought in general, and postmodern nihilism in particular, is very much a result of the bourgeois crisis Barrett describes. At the time of writing *Irrational Man,* Barrett claimed that "what the American has not yet become aware of is the shadow that surrounds all human Enlightenment" (273). Nearly four decades later, with postmodern turns crisscrossing the terrain of academic life and culture in the United States, this is much less the case now and this recognition of the shadow side of Enlightenment reason is an important part of what today is known as "postmodern."

In fact, Barrett's book was very influential during the 1960s and contributed to a sense that things were in change and crisis, a ferment that ultimately fed into the postmodern sensibility. During the 1960s, consequently, a general intellectual mood of change and the dissolution of old paradigms was joined by spectacular political upheaval and struggle throughout the world, along with the emergence of new forms of thought, culture, technology, and life, which would produce the matrix for the postmodern turn.

Thus, one could argue that the postmodern theories of the past two decades exhibit the political and theoretical experiences of the 1960s sublimated into new discourses. The rupture desired and even experienced in the 1960s, a break then described in the discourse of emancipation and revolution, is projected by postmodern discourse onto history itself, but this time to announce the end of lofty visions of historical progress and social transformation. The apocalyptic revolution postulated in the 1960s as the goal of political and theoretical struggle is thus translated into postmodern theory as mutations within society and culture, as the result of new technologies and forms of culture, society, and politics, without the effort of revolutionary struggle. This claim replicates, in effect, the old discourses of technological or cultural determinism, with their pessimistic overtones, while stigmatizing the discourses of emancipation and revolution as obsolete. Once defeated, optimistic hopes for total revolution gave way to cynical dismissals of the project of liberation and the nihilistic world-weariness characteristic of a certain strain of postmodern theory.

Hence, some postmodern discourses bear the marks of defeat in the aftermath of the 1960s. Postmodern doubts concerning the efficacy of modern politics are in part effects of the experience of the dissolution of the political movements and revolutionary politics of the era. Theories of the fragmented and decentered subject might also describe the subjectivities of those involved in the struggles for a new society who found themselves suddenly marginalized and depressed that their hopes were not realized. Some of these individuals were indeed terminally disoriented and confused as a result of drugs, failed dreams, and conflicting impulses during an era of growing conservativism and reaction.

Yet there are more positive versions of postmodern theory that appropriate and expand some of the progressive tendencies and gains of the 1960s into theory and politics. What Hal Foster (1983) and Teresa Ebert (1996) have termed the "postmodernism of resistance" attempts to develop oppositional theoretical and cultural practices against the more oppressive features and practices of modern society. From this perspective, the discourses associated with postmodern emphases on the margins, differences, excluded voices, and new subjects of revolt are related to 1960s attacks on racism, sexism, and other forms of prejudice and to attendant openings to new values, voices, and peoples. Many women, people of color, members of ethnic groups, and gays and lesbians began advocating a politics that directly proceeds from their "subject positions" as oppressed or underprivileged groups and that focuses on their own distinct identities and experiences. Many of these individuals position themselves as "postmodernists" and thus reject modern identities and politics, often for immersion in oppositional subcultures and "identity politics."[6]

Individuals within the new social movements began criticizing the limitations of the university and the academic division of labor. The assault on all academic disciplines, the attack on disciplinarity itself, began in the 1960s and has continued to the present, helping to generate the postmodern turn. In the 1960s, there was contestation of every discipline from philosophy to economics to psychology. The dominant methodologies of these fields were under siege, and their exclusions, silences, limitations, and the ossification of critical inquiry was challenged. The new theories circulated in the 1960s contributed to the vitalization of many disciplines, which were forced to assimilate or otherwise deal with the challenges of Marxism, feminism, structuralism, poststructuralism, and the myriad of new discourses that would nourish the postmodern turn. The questioning of disciplinary boundaries led to calls for multi- and transdisciplinary work of the sort found in the postmodern theories, and this heritage of the 1960s has continued and has directly contributed to the postmodern turn in the university.

The theories and social movements of the 1960s and 1970s made it clear that there were a multitude of forms of power and domination and that there should thus be different forms of struggle and resistance; the essentializing Marxist model, which holds that all forms of oppression can be reduced to exploitation of labor and hence that the working class constituted the privileged social group, was abandoned. As discussed by Foucault, various theorists and activists rejected the hegemony of Marxism as rooted in a totalizing and essentializing logic that subsumed all forms of oppression and resistance to the fulcrum of labor and exploitation. It became widely understood that power had numerous other sources and strategies, working not only in factories but in schools, hospitals, prisons, and throughout culture and everyday life. It was thought, in other words, that there was no center to or essence of power; the focus shifted from exploitation of labor to normalization of identities. The old theme of class solidarity gave way to the new focus on difference and, indeed, disparate political groups and identities fragmented from one another, with each pursuing its own interests and agendas.

Hence, "new social movements" proliferated in the 1970s and 1980s, including feminism, gay and lesbian activism, peace and environmental movements, and various struggles mobilized around race and ethnicity. Although rooted in the social struggles of the 1960s, the new social movements came of age in very different times and in many cases drew sustenance from postmodern theory, Foucault in particular, as validating their specific forms of identity, modes of opposition, and autonomy from class-based politics. By the 1980s, various strains of French poststructuralist theory had mutated into postmodern theory in France and elsewhere. Many of the oppositional postmodern discourses combined sophisticated elements of poststructuralism with impulses from experiences in new social movements and from an academia sharply politicized by the struggles of the 1960s. Academics and activists generated new forms of feminism, queer theory, cultural studies, postcolonial theory, and a wide range of race-based theories. Many of the individuals associated with these trends resisted the limited options of modern or postmodern theory and moved beyond this binary opposition into a new conceptual space. They include such theorists as Richard Bernstein, Homi K. Bhabha, Judith Butler, Teresa de Laurentis, Nancy Fraser, Henry Giroux, Stuart Hall, Donna Haraway, Sandra Harding, bell hooks, Allan Luke, Carmen Luke, Peter McLaren, Linda Nicholson, Steven Seidman, Barry Smart, Gayatri Spivak, Cornell West, and Iris Young who combine themes from both modern and postmodern theory. Many of these theorists attempt to push into a new conceptual space that builds on the strengths of both modern and postmodern theory. Their work is less irrationalist, apocalyptic, and nihilistic than the more extreme versions of postmodern theory and is more radically democratic, attempting to reconstruct rationality, critique, agency, and democracy in the light of poststructuralist and postmodern critiques of modern theory.

Yet many of the theorists just mentioned avoid calling themselves postmodernists, preferring to operate beyond labels, or they distance themselves from some of the exaggerations and extremism of postmodern theory. Some, however, like Steven Seidman (1994), aggressively position themselves as postmodern, albeit of a different mode than Baudrillard, Lyotard, and the proponents of the more exotic European versions of postmodern theory. Indeed, the domain of the postmodern is itself a contested terrain, and there are a variety of positions, often contrasting, that present themselves as constituting a postmodern split from modern theory. We attempt to theorize some of these differences in the following sections. But the explosion of postmodern theories throughout the world in the 1980s and 1990s requires further historical contextualization.

THE POSTMODERN MOMENT: THE 1980s AND BEYOND

Although the initial theorists of the postmodern belonged to the generation of the 1960s, the discourse did not really proliferate and become a dominant intellectual and cultural force until the 1980s. It was primarily younger individuals and groups who picked up on postmodern discourses during the 1980s and 1990s, often in more extreme and aggressive forms, renouncing modern theory and politics en toto. While the first wave of postmodern theorists were influenced primarily by the theoretical and political experiences of the 1960s and the 1970s, the younger postmodern adherents of the past two decades had different cultural experiences. During the 1980s, at the same time that postmodern theories proliferated and circulated, an era of conservative governance unfolded in the United States, Britain, much of Europe, and other parts of the world. Paradoxically, the postmodern discourses were *both* part of the conservative turn toward individualism, local empowerment, and renunciation of the liberal welfare state *and* a form of oppositional discourse that assailed the assumptions of conservative theory and in turn was vilified by conservatives in their culture wars against radical discourses.[7]

In addition, there was something of a generational revolt of the postmodern young against their older modern fathers and mothers. The most vocal and aggressive contemporary champions of the postmodern tend to be younger professors, artists, students, or activists who claim the postmodern as a badge of identity and revolt. Perhaps the intensity of some postmodern discourse has an oedipal dimension, pointing to a revolt by youth and to their radical negation of their patrimony and cultural heritage. This form of the postmodern break also appeals to many women, people of color, and those who have had more marginal positions in the institutions in which intellectual debate and cultural practice takes place. In these cases, the discourse of the postmodern positions its user as oppositional, as avant-garde, as "pomo."

Perhaps only the young could really be postmodern; perhaps the older generation—including ourselves—is so deeply implicated in modern theory and politics that it cannot really make a radical break with its past. Furthermore, the success and currency of the new postmodern discourses are also partly a result of younger individuals who are seeking cultural capital to deploy against the older generation, using postmodern discourses as a badge of revolt, identity, and distinction. As Mike Featherstone reminds us (1989; 1991), journalists, cultural entrepreneurs, and theorists invent and circulate discourses like the postmodern in order to accrue cultural capital, to distinguish themselves, to promote specific artifacts or practices as the cutting edge, and to circulate new meanings and ideas. The discourse of the postmodern especially attracts younger people on the make, or those who wish to distinguish themselves as avant-garde, although it has also allure for many who wish to revive flagging careers or libido with sexy new discourse.

Consequently, we reject Fred Pfeil's description of postmodernism as "a generational 'structure of feeling' rooted in a huge PMC [i.e., professional–managerial class]" (1990: 3, 97ff.).[8] While his baby-boomer/PMC class is part of the consuming public for what he takes to be postmodern culture, it is the so-called Generation Xers who are the more avid producers and consumers of postmodern culture. Moreover, the members of the younger generation who are most aggressively promoting the postmodern discourse today are those marginalized outsiders whose declining job prospects are blocking their entry into the PMC rather than members of the PMC itself. Indeed, many of the most vocal opponents of postmodern theory are precisely those from the baby-boom generation who were socialized into previous modern theoretical discourses and culture, who often find postmodern theory threatening and unpalatable.

Thus, although the initial theorists of the postmodern were 1960s radicals who by the 1970s and 1980s had become successful academic intellectuals (i.e., Baudrillard, Foucault, Jameson, Laclau, Lyotard, Mouffe, etc.), the discourse and cultural forms seem to have resonated most fulsomely with those coming of age in the 1980s and 1990s, for whom the postmodern became a sign of identity and a weapon against their elders. Yet all too often, the younger Generation X is drawing on and replicating postmodern discourse without an informed grounding in the modern tradition, not having read Voltaire, Diderot, Kant, Hume, Hegel, Durkheim, Dilthey, Weber, or even Marx; it is as if the only books in one's library should be published by Semiotext(e). They therefore unavoidably construct superficial, stereotyped, and totalizing models of modern theory and the Enlightenment, setting up straw targets to blow over with an enthusiastic gush of hot air. Ignorance of the modern tradition inevitably entails abuse of postmodern theory itself, since it leads to exaggerating the novelty of postmodern breaks with earlier theories and fails to see how a legion of modern theorists themselves challenged ahistorical, essentializing, and totalizing tendencies within modern theory (see Antonio and Kellner, 1994; Best, 1995).

But the new postmodern theories have gained a following in large part because there have been significant changes throughout the world, changes that they attempt to articulate. In recent decades, new technologies have emerged that have altered the pattern of everyday life and have powerfully restructured work, leisure, education, communication, politics, and personal identities. New computer technologies have replaced many jobs and created new ones, providing as well novel forms of accessing information, communicating with other people, and plugging into the joys of a computer-mediated public sphere. New media technologies are bringing an unprecedented flow of sights and sounds into people's homes, and novel virtual worlds of entertainment, information, sex, and politics are reordering perceptions of space and time, erasing firm distinctions between reality and artificiality while producing new modes of experience and subjectivity and dramatically changing fields as disparate as philosophy, architecture, science, war, and law—as we shall document in this book and forthcoming studies.

We believe that the experiences and effects of new technologies and the transformations in global capitalism provide the most important factors in creating shifts and changes in the present moment, which helps account for the rise of postmodern discourse. Political economists have been arguing that we are entering a new "postFordist" society in which the modern, Fordist form of capitalist society, marked by mass production and consumption, state regulation of the economy, and a homogeneous mass culture, is being replaced by "more flexible" modes of sociopolitical and economic organization (see Harvey, 1989, for summaries of this position, as well as Lash and Urry, 1987, 1994).[9] These new forms of postmodern economy and society are produced by transnational corporations replacing the nation-state as arbitrators of the economy in an emergent stage of transnational capitalism that erases previous boundaries of space and time and that produces an ever-expanding global marketplace and division of labor, with novel forms of speculative capital, new forms of production and distribution, rapidly expanding emigration and class restructuring, and a cornucopia of new consumer goods, information technologies, and services.

As we will see throughout our studies, the postmodern turn is intimately bound up with globalization and the vicissitudes of transnational capitalism. The expansion of the capitalist world market into areas previously closed off to it (i.e., into the communist sphere or various Third World countries) is accompanied by the decline of the nation-state and its power to regulate and control the flow of goods, people, information, and various cultural forms. Transnational organizations like the United Nations, the World Trade

Organization, and the World Bank and new arrangements such as the North American Free Trade Agreement (NAFTA) and the General Agreement on Tariffs and Trade (GATT) are assuming global power and taking away some of the prerogatives of the nation-state, which was a major institution of modernity. Moreover, transnational corporations are growing in power and wealth, with a global market economy disseminating throughout the world fantasies of happiness through consumption and products that allow entry into the phantasmagoria of consumer capitalism. A world financial market circulates capital in international exchanges that bind together markets and investors. The capitalist mode of production encircles the earth, penetrating ever-new markets while producing innovative products and fashions and eroding tradition and national economies and identities.

Transnational economic change often has tremendous local impact. Whole regions are devastated when industrial production shuts down and moves to areas with lower wages and less government regulation (Bluestone and Harrison, 1982). Such "deindustrialization" has created huge "rust belts" of previously prosperous industrial regions, as in the case of Flint, Michigan, which suffered major economic decline with the closing of its General Motors automobile plants, an episode poignantly documented in Michael Moore's film *Roger and Me* (1989). Automation, computers, and new technologies have axed entire categories of labor while corporate reorganization has eliminated layers of management, producing growing unemployment. More than ever, the world economy is bound together so that typhoons in Japan or financial irregularities in Britain influence the entire world.

As a reaction to the homogenization and commodification of a globalized economy and culture, there has also been a significant eruption of subcultures of resistance that have attempted to preserve specific forms of culture and society against transnational media and consumer culture (see Grewal and Kaplan, 1994). This reaction is evident in phenomena as disparate as the explosion of nationalism in the former Soviet bloc and elsewhere, the rise of religious fundamentalism, expressions of tribalism and traditional cultures in various parts of the world, and the emphasis on the local, the particular, the marginal, and the heterogeneous in some versions of postmodern theory.

The postmodern focus on otherness, difference, and heterogeneity is also in part a function of decolonization and of the immigration of people of color all over the earth. Vast migrations and diasporas of people of color mostly to the metropolitan cores of the more developed countries, have created new fusions of cultures and hybrid identities. Globalization thus involves the circulation of difference and otherness as well as homogenization (on the patterns of immigration see Hall, 1991; Lash and Urry, 1994; Hamamoto, 1996).

We might also point to the worldwide economic crisis that, since the

1970s, has put an end to the post-World War II era of prosperity and has created a palatable sense of a different set of life possibilities. Older expectations are dissolving in a new world of corporate restructuring and downsizing, and deindustrialization and automation are eliminating entire job categories and making employment increasingly uncertain in all sectors, including management (Gorz, 1982; Shaiken, 1984; Aronowitz and de Fazio, 1994; Rifkin, 1995). Cumulatively, the enormous transformations in the economy, politics, society, culture, and everyday life have nourished a sense that a rupture with the past has occurred, which in turn feeds the production and circulation of postmodern discourses.

For these reasons, there has been a significant shift toward postmodern theories in the Western world during the past two decades. Indeed, postmodern discourse has now circulated throughout the world, as the translation of the key works of postmodern theorists into every major language signifies. Debates concerning the postmodern are raging everywhere, and postmodern discourses have themselves become hot commodities in the global marketplace, making the postmodern turn a global one.[10] Postmodern discourse is, to be sure, a Western discourse, but one of the claims of many versions of postmodern theory, as first anticipated by Marx and Engels, concerns the globalization of the modern world. Contemporary postmodern theory stresses a growing interconnectedness of markets, politics, and culture in a high-tech global village where information and images simultaneously penetrate the entire world (Harvey, 1989; Jameson, 1991). In this new world (dis)order, in which economic developments in Japan influence markets in Europe and the United States or a war in the Middle East affects the entire world, global developments penetrate local ones, traditional cultures are bombarded with new ideas and phenomena, and cultural and political wars between old and new ideas and forces explode on a planetary scale (see Cvetovitch and Kellner, 1997).

In addition, postmodern discourse articulates fin-de-siècle anxieties concerning the end of an era and the demise of the certainties, orthodoxies, and positions that have sustained thought and politics over the past three centuries. Shedding of old habits of thought and action is often difficult; thus the postmodern turn evokes threats and challenges that are often anxiety producing, although, as noted, it also contains exciting challenges and experiences. It is curious that apocalyptic thought frequently erupts at the close of a century, and as the 20th century comes to an end there are many who see an entire world order—modernity—dissolving as an uncertain future quickly approaches (or is already here).

This dialectic of reorganization and disorganization in the global restructuring of capitalism, the dramatic impact of new technologies, and the accompanying changes throughout social and personal life suggest that we are entering an especially complex period that requires fresh theories and politics. The discourses of the postmodern, at their best, address these novel experiences and conditions and help illuminate contemporary realities. Important forces behind the need for critical postmodern theories are, therefore, that new technologies and an emergent global reorganization of transnational capitalism are producing novel forms of society, culture, politics, and identities and that we need new theories to chart and map these developments. Yet whether existing postmodern theories provide the best theoretical resources to conceptualize, contextualize, and evaluate the very phenomena on which we believe its currency depends is an open question that we raise throughout our studies.

FROM THE MODERN TO THE POSTMODERN

> I believe we are coming to a watershed in Western society: we are witnessing the end of the bourgeois idea—that view of human action and social relations, particularly of economic exchange—which has molded the modern era for the last 200 years. And I believe that we have reached the end of the creative impulse and ideological sway of modernism, which, as a cultural movement, has dominated all the arts, and shaped our symbolic expressions, for the last 125 years.
>
> —Daniel Bell

The postmodern turn contains a mutating mixture of risks and excitement, losses and gains, resulting from destruction of the old and creation of the new. Many individuals, such as Bill Gates (1995), president of Microsoft, are celebrating the emerging technological society as a new era of job and profit possibilities, with exciting new forms of culture and communication, promising a technological utopia. Others stress the downside, emphasizing in apocalyptic fashion the collapse of the old modern society in a new postmodern scene of "panic," "spasm," and "crash" (Kroker et al., 1989; Kroker, 1993; Kroker and Weinstein, 1994). Certainly, there will be enough crises and catastrophes to support a cottage industry of postmodern gloom and doom for a good many years to come, but there are also new possibilities for a better life. Our position toward the novelties of the present is precisely the dialectical optic developed by Marx and Engels toward the new modern capitalist society, which was the most dynamic and revolutionary social order in the history of the world, as well as one of the most destructive (see Marx and Engels, 1978; Jameson, 1991). Marshall Berman (1982) has argued that the best modern theories shared this dialectical vision that sorted out the positive from the negative aspects of modernity, and he criticized a flattening out of vision in the 20th century that took either wholly positive or wholly negative positions toward the modern order. We agree with him that what is needed today is a critical and dialectical vision that renounces both cheery celebration of the new technological utopia speeding down the information superhighway and wholly negative jeremiads against the present age, which are dispiriting and disempowering unless accompanied by a positive vision of hope and possibility.

In this turmoil of dramatic social and cultural transformation and in the proliferation of discourses of the postmodern that purport to explain it, there is an urgent need for critical analysis of contemporary trends, developments, and discourses. Charting the postmodern turn requires compasses and maps to stake out the terrain traversed, and in the following studies we will attempt to supply our readers with some conceptual markers, as we ourselves enter into some of the controversies of the present and attempt to break fresh ground in analyzing salient aspects of the present age. Our project in this book is thus to delineate the postmodern paradigm as it travels through theory, arts, culture, and the sciences and to use the discourse to illuminate the present era. The work of genealogical and conceptual delineation is difficult in that there are fundamentally different definitions of the postmodern and disparate emphases and positions among the various fields of theory, culture, science, and the specific arts such as architecture, painting, or literature. Moreover, there is an anti-essentialist and formalist attitude within the postmodern sensibility that eschews fixed definitions. Against postmodern nominalism, which renounces general categories in favor of fragments, particulars, and chaotic heterogeneity, we wish to elucidate the postmodern turn in various fields and to discern its overall significance, form, and effects as an important emergent paradigm shift.[11]

In order for postmodern discourse to have substantive cognitive content, certain distinctions need to be made, and the family of terms of the postmodern must be distinguished from the discourses of the modern. In our previous book, *Postmodern Theory*, we distinguished between modernity and postmodernity as two different historical eras; between modernism and postmodernism as two conflicting aesthetic and cultural styles; and between modern and postmodern theories as two competing theoretical discourses (see Best and Kellner, 1991). Building on these analyses and historical genealogy, we wish here to offer some further conceptual clarifications to try to illuminate the complex field of modern and postmodern discourse.

Notions of a postmodern break with the modern era and modern culture and society appeared regularly from the 1940s to the present in the fields of history and social theory (see the genealogy in Best and Kellner, 1991); in the arts, a current of postmodernism emerged that rejected modernism, and postmodernism's practices and forms eventually erupted in every artistic field during the 1960s (see Chapter 4). Modern paradigms in philosophy, literature, science, and the arts were decisively formed in the social context of modernity, which involves a shattering of tradition, the replacement of a religious worldview by a secular worldview, and the instantiation of a social-institutional logic based on incessant change and progress (Berman, 1982). Science, capitalism, and technology were the driving forces behind rapid social development, while theorists and artists affirmed change and novelty and produced new theories and visions. Despite the specialization of disciplines, an overarching *modern paradigm* emerged in society, beginning perhaps in the 15th century and continuing strongly through the end of the 19th century, organized around mechanical metaphors, deterministic logic, critical reason, individualism and humanist ideals, a search for universal truths and values, attempts to construct unifying and comprehensive schemes of knowledge, and optimistic beliefs in progress and the movement of history toward a state of human emancipation. Through a series of revolutions—geographic (colonialism), intellectual (the Renaissance, modern science, and the Enlightenment), economic (capitalism), political (bourgeois democracy), technological (the Industrial Revolution), and artistic (modernism)—the world of Newton, Kant, and Marx became fundamentally different from the premodern world of Dante, Aquinas, and Augustine.

Of course, the modern paradigm is highly complex, diverse, and contested. The Enlightenment, for example, although comprising a general belief in reason and critical thought as the key to human emancipation, was divided within itself on major issues such as religion, politics, and human nature (see Antonio and Kellner, 1994; Best, 1995). Dissent from the modern paradigm began at the moment of its formation in science, the emerging capitalist economy, the bourgeois political revolutions, and the first flowerings of the Enlightenment. Throughout its entire development, from Burke to Blake to Bakunin and Baudrillard, the values of modernity were hotly contested, as modern paradigms in theory, the arts, and science began to take shape, were criticized, and developed further. In the realm of theory, sustained and powerful critiques of totalizing modes of rationality, a major complaint within dissenting traditions, was begun in the mid-19th century by European existentialists, was continued through the early 20th century by U.S. pragmatists, and culminated in the second half of the 20th century with poststructuralism and New French Theory, all helping to generate a postmodern turn.

By the 1960s, with their complex lineage of anticipations, postmodern paradigms in various fields began to take shape in reaction to the orthodoxies and problems in modern paradigms. But the sea change that has taken place in virtually every discipline since the 1960s has *also* been influenced by economic and technological changes occurring during the whole post-World War II period and, as we have already argued, was decisively influenced by the political turmoil of the 1960s. The modern paradigm, which began to emerge at various levels in the 15th and 16th centuries and was clearly dominant by the time of the American and French Revolutions, is now in great crisis at all levels—from the values of humanism to the deterministic and mechanistic logic of the sciences to the ideology of growth and progress. Consequently, there is today an *emerging postmodern paradigm* organized around a family of concepts, shared methodological assumptions, and a general sensibility that attack modern methods and concepts as overly totalizing and reductionistic; that decry utopian and humanistic values as dystopian and dehumanizing; that abandon mechanical and deterministic schemes in favor of new principles of chaos, contingency, spontaneity, and organism; that challenge all beliefs in foundations, absolutes, truth, and objectivity, often to embrace a radical skepticism, relativism, and nihilism; and that subvert boundaries of all kinds (see Chapter 6).

The postmodern paradigm that we see emerging in society and culture as a cumulative result of paradigm shifts in specific disciplines must not be understood in terms of a closed, unanimous, or finalized framework that rigidly and inexorably informs all thought today, nor should it be seen as the new "cultural dominant" as Jameson (1984, 1991), for example, argues. In the current conditions of crisis and ferment, the postmodern paradigm is only emergent and is strongly resisted by modernist orthodoxy, as well as being conflicted among competing tendencies, as we will show throughout our studies. As we write, modern neo-positivist approaches still prevail in the social sciences, behaviorism dominates psychology, and the Anglo-American analytic tradition still governs philosophy.

Yet in the past decade, postmodern interventions have taken place in a wide range of academic disciplines, challenging basic assumptions, practices, and the very division of intellectual inquiry and education into separate fields (see the studies in Rosenau, 1992; Seidman and Wagner, 1992; Dickens and Fontana, 1994; Hollinger, 1994; Seidman, 1994; Ritzer, 1997). Postmodern theory has penetrated almost all academic fields, producing critiques of modern theory and of alternative theoretical practices in philosophy, politics, economics, anthropology, geography, environmental theory, education, and just about every domain from the humanities to the sciences (see Chapters 5 and 6). Consequently, nearly every academic discipline and profession have been challenged and confronted with alternative approaches that often fly the banner of the postmodern, and every domain of society is undergoing transformations to which the term "postmodern" is applied.

In different fields, the postmodern turn takes distinct forms, and throughout this book we shall examine a series of postmodern interventions in various domains and topics. What is at stake is not only standard methods and ideas within each academic field but the concept of disciplinarity itself. As our own studies indicate, the postmodern turn comprises border crossings that traverse standard academic domains and boundaries. Postmodern theory—along with the critical theory of the Frankfurt School, feminism, British cultural studies, and other contemporary approaches to society and culture—is transdisciplinary, exploding artificial academic divisions of labor and creating new forms of discourse, critique, and practice.

Yet the discourse of the postmodern has become more than just an academic affair. It has filtered down into journalism and common usage, often ludicrously. MTV had a "postmodern" segment of music videos, though the products displayed under this rubric were little different from the regular fare, and it is now popular to describe MTV in its totality as postmodern (Kaplan, 1987; Grossberg, 1993; Lewis, 1990). Newspaper articles regularly use the word "postmodern" to describe a broad array of political, social, and cultural phenomena; *The New York Times* (May 12, 1993), for example, published a story with the headline "Forget the Bologna on White, Here Comes the Post-Modern Sandwich." *Playboy* magazine had an article on postmodern comic books in the July 1995 issue and during 1995, *NBC Nightly News* began featuring oddities of the present under the rubric of "postmodern," such as presenting pop music stars and comic book characters on U.S. postage stamps. In February 1996, *Harper's* featured a story on the "postmodern kiss," in September/October 1996, *Foreign Affairs* had a story on "postmodern terrorism," and in January 1997, Barnes and Noble advertised a new "postmodern diet book."

One may laugh at muddled pop media usages of the discourse, but things are often not much clearer when we turn to the theoretical discourses of scholarly books or presentations. Frequently, academic commentators simply assume that we are in a postmodern age without any specific analysis. Often the term "postmodern" is used for phenomena that are arguably modern, and the discourse is merely employed to indicate the contemporary moment in which we live, or contemporary novelties, without substantive analysis.[12] Our favorite example of academic abuse of the concept concerns a sociology professor whom we asked to describe more clearly what he meant by "postmodern" at a 1992 conference in which he continually described the current social order as "postmodern." He answered that the best description of "our postmodern society" was found in the passage in "The Communist Manifesto" where Marx and Engels describe a state "where all that is solid melts into air." Of course, as Marshall Berman (1982) has shown, the "Manifesto" is a virtual hymn to modernity and is a key text of modern theory.

We have found that many books and articles that use the discourse of the postmodern either do not really theorize it or they provide inconsistent and confusing accounts. In his many books of the late 1980s and early 1990s, Zygmunt Bauman seems to offer a different characterization of the postmodern in every book and sometimes within a single book, as in *Intimations of Postmodernity* (1992), where he provides a list of allegedly postmodern phenomena and then defines postmodernity as "more than anything else—a *state of mind*" characterized by "its all-deriding, all-eroding *destructiveness*" (viii). He next describes its dogmatic modes, and its hypercritical and self-reflexive aspects, but he then defines it "as a *re-enchantment* of the world that modernity tried hard to *dis-enchant*" (x). Lest one think that Bauman's constantly expanding and mutating features of the "postmodern mind" is inconsistent, he concludes that "incoherence is the most distinctive among the attributes of postmodernity (arguably its defining feature)" (xxiv).

Bauman's problem is that he veers between essentializing the postmodern and declaring it to be incoherent and without clear identity. Throughout his studies, he reduces a complex phenomenon to a monolithic entity as when he collapses "postmodernity" into a "postmodern mind" and culture:

> Postmodernity (and in this it differs from modernist culture of which it is the rightful issue and legatee) does not seek to substitute one truth for another, one standard of beauty for another, one life ideal for another. Instead, it splits the truth, the standards and the ideal into already deconstructed and about to be deconstructed. It denies in advance the right of all and any revelation to slip into the place vacated by the deconstructed/discredited rules. It braces itself for a life without truths, standards and ideals. It is often blamed for not being positive enough, for not being positive at all, for not wishing to be positive and for pooh-poohing positivity as such, for sniffing a knife of unfreedom under any cloak of saintly righteousness or just placid self-confidence. The postmodern mind seems to condemn everything, propose nothing. (1992: ix)

In this passage, note how Bauman essentializes the "postmodern mind" as an actor, as a subject that engages in multiple forms of activity. But his hypergeneralizations create a model of what this mind does without mentioning various postmodern thinkers who actually make this move. Note here also the litany of "its" that supposedly describe "*the* [our emphasis] postmodern state of mind." But in fact there is no "postmodern mind"; rather, there is a complex set of postmodern perspectives that sometimes coalesce into distinct paradigms and often coexist uneasily with each other and with modern perspectives. Reducing "the postmodern mind" to a simple set of defining characteristics (relativism, irrationalism, nihilism, destructiveness, incoherence, or whatever) makes it easy for its critics to dismiss postmodern discourse, but in fact, as we are trying to argue, the postmodern turn is more complex and differentiated than many of its advocates and critics realize.

Indeed, opponents of postmodern theory also reify it, as do some of its proponents. In an attack on postmodern theory, Barbara Epstein, for instance, writes:

> As a campaign for dominance, postmodernism belies its claim to value diversity; it treats other perspectives with scorn, often dismissing them as outmoded and conservative. It has created an arena in which statements are all too often grandiose and self-referential, in which participants congratulate themselves on their radicalism or "transgression," and in which ideas are judged by their conformity to the shifting ideological trends

> within this world rather than in terms of their ability to shed light on any external reality. It is difficult to debate postmodernism because it is not a set of claims. It is a set of attitudes, the central one being the suspicion of all claims—and the concomitant effort never to make any claims, or at least, never to acknowledge them. Postmodernism sees itself as the unique possessor of a critical perspective, with a foothold above the fray, a perspective from which it may judge the claims of others without exposing itself to being judged in return. (1996: 57–58)

Note how both Bauman, who advocates postmodern discourse, and Epstein, who opposes it, use the term "it" to reify complex postmodern discourse into an essentialized and active subject that carries out specific effects, which are either affirmed or attacked. It is not the "postmodern mind" or "postmodernism" that does anything, but rather specific theorists who use the discourse, and different theorists use it in significantly different ways, rendering illicit attempts to excessively generalize or essentialize postmodern discourse.

Curiously, this setting up a hypostatized and reified "postmodernism" replicates the same strategy and discourse often used to dismiss Marxism and feminism, a strategy in which their opponents argue: "Marxists ______ [fill in the reified blanks]," or "feminists ______ [fill in another set of theoretical sins]." Such discourse, however, creates a straw model, covers over differences within the target of critique, and thus thoroughly misrepresents that polemical target. To be sure, some proponents of postmodern theory suffer precisely from the problems that Epstein enumerates, but her critique is vitiated by its generality, polemical bias, and oversimplified reductionism.

There are, in fact, many different versions of postmodern theory, and in *Postmodern Theory* we examined some of their shared themes and often striking differences (Best and Kellner, 1991). In a strict sense, then, there is no such thing as "postmodern theory"; rather, there are a diversity of postmodern *theories*. We warn against two extreme reactions to the discourse of the postmodern: uncritical embrace, as if postmodern theory was *the* key to the contemporary universe, or total rejection, as if it were just a fad or novelty of no real significance, or a malignant virus infecting modern culture. Both reactions assume that there is a coherent and unified "postmodern theory" that consists of positions, ideas, and politics that one can either happily affirm or contemptuously reject. Rather, there are a wide range of postmodern theories with a tremendous complexity of often-conflicting positions. There are significant differences between thinkers like Baudrillard, Derrida, Foucault, Lyotard, and Rorty, with many fierce polemics among them. Moreover, theorists like Foucault, Lyotard, and Baudrillard have evolved through several different stages of development, often with serious disparities and conflicts among their positions in different periods. Thus the only way to appraise postmodern theories is to actually engage their full complexity, to chart their diverse effects, and to see how they function in a variety of disciplines, contexts, topics, and texts—as we attempt in the following studies.

In general, we have found that the term "postmodern" functions in many contemporary discourses as a marker for the new, for that which does not fit into old paradigms, or, in some cases, for those novelties that the theorist is unable or too lazy to theorize. Many who use the terminology just list a set of things that are said to be "postmodern," sometimes engaging in overkill and calling everything contemporary "postmodern," or they simply offer an arbitrary list of characteristics, many of which are arguably modern. Often the concept of the postmodern is simply used as a slogan or buzzword without any analysis at all, as in the case of any number of books that have the term "postmodern" in the title without engaging the concept, such as *The Postmodern President* (Rose, 1988). Moreover, it seems that almost everyone who uses the term "postmodern" has a different definition; thus it is not at all certain what the term means or how it is being used.

Although the discourse of the postmodern frequently muddles more than it illuminates, it has a symptomatic value. While the concept "postmodern" is often an empty signifier and a sign that more concrete theorization is being avoided, it is also a sign that something is new and needs to be theorized, that something is bothering us and requires further thought and analysis, that new and perplexing phenomena are appearing that we cannot yet adequately categorize or get a grip on. Thus the nomenclature "postmodern" is often a placeholder, or semiotic marker, for novel phenomena that deserve our critical attention. In this book, we argue that if the discourse of the postmodern is to be more than just a buzzword for pop journalism and lazy or muddled theory, it must be clearly analyzed, distinguished from the modern, and provided with specific content and force. It will therefore be one of the purposes of our studies to interrogate some contemporary discourses of the postmodern, to demonstrate both illuminating and significant uses of the concept and confusion, sloppiness, and abuse. But first we need to make a few further key distinctions.

THE DISCOURSE OF THE POSTMODERN: SOME CONCEPTUAL CLARIFICATIONS

> An ambivalent response to what Barthes might have called the "happy babel" of the Post seems on the whole more honest and in the long run more productive than a simple either/or. Viewed benignly, the degree of semantic complexity surrounding the term might be seen to signal the fact that a significant number of people with conflicting interests and opinions feel that there is something sufficiently important at stake here to be worth struggling and arguing over.
>
> —DICK HEBDIGE

One should be clear that the concept of the postmodern is a cultural and theoretical construct, not a thing or state of affairs. That is, there are no phenomena that are intrinsically "postmodern." The concepts are generated as theoretical constructs used to interpret a family of phenomena, artifacts, or practices. Thus, the discourses of the postmodern produce their objects, whether a historical epoch of postmodernity, postmodernism in the arts, or a new type of postmodern theory. This is not, however, to affirm the "theoreticism" of Hindess and Hirst (1975), who claim that there is no "history" or "reality" independent of theory. Obviously, there are social and historical phenomena from which theorists derive concepts like postmodernity, just as there are practices, artifacts, and authors in the field of culture from which theorists derive the term "postmodernism." Yet which phenomena, practices, artifacts, and so on are seen as "postmodern" is itself a function of the theoretical assumptions that denominate some things as "postmodern" and others not.

Consequently, postmodern concepts are primarily conceptual constructs meant to perform certain interpretive or explanatory tasks and are not neutral descriptive terms that define preestablished states of affairs. This is an obvious point, but an important one; it explains why there are so many conflicting postmodern theories circulating: Each discourse constructs its own objects, calling various phenomena "postmodern," having different genealogies, and contesting other theories. Thus, one should be clear that when we are dealing with postmodern discourse, we are operating on the level of theory and need to make appropriate clarifications and distinctions.

It seems sensible to distinguish between two quite different types of postmodern theory. At one pole, there is what might be called extreme postmodern theory, which posits a radical rupture between modernity and postmodernity and between modern and postmodern discourse and practice. Such extreme, or "strong," postmodern theory puts its emphasis on the *post* and its break from the modern and finds its ideal type in the texts of Baudrillard and his followers, as well as in some subcultures that use the pathos of the postmodern as a sign of identity and distinction. Those who claim that we are now in a new era of postmodernity that is fundamentally different from the modern era are examples of what we are calling extreme postmodernists—as are those who call for a complete breach with modern theory, culture, or politics and new postmodern ways of thinking, writing, and living. This tendency usually attacks modern discourse or politics as being obsolete and calls for new beginnings and a break with the past. Modern theory is rejected in toto as a set of foundationalist, universalist, essentialist, and totalizing practices that must be renounced for a wholly new mode of postmodern theory, ethics, and politics. Strong postmodernists who take this position tend to be true believers and often ideologues who totally reject the theories, discourses, and politics of the past modern era. They tend to engage in polemical assaults on modernity and modern theory and politics, and they often fervently champion postmodern discourse, rhetoric, and style. Such adherents are often found among the younger generation of pomos, who are able to jettison the baggage of modern theory more easily than some of their elders, who were socialized within the modern tradition.

Extreme postmodernists should be distinguished from those who use a more qualified and modest mode of postmodern discourse, who do not advocate a fundamental break either with modernity or with modern theory.[13] The more moderate versions put their emphasis on the *modern* and interpret the postmodern merely as a mutation of the modern, as a shift within modernity. Moderate postmodern theorists like Lyotard (sometimes), Foucault, Laclau, Mouffe, Harvey, Rorty, and others combine modern and postmodern discourses and interpret the postmodern primarily as a modality of the modern rather than as its radical other. They do not usually carry out extreme attacks on modern theory or make extravagant claims for major historical ruptures but, instead, simply use certain versions of postmodern theory for specific tasks.[14] Other less extreme forms of the appropriation of postmodern positions include the work of theorists such as Bernstein, Fraser, Harding, Nicholson, Seidman, Smart, and West, who use postmodern categories and insights to rethink modern theory without abandoning its core features, such as concern for truth, objectivity, ethics, and normative critique. Such theorists combine modern and postmodern perspectives, drawing on both traditions as providing resources to do theory and critique in the present age—a position with which we identify.

Indeed, we choose to deploy the discourse of a postmodern "turn" or "shift" instead of "rupture" which stresses extreme breaks, discontinuities, and an apocalyptic sense of ending and completely new beginnings. Rather, for us the notions of "shift" and "turn" signify novel developments, yet also retain continuities with modernity and modern theory, pointing to shared assumptions, presuppositions, modes of thought and discourse, practices and strategies, and vision. Seeing the postmodern as continuous with the modern eschews postulating a radical rupture in theory and history. Such moderate versions describe postmodern phenomena more modestly and make less dramatic claims for the radicality of their theoretical positions while appreciating the novel qualities of the contemporary era. Thus, from the moderate perspective, postmodernism in the arts would be another mode of modernism, another avant-garde avatar of the new, another search for a contemporary mode of cultural expression that has always been the goal of the modern.[15] Likewise, taking postmodern theory as continuous with and supplementary to modern theory interprets postmodern discourse as an intensification of critical tendencies of such modern figures as Nietzsche, Heidegger, or Dewey rather than as a leap into a whole new mode of discourse altogether.[16]

One could indeed argue that the modern itself is highly polysemous and unstable, constantly undergoing change and development. On this reading, the modern is always articulating itself in novel or radically new forms, and the "postmodern" can thus be interpreted as a form of the modern. Indeed, throughout this book we will argue that the postmodern is a radicalization of the modern, which intensifies modern phenomena like commodification, massification, technology, and the media to a degree that generates genuine discontinuities and novelties from the modern world. The virtue of this position is that it avoids making exaggerated claims of discontinuity and rupture and that it stresses important lines of continuity between the modern and the postmodern.

Yet in some ways the more extreme versions of postmodern theory and practice are more interesting than the qualified versions in that they more radically challenge existing theories, practices, and realities. One of the virtues of strong postmodern theory is that it forces us to rethink our basic presumptions, methods, and modes of practice. Indeed, it is precisely the putting in question of modern discourses, the development of novel perspectives, and the rethinking of theory, culture, and politics that make the postmodern turn an exciting and challenging one. At their best, therefore, postmodern theories have forced many to rethink their theoretical and political assumptions and have generated a wealth of new theories and practices during the contemporary moment.

Yet many versions of strong postmodern discourse are arguably too extreme, making unsubstantiated claims for the radicality of the rupture, failing to adequately thematize the alleged break or to provide empirical evidence for their claims. Many moderate postmodernists, who see the postmodern simply as a modality of the modern, also often undertheorize it or provide weak and confusing analyses. Moreover, there is often slippage between the stronger and weaker versions. Although Baudrillard is arguably the ideal type of an apocalyptic postmodernist, in some interviews he distances himself from the concept (while in other interviews and published writings he uses the term positively).[17] Lyotard too, who with his *Postmodern Condition* has done as much as anyone to popularize the notion of a new postmodern condition, often rejects the concept's extreme use, defining it more modestly as a "mood" or "state of mind" or "modality" of the present.[18] Baudrillard and Lyotard, however, want to have it both ways: to be avatars of the postmodern when it suits them (or their audience of the moment) and to distance themselves from the discourse when they wish. Such theoretical opportunism is indeed characteristic of a postmodern mood that eschews being pinned down to positions, that revels in irony and the indeterminacy of meaning, and for whom theoretical discourse is not a search for truth but a form of play, indifferent to old-fashioned matters such as logical consistency.

There is also an important conceptual distinction to be made between "ludic" and "oppositional" postmodern discourse. An oppositional postmodernism, as we described it above, is a product of new social movements and the impulse to oppose and resist existing society. Oppositional postmodernism strongly opposes the established society and culture and seeks new forms of critique and opposition. Distancing itself from modern theory and politics, oppositional postmodernism seeks new forms of resistance, struggle, and social change. In this sense, it is continuous with modernism in its seriousness and commitment to critique, struggle, and opposition.

Ludic postmodernism, by contrast, is highly ironic, playful, and eclectic, advocating a pluralism of "anything goes" and excessive relativism and subjectivism. Some ludic postmodernism is affirmative toward existing reality, whereas other forms are more cynical, advocating in some cases an intense nihilism and pessimism. The element of play in ludic postmodernism constitutes some of the attractions and pleasures of the postmodern turn, although as Ebert (1996) notes, it can also foster a hostility toward theory and politics, legitimating an escape from theory and politics rather than serving to produce their reconstruction.[19]

For some, the postmodern turn simply provides some new ways of doing theory and engaging contemporary society and culture. For others, however, the postmodern turn constitutes a new mode of existence, and these pomos live the postmodern with a certain style, attitude, and élan. For them, the postmodern turn is a way of life, permeating everything from their ideas to fashion. By contrast, some academic postmodernists are using new categories merely to improve contemporary theoretical discourse, or perhaps just plugging in to a fashionable discourse, attempting to keep up with the latest trends.

Indeed, as noted, the emergence of the postmodern has much to do with battles for cultural capital in the present age. One way of contesting previous theories, canons, and models is to declare their obsolescence or to radically negate their claims to truth, virtue, utility, or whatever. One accrues cultural capital by distinguishing one's work and positions from that of others, by attaching oneself to popular phenomena, and by adopting new theoretical and cultural styles that enable one to identify oneself as hip, cool, with it, and *au courant.* Making the postmodern turn in this sense is fun, allowing one to engage in promiscuous excursions beyond the rules and conventions of standard discourse and practice. But the struggle for cultural capital makes the debates over the postmodern highly serious as well, with those invested in modern discourses often sharply denouncing those who have made the postmodern turn, who in turn frequently respond by attacking as outmoded or obsolete the discourses of their modern critics.

In addition to distinctions between extreme and moderate versions of postmodern theory and between ludic and oppositional postmodernism, we can distinguish between "postmodern" as a temporal marker, designating that which comes *after* the modern, and as a philosophical marker, emphasizing that which is *against* the modern and is thus *anti*-modern. The former, temporal sense may simply depict changes in contemporary society and culture without normative discrimination, whereas the latter carries out a critique of the modern in favor of new, postmodern positions. Yet, as Habermas (1987) has fruitfully suggested but has inadequately theorized, there are distinct continuities between some postmodern theories and the entire counter-Enlightenment and conservative tradition of modern theory.

Thus, some anti-modern discourse is aimed toward producing a radically new postmodern future, while other critics of the modern seek a resurrection of tradition and a return to the past. On our analysis, this backward-looking anti-modern tradition stretches from early conservatives such as Edmund Burke, Joseph de Maistre, and Hippolyte Taine, to romantics like William Blake, Ralph Waldo Emerson, and William Wordsworth to existentialists such as Kierkegaard, Nietzsche, and Heidegger. Beginning with Burke's celebrated *Reflections on the French Revolution,* written just one year after the 1789 French Revolution, the institutions and values of modernity—Enlightenment, secular rationality, criticism, individualism, capitalism, and revolution—were sharply contested from various quarters. Burke set the tone for the conservative reaction in his defense of monarchy, the church, nobility, and community against what he and others saw as the corrosive influences of modernity that created historical regression rather than progress. Burke and others denigrated modern notions of universal values as conflating crucial differences among various nationalities and local traditions. They ridiculed the Enlightenment notion that individuals can be governed by reason, and they stigmatized the "masses" as being ruled by base desires and appetites. For conservatives, the transformation of "liberty, equality, and fraternity" into the Terror was an inevitable effect of disruptive modernizing influences. They believed that social stability could only be maintained through the time-honored wisdom of tradition and the rule of the church and nobility.

The romantics cannot easily be assimilated to the conservative tradition. Some rejected Christianity, abhorred stale dogmas and tradition, and embraced the modern ethos of individualism and freedom. But, like the conservatives, many were deeply religious in their outlook and repudiated the secularism and empiricism of the Enlightenment along with the mechanical worldview of modern science. In their search for divine communion, for the unity of the individual self with the cosmic Self, romantic philosophers and artists embraced mysticism, pantheism, and exoticism. They argued that both reason and scientific empiricism were limited to surface appearances of reality, beyond which lies the infinite. Although some romantics supported the French Revolution in its early phase, they all rejected the Enlightenment emphasis on reason and logic as sterile and inhibiting and championed passion and imagination as the true liberating forces of human beings.

Like the romantics, the 19th- and 20th-century thinkers associated with existentialism affirmed passion, instinct, and spontaneity over reason, logic, and deliberation. Some, like Kierkegaard, Gabriel Marcel, and Paul Tillich, remained Christians, but others, like Nietzsche, Sartre, and Albert Camus, adopted an aggressive atheism. Existentialists criticized mass society as a deadening, homogenizing force and advanced highly individualistic positions. With some exceptions, such as Sartre, they took regressive (Nietzsche) or even fascist (Heidegger) political positions that renounced Enlightenment emphases on equality and democracy, and they rejected social ethics and modern forms of society altogether.

In one way or another, 18th- and 19th-century conservatism, romanticism, and existentialism form the backbone of a counter-Enlightenment tradition and are important influences on some versions of postmodern theory.[20] Kierkegaard, for instance, anticipated postmodern theory with his critiques of reason and the Enlightenment and his insights into how the emerging mass media were producing a new realm of social experience and a "phantom public" that lost its individuality and capacity for critical thought. Like Kierkegaard, Nietzsche carried out a radical critique of reason, the Enlightenment, and modern philosophy, a critique that anticipated the postmodern turn in thought and also argued that modern society had become so chaotic, fragmented, and devoid of "creative force" that it had lost the resources to create a vital culture, thus demanding a break with modernity and the creation of a new society and way of life—a radical rejection of modernity repeated by Heidegger (see Chapter 2).

There are obviously major differences among the precursors of the postmodern turn: Conservatives, for example, reject the individualism endorsed by romantics and existentialists; atheistic existentialists articulate a sense of cosmic absurdity that is felt by neither conservatives nor romantics but that is taken up by some postmodernists. Yet there are significant similarities uniting all three movements. Perhaps most important, each camp attacks reason as a largely repressive or regressive force, and with this they stand in solidarity with numerous postmodernists. Each has helped to shape the aestheticist motifs of some versions of a postmodern theory that privileges nonrational over rational forces. Postmodern discourse, therefore, is often *anti-* rather than merely *after* modern theory, and in some contemporary forms it assaults theory itself, stigmatizing it as reductive and totalizing, while favoring a fragmented, aphoristic, or essayistic mode of writing.

BETWEEN THE MODERN AND THE POSTMODERN

The most extreme postmodern discourse holds that we have entered a dramatically novel era that requires entirely new theories and politics. Such postmodern theories posit a decisive break with modern culture and society, assuming that the historical epoch of modernity is over and that we are living in an entirely new social order, a postmodernity. The claim is analogous to the earlier modern arguments that the Enlightenment broke with the unreflective, childlike past (Kant) or that the French and Industrial Revolutions produced an altogether new modern society radically different from traditional society. Modern social theory arose as an attempt to describe the emergent industrial–democratic societies. Classical modern theorists like Marx pointed out that for the first time in history, life was organized around the production (and later consumption) of commodities. Industrial men, women, and children were becoming species of a new type of *homo faber* (literally, humans who make) who were condemned to novel modern forms of misery and servitude. Nietzsche depicted in turn the innovative social and cultural forms of modern society and the rise of a modern state and mass society, while classical social theorists like Durkheim and Weber analyzed the forms of social differentiation, rationalization, and secularization characteristic of modernity (see Antonio and Kellner, 1992, 1994).

Some postmodern theorists claim that a rupture has taken place in history every bit as great as the rupture between modern and premodern societies. The break is described by postmodernists as the transition from modernity to postmodernity, by Marxists as the restructuring of global capitalism and the emergence of a new regime of post-Fordist accumulation and transnational capitalism, and by sociological theorists as the move to a postindustrial or information society. A wide range of theorists interpret the contemporary moment in terms of dramatic mutations of the economy, polity, society, and culture, characterized in a variety of competing vocabularies. There is thus a multiplicity of new discourses that have attempted to capture the novelty of the present moment, of which postmodern theories are among the most prominent.

Although it is prudent to be skeptical of extreme postmodern claims that would render obsolete the assumptions, values, categories, culture, and politics of the modern era, it must be admitted that significant changes are taking place and that many of the old modern theories and categories can no longer adequately describe our contemporary culture, politics, and society. Whereas the modern era swept in unprecedented forces of secularization, rationalization, commodification, individualization, urbanization, nationalism, bureaucratization, and massification, since the 1960s we have seen the decline of the nation-state, a tumultuous process of decolonialization, explosions of ethnicity and fundamentalism, cultural fragmentation, and the erosion of belief in progress and Enlightenment values. In addition, revolutionizing new phenomena have appeared, such as automation, robotics, and computers and huge data banks; media culture has risen to unparalleled power; and new forms of virtual reality and hyperreality have spread, challenging our definitions of subjectivity and objectivity.

And yet the extreme claims for a postmodern break and rupture do violence to our sense of enduring continuities with the past and to the fact that many ideas and phenomena that are claimed to be "postmodern" have their origins or analogues precisely in the modern era. Many of the cultural and technological phenomena described as "postmodern" are produced by the classical dynamics of commodification logic and cannot be separated from the developments of a new global capitalism (e.g., the technologies of the information society). Often what is described as "postmodern" is an intensification of the modern, a development of modern phenomena such as commodification and massification to such a degree that they appear to generate a postmodern break (see Chapters 2 and 3).

Consequently, we will argue throughout this book that we are living between a now-aging modern era and an emerging postmodern era that remains to be adequately conceptualized, charted, and mapped. In Richard Bernstein's (1991) appropriation of Walter Benjamin and Theodor Adorno, we are living in a "new constellation" of changing elements that are irreducible to a common denominator, in a "force-field" of dynamic interplay between the old and the new. Living in the borderlands between the modern and the postmodern means negotiating constant conflicts between the old and new and confronting perplexing and often disturbing change. Condemned to a seemingly unending state of transition, permanent tension and strife appears to be a defining modality of the between. Moreover, the discourses that strive to describe this condition are also in conflict and at odds with each other, condemning us to unending theory and culture wars with no truce in sight.

Historical epochs do not rise and fall in neat patterns or at precise chronological moments. Perhaps our current era is parallel in some ways to the Renaissance, which constituted a long period of transition between the end of premodern societies and the emergence of modern ones. Such periods are characterized by unevenly developing multiple levels of change and the birth-pangs associated with the eruption of a new era. In fact, change between one era and another is always protracted, contradictory, and usually painful. But the vivid sense of "betweenness," or transition, requires that one grasp the connections with the past as well as the novelties of the present and future. Thus, it is important to capture both the continuities and discontinuities of the postmodern with the modern, in order to make sense of our current predicament.

Living in a borderland between the modern and the postmodern creates tension, insecurity, confusion, and even panic, as well as excitement and exhilaration, thus producing a cultural and social environment of shifting moods and an open but troubling future. The concept of a postmodern turn includes a recognition of the risks and dangers in the current social constellation, as well as the hope of new possibilities and excitement. The postmodern turn is thus deeply implicated in the moods and experiences of the present and is an important component of our contemporary situation. The very ubiquity of the discourse of the postmodern, its constant proliferation, its refusal to fade away, and its seeming longevity—several decades is a long time for a mere "fad" in our rapidly changing world—suggest that it is addressing current concerns in a useful way, that it illuminates salient present-day realities, that it resonates with shared experience, and that it is simply an ingrained part of the current critical lexicon that one has to come to terms with, one way or another.

Consequently, it would be a mistake to dismiss the discourse of the postmodern out of hand as an ephemeral fashion, as if it were a theoretical hula hoop. Although many predicted that the phenomenon would soon be over, there continue to be waves of books, articles, and conferences and a proliferation of postmodern discourses in every academic field, as well as in the popular media.[21] People continue to feel passionately about the drama of the postmodern, and the discourse obviously speaks to important changes in our culture and society and by now has acquired a certain weight and substance. Groups and individuals marginalized in the society, culture, and university have taken the concept as their own and use it as a banner to oppose the established order of things. Since many of these individuals are younger, one expects that the discourse will continue to be used for some time to come.

In addition, the discourse is remarkably flexible and open, and individuals can use it to promote a lot of different agendas, as well as a lot of gibberish. Although some discourse of the postmodern (i.e., that of Baudrillard and his followers) is exceedingly cynical and skeptical, those who wish to promote religion have also used the discourse to attack modernity and Enlightenment rationalism (e.g., Smith, 1982). Individuals have used it to advance a tremendous variety of theoretical and political agendas, producing a bewildering cacophony of postmodern discourses. Many have invested cultural capital in promoting the turn, while others are invested in attacking it. The discourse of the postmodern is thus an integral part of the contemporary scene and will, we believe, be with us for a long time to come. Therefore, we must take it seriously and engage it critically.

We conclude that we are currently forced to live in parentheses, between the old and the new, in a borderland between the modern and the postmodern, in an interregnum period in which the competing regimes are engaged in an intense struggle for dominance. In this situation, it is necessary to engage the discourse of the postmodern, its differences with the modern, and the ways they intersect as well as conflict in the present situation. In the following chapters, we will accordingly trace some of the circuitous pathways to the postmodern, indicating their multiple origins, genealogies, and contemporary manifestations, suggesting to us in their variety, richness, and shared features that we are currently undergoing a paradigm shift and thus exist in a field between the modern and the postmodern that we need to chart and navigate.

NOTES

1. See the books on postmodernity and the postmodern turn listed in the Preface, Note 1, for further sources on these themes.

2. On the events of May '68 in France, see Cohn-Bendit et al. (1968) and Lefebvre (1969).

3. On the experiences of the 1960s, which generated an aura of apocalyptic break with the past, see Katsiaficas (1987), Gitlin (1987), and Caute (1988).

4. "Structuralist Marxism" was developed in the 1960s by Louis Althusser and his followers in France and quickly spread throughout the world. It maintains the primacy of wholes over parts, privileging social structures and institutions over individual and group struggle and revolt, while attacking previous versions of "humanism." For overviews and appraisals, see Clark, et al. (1980), Benton, (1984), Resch (1992), and Kaplan and Sprinker (1993). Postmodern theory would attack Structuralist Marxism's totalizing gestures, its concepts of sharply delineated economic, political, and social structures, and its lack of adequate perspectives on subjectivity, history, and political resistance.

5. See Derrida (1994) for a discussion of how the Marxian critique of ideology marked the new French theories of the 1960s and how the critical spirit of Marxism provided an ambience that leading thinkers appropriated, rejected, and responded to in complex ways.

6. The concept of "identity politics" refers to a politics in which individuals construct their cultural and political identities through engaging in struggles that advance the interests of the groups in which they identify. Identity politics emerges from the splintering of the social movements of the 1960s and 1970s, in which, for example, some black groups pursued a solely black nationalist and separatist program, and some women pursued a strictly feminist agenda. Many other groups also pursued their own agendas, eschewing the movement and alliance politics that had emerged during the mid-1960s to bring people together to oppose the Vietnam War and to transform United States society. By the 1980s and 1990s, the now often warring and highly fragmented political groups were pursuing a politics of identity marked by cultural politics, single-issue group focus on the concerns of one's chosen social group, and rejection of alliance politics. Such versions of identity politics advocate the interests of specific groups over alliances and coalitions, following upon earlier forms of cultural nationalism and separatism. See the discussion of the politics of identity and difference in Best and Kellner (1991) and on identity politics, see Rutherford (1990), Keith and Pile (1993), Gitlin (1995), and Best and Kellner (forthcoming).

7. For examples of conservative attacks against postmodern discourses, see

Bloom (1987), Kimball (1990), D'Souza (1991), and Bernstein (1994). These conservative attacks suggest that it is a mistake to dismiss postmodern theory as itself a reactionary discourse as do Habermas (1987), Callinicos (1989), Norris (1990), and many on the left. In fact, postmodern theory is a highly complex synthesis of a wide variety of ideological impulses. As we will argue, we also believe it is a mistake to reduce the postmodern to one defining position, as such one-sided characterizations inevitability distort this complex and protean phenomenon.

8. "PMC" refers to an allegedly new "professional–managerial class" that appeared in the 1950s and 1960s, forming a major new class stratum between the industrial working classes and the capitalist class. We are skeptical of Pfeil's claim that it is the baby-boomer PMC class that is the audience and promoters of postmodern culture. Rather, the audiences for the postmodern texts that Pfeil cites (i.e., the music of the Talking Heads, *Repo Man*, Laurie Anderson, Philip Glass's opera *Einstein on the Beach*) are some intellectual baby-boomers and the younger Generation Xers who like the irony and nihilism of the texts Pfeil takes as symptomatic of postmodern culture. Thus it is segments of a university-based baby-boomer generation *and* a younger Generation X that we find to be the basis for the postmodern turn, and not just a "professional–managerial class," which we imagine to be more pedestrian in its cultural tastes; see our further discussion in Best and Kellner (forthcoming).

9. Harvey (1989) claims that the postmodern turn emerges as a result of post-Fordist developments in the economy and subsequent shifts in "space–time" experiences involving a sense of rapid change and an overcoming of spatial distance. But Harvey's analysis is too economistic and ignores a wide range of political, cultural, and theoretical influences on postmodern discourse, influences that we examine in this book. Jameson (1991) describes postmodernism as "the cultural logic of late capitalism," arguing that a new postmodern culture of depthless, fragmented, and disconnected images, artifacts, and selves characterizes the dominant mode of contemporary culture and society, but he does not adequately analyze the complex mediations among new forms of economy, society, and culture, nor does he show how the economy produces new postmodern forms of culture and society. Lash and Urry (1987, 1994) argue that a new "disorganized capitalism" has produced a postmodern culture, but their analysis of the postmodern is undertheorized, and against their claim that global capitalism is undergoing disorganization, we believe that it is also carrying out a reorganization as well.

10. The translation of our book *Postmodern Theory* into Chinese, Korean, and Japanese attests to the global interest in the discourses of the postmodern. Yet globalization alone is not a mark of the postmodern, as a passage in the prototypically modern text of Marx and Engels, "The Communist Manifesto," should remind us: "The needs of a constantly expanding market for its products chases the bourgeoisie over the whole surface of the globe. It must nestle everywhere, settle everywhere, establish connections everywhere" (1978: 476).

11. We thus partially disagree with Steven Connor, who writes:

> Instead of asking, what is postmodernism", we should ask, where, how and why does the discourse of postmodernism flourish?, what is at stake in its debates?, who do they address and how? This series of questions shifts attention from the meaning or content of

> the debate to its form and function, so that, to borrow Stanley Fish's formula, we ask, not, what does postmodernism mean", but, what does it *do*? (1989: 10)

We see *both* of these concerns as important and hence try both to elucidate the defining features of postmodern discourse and to note its uses and effects in various fields.

12. For example, Steven Seidman (1994) gives to his book *Contested Knowledge* the subtitle *Social Theory in the Postmodern Era*, but without any analysis of what constitutes the contemporary era as "postmodern." The only place that he explicates his notion of postmodernity is his epilogue, where he defines a "postmodern culture" in terms of "acceptance of the irreducible pluralistic character of social experiences, identities, and standards of identical selves" (324). But liberal pluralism is a feature of modern culture, and there is no evidence that it has triumphed among the population at large to constitute a new social system. Seidman then suggests that "postmodernity underscores a process of dedifferentiation or the collapse of boundaries" of modern binary oppositions and hierarchies of knowledge (324), but there is no evidence that this is indeed a defining feature of a new social order. Actually, Seidman is contrasting what he sees as modern and postmodern mind-sets, or paradigms of theory, and is not really analyzing a "postmodern era"; he is thus merely assuming that we are in a new social space without mapping its postmodernity.

13. Lyotard, for example, provides some texts in which he interprets the postmodern merely as a modality of the modern, as in the appendix to *The Postmodern Condition*, where he writes that the postmodern "is undoubtedly a part of the modern" (1984: 79) and then contrasts the modern and the postmodern as various moves within the modern. On other occasions, Lyotard deploys more extreme rhetoric, declaiming the novelty of the postmodern and the need for a break with the modern. Categories of the postmodern are ideal types that more or less fit some thinkers, texts, and positions, although many theorists and some texts share qualities of both the moderate and extreme versions. Yet it is not possible to always neatly apply these categories to actual empirical instances. Consequently, postmodern categories are really hermeneutical devices that are used to illuminate certain features of contemporary society and culture.

14. Jameson (1991) postulates an extreme break, or *coupure*, in culture and history but generally uses modern theories to analyze the situation, although he also puts in question many modern positions. Some theorists, like Rorty, are hard to pin down because they combine modern and postmodern insights and concepts. Against the foundationalist tradition of Western theory, Rorty argues for the historical embeddedness and contingency of all thought and values. Proclaiming the end of philosophy as a culturally authoritative discipline able to ground or adjudicate competing value claims, he adopts a postmodern aestheticism that embraces art and literature over theory and reason. Yet Rorty also clings to the modern values of liberalism, humanism, reformist politics, and the idea of progress. Thus, we find Rorty, the self-described "postmodern bourgeois liberal," straddling both worlds (see Rorty, 1991: 197ff.).

15. This is the sense that Calinescu (1987) ascribes to postmodernism when he identifies it as one of five faces of modernity rather than as a complete rupture with the modern. Some theorists, however, postulate a more radical break between modernism and postmodernism in the arts, such as those theorists of postmodern archi-

tecture who define it as a successful rebellion against the modern and reject the project of modern architecture altogether (see our discussion in Chapter 4 of the different theories of postmodernism in the arts).

16. See Antonio and Kellner (1994) for indications of how many of the critiques of modern theory associated with postmodern theory were anticipated and developed by modern theorists themselves to create models within modern theory that did not suffer from its alleged faults and deficiencies. Thus, the postmodern critique of modern theory is not as novel as some of its champions proclaim, and many postmodern criticisms were anticipated within modern theory itself, which since Descartes and through Kant, Hegel, Marx, Nietzsche, and others was constituted as a perpetually critical discourse that would put everything in question, including its own theoretical premises and positions, and that would constantly generate new positions.

17. For documentation of Baudrillard's positive uses of the term "postmodern," when he identifies with the discourse, see Kellner (1989b, 1994). Gane (1991) argues that Baudrillard should not be read as a postmodernist, and then supplies an interview in which Baudrillard distances himself from the discourse (Gane, 1993: 21)—but in subsequent texts Baudrillard continues to use the discourse of the postmodern (Baudrillard, 1994), as Gane himself (1995) is forced to admit. In fact, Baudrillard is *the* theorist of the postmodern rupture, even when he does not specifically use the discourse of the postmodern. Yet in some works, he takes pleasure in generating uses of the terms and distinctions between the modern and the postmodern, but distances himself from the discourse when interviewers and critics confront him with certain postmodern positions with which he doesn't identify.

18. Lyotard, for example, writes that "'postmodern' is probably a very bad term because it conveys the idea of a historical 'periodization.' 'Periodizing,' however is still a 'classic' or 'modern' ideal. 'Postmodern' simply indicates a mood, or better a state of mind" (1986–1987: 209). Yet, in other texts (*The Postmodern Condition*, for example), he makes much stronger claims.

19. Following Foster (1983), Ebert (1996) distinguishes between a ludic postmodernism and a postmodernism of resistance, valorizing the latter. While this distinction is useful strategically, as is Ebert's critique of ludic postmodernism, some ludic postmodernism has progressive moments, and the postmodernism of resistance could be reactionary and highly problematical (e.g., fascism is an obviously reactionary resistance to modernity); thus, it is problematic to distinguish between "good" and "bad" postmodernism in general categories. Instead, one must make such discriminations in concrete contexts, as we shall do in our studies.

20. Though there are pronounced anti-Enlightenment and anti-modern tendencies in some versions of postmodern theory, as well as strong forces of irrationalism and radical individualism, one cannot, with Habermas (1987) and Callinicos (1990), tar all postmodernists with the same brush as modalities of irrationalist conservatism since there are conflicting tendencies among them and within the works of individual theorists themselves, mixing conservative and radical elements.

21. Predictions that the postmodern turn is over appeared periodically throughout the 1990s (see Frow, 1991; Rosenthal, 1992; Alexander, 1995; Waters, 1995). Yet conferences, books, and articles continue to engage the phenomenon. Computer database searches for uses of the term "postmodern" that we have done

from 1994 through May 1997 indicate that hundreds of articles appear each month with the term and attest that the use of the concept of the postmodern is steadily growing and is probably here to stay. In this situation, it is better to critically engage it and intervene in the debates over the postmodern than to ignore it or to merely denounce it in a one-sided and polemical fashion Moreover, as Graff suggests (1985: i–ii): "For to the intellectual historian, the law of mind over matter must be honored: once a certain number of people *believe* that a concept like the Post-Modern marks a real change in the cultural climate, that change *becomes* a reality to be reckoned with, even if the reality is not exactly what most users of the term think it is."

CHAPTER TWO

Paths to the Postmodern

Kierkegaard, Marx, and Nietzsche

> [Money is] the visible divinity—the transformation of all human and natural properties into their contraries, the universal confounding and overturning of things: it makes brothers of impossibilities. It is the common whore, the common pimp of people and nations.
>
> —KARL MARX

> What I relate is the history of the next two centuries. I describe what is coming, what can no longer come differently: *the advent of nihilism.*
>
> —FRIEDRICH NIETZSCHE

> No one knows who will live in this cage in the future, or whether at the end of this tremendous development entirely new prophets will arise, or there will be a great rebirth of old ideas and ideals, or, if neither, mechanized petrification, embellished with a sort of convulsive self-importance.
>
> —MAX WEBER

In the realm of philosophy and social theory, there are many different paths to the turn from the modern to the postmodern, representing a complex genealogy of diverse and often divergent trails, as the postmodern turn winds and twists through different disciplines and cultural terrains. One pathway moves through an irrationalist tradition from romanticism to existentialism to French postmodern theory, via the figures of Nietzsche, Heidegger, and Bataille. This is the route to the postmodern charted by Habermas in *The Philosophical Discourse of Modernity* (1987), a road that ultimately leads for him to the dead end of irrationalism and the catastrophe of fascism.

More positive narratives of the postmodern turn in theory include Richard Kearney's (1988) journey through the progression of premodern, modern, and postmodern modes of thought to the triumph of a new postmodern imagination and vision. Ihab Hassan (1987) describes the outlines of a postmodern culture of "unmaking" that emerges out of modernism, pragmatism, and changes in modern science that, at its best, will help advance William James's vision of an "unfinished pluralistic universe." John McGowan (1991) in turn tells the story of the emergence of poststructuralist, neoMarxist, and neopragmatist postmodern theories arising out of the tradition of Kant, Hegel, Marx, and Nietzsche, building on but overcoming the limitations of their predecessors. Many accounts privilege Nietzsche and Heidegger as key progenitors of the postmodern turn who generated new modes of thought, new forms of writing, and new values (Vattimo, 1988; Kolb, 1990).

Our story differs from these accounts, as we argue that key predecessors of postmodern theory analyze the fundamental trends of modernity whose intensification helps produce a new postmodern society and culture, as well as generate critiques of modern thought and society that anticipate the postmodern turn. We trace in this and the following chapter a path from Kierkegaard to Marx and Nietzsche through Debord and the early Baudrillard to the proliferation of postmodern theory in the 1970s and 1980s. We are aware that there are other possible narratives of the trajectory to postmodern theory, but we believe that our account avoids both the polemical biases of one-sided attacks on postmodern theory and a triumphalist narrative that celebrates an alleged postmodern victory over modernity and modern theory.

We are also concerned to criticize discourses that either exaggerate the postmodern turn or deny that significant changes have taken place. The former decontextualize the transformations that the discourse of the postmodern alludes to and occlude the continuities with the core dynamics of modern culture, society, and theory. The latter deny that there are seriously novel phenomena that warrant the label "postmodern." In contrast to both approaches, we will argue that the defining features of what has been identified as postmodernity in recent decades began emerging in the 19th century and that therefore the postmodern turn in society and culture represents a radicalization and an intensification of the modern that generates genuinely novel social and cultural phenomena to which the term "postmodern" has been applied. In particular, we will argue that the phantom public identified by Kierkegaard as a product of the mass media, the processes of commodification and abstraction theorized by Marx, the trends of massification analyzed by Kierkegaard, Nietzsche, Heidegger, and others, and the emergence of a new, highly rationalized society and technological world described by Marx, Nietzsche, Heidegger, Weber, and the Frankfurt School were all intensified and radicalized in the 20th century, helping to produce postmodern social and cultural forms.

We are also concerned to demonstrate how critiques of the basic assumptions of theory within the modern tradition generated a postmodern turn, creating new modes of discourse, writing, and critique. Modern theory was secular and humanistic, focusing on the abilities of human beings to discover the truth and to reconstruct their worlds. It was assumed that there were order and laws in the natural and social worlds that reason could discern and that reason was able to represent and control the world. Reason was deemed the distinctive human faculty, the cognitive power that would enable humans to dominate nature and create moral and just societies. Faith in reason resurfaced in the Renaissance and the scientific revolutions of the 16th and 17th centuries, was enthroned in the 18th-century Enlightenment, and was triumphant, though challenged, by the 19th century. In the following discussions, we will see how key 19th-century thinkers both questioned the pretensions of reason and modern theory and critically analyzed the ideologies of modern society, thus opening the way for a postmodern rejection of the modern and the search for new paradigms.

Accordingly, in this and the next chapter, we will show both how certain trends in modernity generated the transition to an alleged postmodernity and how 19th-century critiques of modern theory helped engender postmodern theory. On this account, postmodernity emerges out of modernity, out of a magnification of the dynamics of commodification, massification, and technification, while postmodern theory originates from modern theory's critiques of some of its own presuppositions and values.

KIERKEGAARD, MASS SOCIETY, AND THE ROAD TO THE POSTMODERN

> The age of the encyclopedists, the men who indefatigably wrote folios, is over; now it is the turn of the lightly equipped encyclopedists who dispose of the whole of existence and all the sciences *en passant*.
>
> —Søren Kierkegaard

The Danish religious philosopher Søren Kierkegaard is an important predecessor of the postmodern turn not only in his much-discussed attacks on Enlightenment reason but also in his critique of societal massification and the role of the press in the creation of a "phantom public." His vision of massification and of individuals lost in the world of the media and public opinion provides an uncanny anticipation of the later postmodern theories of Baudrillard and others. We argue that Kierkegaard can be read as the first existential critic of modernity who attacked the modern age for its crippling effects on the individual. He also anticipated theories of postmodernity that specify the ways that the intensification of modernity has changed the very nature of the individual, society, politics, and experience. Consequently, rather than perceiving Kierkegaard solely as a religious thinker who had little to say about modern culture and society, we suggest that he had some incisive insights into the fate of the individual in the modern age and produced a provocative critique of modernity.

Kierkegaard and the Critique of the Press

Critiques of mass culture and the press began emerging during the late 18th century. Goethe argued that the press constitutes a squandering of time wherein the reader "wastes the days and lives from hand to mouth, without creating anything" (quoted in Lowenthal, 1961: 20). Anticipating Kierkegaard, he criticized the ways that modern entertainment and the press promoted passivity and conformity, noting in a ditty how the press is eager to provide its readers with almost anything except dissenting ideas:

> Come let us print it all
> And be busy everywhere;
> But no one should stir
> Who does not think like we. (quoted in Lowenthal, 1961: 20)

Others had more optimistic appraisals of the impact of mass media, and particularly the press. Karl Marx, for instance, had an especially high opinion of the press in the promotion of democracy and civil liberties, writing in 1842:

> The free press is the ubiquitous vigilant eye of a people's soul, the embodiment of a people's faith in itself, the eloquent link that connects the individual with the state and the world, the embodied culture that transforms material struggles into intellectual struggles and idealizes their crude material form. It is a people's frank confession to itself, and the redeeming power of confession is well known. It is the spiritual mirror in which a people can see itself, and self-examination is the first condition of wisdom. It is the spirit of the state, which can be delivered into every cottage, cheaper than coal gas. It is all-sided, ubiquitous, omniscient. It is the ideal world which always wells up out of the real world and flows back into it with every greater spiritual riches and renews its soul. (in Marx and Engels, 1975: 165)

By the 1840s, the press was thus a contested terrain with fervent defenders and critics. Some saw it as an instrument of progress and enlightenment, whereas others saw it as a vehicle of distraction and banality. Different political groupings were developing their own distinct presses and attempting to shape public opinion in different ways. Though Goethe and others made some critical remarks concerning the press of the day, one of the first systematic and sustained philosophical attacks on the press is evident in Kierkegaard's polemic with the satiric Danish review *The Corsair*, which published articles making fun of him in late 1845, inciting him into a literary duel with the journal. Kierkegaard's efforts constitute one of the first critiques of the press as an instrument of massification and manipulation, anticipating later critical theories of mass media and society.

Kierkegaard published two rather short criticisms of the Danish satirical journal *The Corsair* and of a journalist, P. L. Möller, who had offended him

(Kierkegaard, 1982: 38–50). Möller and *The Corsair* countered with frequent attacks on the Danish writer over the course of many months, which prompted from Kierkegaard a series of unpublished reflections in his journals.[1] Standard biographies of Kierkegaard and interpretations of his work perceive the affair as decisive in leading him to give up his idea of becoming a country pastor and in driving him to continue his prodigious literary output, which proceeded at its typical ferocious pace until his death in 1855 (see Lowrie, 1962: 347ff.; Hong and Hong, 1982: xiiff.). In our reading, the affair led Kierkegaard to sharpen his critique of "the present age," which henceforth would become a central feature of his existential philosophy and of the German and French existentialism that followed him. We thus see the issues Kierkegaard raised vis-à-vis *The Corsair* to have their ultimate significance in a critique of the press, mass society, and modernity itself—themes that would be of crucial importance as well for postmodern theory.

Kierkegaard's assault on the press began in December 1845 with specific criticisms of *The Corsair*'s personalities, practices, and effects, and it was extended in his own journal reflections and later writings into a polemic against the press and "the present age." Kierkegaard invited further attacks with his intervention, and *The Corsair* responded with a series of sharp polemics and satires on his personality and caricatures of his appearance (Kierkegaard 1982: 105–137), especially his thin legs and uneven trouser lengths (one leg was apparently shorter than the other). The result was a general ridicule of Kierkegaard in Copenhagen that made it difficult for him to take his customary walks without attracting attention and mockery, which, as his journal entries testify, hurt him deeply.

Characteristically, Kierkegaard began working out his reflections on the affair in his journal. In an 1846 entry, "The Dialectic of Contemptibleness," he attacked the scurrilous journalistic practice of slander and gossip and suggested that it was really not shameful to be attacked by something, or someone, contemptible. In a lengthy 1846 entry titled "Some Instructive Comments on *The Corsair*'s Drastic Errors," Kierkegaard reflects in a more social and political vein on journalism in which "personal attacks are made on private individuals," writing:

> In everyday life, when suddenly there is too much fresh fish on the market, the police ban it; but when tainted wit is offered for sale in such a mass that even the rare good witticism is spoiled by the encompassing mass, when it is offered for sale in such quantity that it is like everyday bread, when nothing happens. . . . But when a private citizen is assaulted in print, at times regularly every week, by a malinger setting the dog, so to speak, of irresponsibility, coarse brutality, and passion on him, nothing happens. No, nothing happens; the only thing that happens is that the business flourishes, that few civil employees, even in the higher positions,

> and no artists and scholars in Denmark are paid on the same scale. (1982: 166–167)

In this passage, Kierkegaard's anger is still focused on *The Corsair* itself, and the implied solution to the problem seems to be government regulation that would take action against a publication as vile as Kierkegaard perceived *The Corsair* to be. The interesting conclusion that Kierkegaard draws, however, is that when "passion and commercial interest determine the issue," when "the rattle of money in the cashbox" is at stake, the propensity for corruption increases (172). Kierkegaard reveals an insight here into the economic roots of the features of the press that he finds scandalous, and another journal entry, "A Personal Statement in Costume," considers the responsibility of the press and writer, sharpening his attack on *The Corsair* while extending his critique to the press in general and to the present age. Immoral slander, such as he finds in *The Corsair*, "is of no benefit whatsoever" and "does great harm because it seduces the unstable, the irresponsible, the sensate, those who are lost in earthly passions, seduces them by means of ambiguity, lack of character and the concealment of brash contempt under the pursuit of the comic" (179–180).

Interestingly, Kierkegaard's favorite metaphor for the press is that of a vicious attack dog (1978: 95f., 136–137). He does not theorize the press as a guardian of the public's interests, as it was initially conceived to be,[2] but, rather, as a predator that goes after individuals in a contemptible way. Moreover, in terms of Kierkegaard's theory of the stages of existence, the press pertains to the *aesthetic* stage and lacks any ethical scruples whatsoever. In his theory of "stage of life's way," which we discuss below, Kierkegaard contrasts the aesthetic stage, dedicated to sensuous pleasures including art and eroticism, with the ethical stage of duty and commitment and with the religious stage, which he conceived to be the highest level, concerned with the question of eternal life and the salvation of the soul (see Kierkegaard, 1987, 1988). Kierkegaard attacks the press both as an aesthetic phenomenon that manipulates passions and as an entity devoid of ethical scruples and responsibilities, thus posing the question of the effects and ethics of the press in the present age.

The press, he argues, is fundamentally irresponsible because its writers are anonymous and do not assume responsibility for what they print. In addition to undertaking an ethical critique of the press, Kierkegaard was one of the first to see that the press is a mass medium that addresses its audience as members of a crowd and that itself helps massify society. The press plays a fundamental role, Kierkegaard suggests, in producing a public, a crowd devoid of individuality and independent judgment, their thought determined by the authority of printed words and editorial fiat. The average man in the street, Kierkegaard suggests, "believes that what appears in the newspapers is public opinion, the voice of the people and of truth" (1982: 186). For this mentality, "anything that appears in print" is regarded "as infallible" by the average reader who asks: "Is it possible that anything can be a lie which is printed in countless copies, is read all over the country, and, from what I hear, no one yet has ventured to refute!" (186). This complaint is still frequently registered today by individuals who have been wrongly slandered by newspapers yet who only have "retractions" and admissions of error printed later, after the damage has been done and often in fine print near the back of the paper.

Kierkegaard points to the ways that the press simulates authority and objectivity and can thus make a lie appear as truth or an opinion as fact. Repeating and circulating stories in the press provides an aura of fact, and opinions expressed appear important and substantial. Kierkegaard shows how the press in this way manufactures public opinion and promotes social untruth under the guise of truth and objectivity. Indeed, the history of journalism has many examples of lies that destroyed individual reputations, formed public opinion, and promoted events like wars.[3]

Kierkegaard also notes how the press delights in scandal and in inflating

trivialities into significant events, thus obscuring and ignoring more important and significant issues—much as it continues to do today. Indeed, Kierkegaard saw this trivialization of personality and issues as his own personal fate while the gossip and public response to the *Corsair* affair unfolded:

> Petrarch believed that his Latin writings would make him immortal, and his erotic poetry did just that. Fate is even more ironic toward me. Despite all my diligence and efforts, I have not been able to fathom what the times demanded, and yet it was so very close to me; how incredible that it did not occur to me, that someone else had to say that it was my trousers. What rare luck, for my life acquires exceptional significance; to my contemporaries I acquire significance by means of my trousers, but for a later generation perhaps my writings will help a little. (1982: 187)

In Kierkegaard's view, the press "gives rise to the abstraction's phantom, 'the public,' which is the real leveler" (1978: 93). The public produced by the press is an abstraction, but it is one to which individuals conform, fearing to be different and wanting to fit in with the crowd. The press thus helps produce uniformity of thought, contributing to a general leveling process, to producing a public, a crowd, that is little more than a "rabble," devoid of character and individuality. The press appeals to base desires and curiosity, encourages idle chatter and gossip, empties events of real significance by focusing on trivialities, and in general deflects public attention from the important issues of the day. Consequently, the press transforms individuals into passive objects and "actually destroys all personality" (1982: 218).

The Public, Massification, and Reflection

Kierkegaard thus inverts the liberal theory of public opinion (which is supposed to protect the interests of the public against corrupt authority) by claiming that the press creates a phantom public devoid of character and individuality. Consequently, Kierkegaard, like later postmodern theorists, ascribes to the communications media a tremendous role in producing a mass society without distinction, individuality, or conviction. Devoid of individuality, the masses themselves are an abstraction, and the main effect of modern society is a leveling of the population into a mass. Leveling is twofold: It prevents inward depth in the individual, and it levels all individuals to the same plane of debased mediocrity. Leveling involves the ascendancy of the "public" over the (differentiated) community and the individual. The public is

> made up of unsubstantial individuals who are never united or never can be united in the simultaneity of any situation or organization and yet are claimed to be a whole. The public is a corps, outnumbering all the people

> together, but this corps can never be cleaned up for inspection; indeed, it cannot even have so much as a single representative, because it is itself an abstraction. (1978: 91)

Within the public, individuals are simultaneously massified and serialized, amalgamated and fragmented.

As both a leveled and a leveling force, the public is an abstraction that "annihilates all the relative concretions of individuality" (Kierkegaard, 1978: 92). But, because it is an abstraction, the public really doesn't exist. It is a "monstrous nonentity," a modern fiction that lacks reality because it lacks concretion. Articulating the problem of "the silent majority," Kierkegaard says:

> The public is a concept that simply could not have appeared in antiquity, because the people were obliged to come forward *en masse in corpore* in the situation of action, were obliged to bear the responsibility for what was done by individuals in their midst, while in turn the individual was obliged to be present as the one specifically involved and had to submit to the summary court for approval or disapproval. (93)

Crucially, anticipating later postmodern arguments (see Chapter 3), Kierkegaard argues that the derealization and desocialization of the community is created through mass media and, specifically, the press: "Only when there is no strong communal life to give substance to the concretion will the press create this abstraction" (1978: 91). The press, therefore, not only helps to create the masses, it simulates the fictive "public" as a real referent to mask the fact that the individuated community it addresses doesn't exist: "If the age is reflective, devoid of passion, obliterating everything that is concrete, the public becomes the entity that is supposed to include everything" (91). Kierkegaard thus saw and understood, in its earliest form, the "mighty power" of the mass media: its biases disguised as "objectivity" and its role in *creating* the public it claims to "represent" or address. For the press absolves the individual of the responsibility of thinking and gaining knowledge. The individual who reads the press thinks he or she is in the know, is informed, but really is just absorbing ideas and opinions that are manufactured for public consumption and that in turn create public opinion and a massified public.

Verging on a materialist perspective, Kierkegaard also understood the press as a monopoly power under the control of capital: "A newspaper's first concern has to be circulation; from then on the rule for what it publishes can be: the wittiness and entertainment of printing something without any relation to [genuine] communication" (1982: 220). Kierkegaard interpreted this economic determination in terms of the form and content of "information," now a salable commodity, and described its leveling effects on the individual, who consumes it for entertainment rather than information value. In passages that sound much like the work of the young Marx, Kierkegaard sees the modern age as a greed-ridden and narcissistic era that sacrifices "the resource of a consistent and well-grounded ethical view" (1978: 74) to a hedonistic selfishness that fetishizes money as an ultimate source of value in an abstract and "unreal" world: "So ultimately the object of desire is money. A young man today would scarcely envy another his capacities or his skill or the love of a beautiful girl or his fame, no, but he would envy him his money. Give me money, the young man will say, and I will be all right" (1978: 75). Money usurps the place of God in the secular world of modernity and is worshipped by nearly all—a position also dramatized by the young Marx:

> The stronger the power of my money, the stronger am I. The properties of my money are my, the possessor's, properties and essential powers. Therefore what I am and what I can do is by no means determined by my individuality. I am ugly, but I can buy the most beautiful women. Which means to say that I am not ugly, for the effect of ugliness, its repelling power, is destroyed by money. (in Marx and Engels, 1975: 377)[4]

Marx and Kierkegaard thus describe an inversion process. For Marx, the possession of money transforms ugliness into beauty, stupidity into wisdom, infirmity into power. For Kierkegaard, the essential and inessential are inverted, with the press elevating the inessential into the essential and modernity putting secular values in the place of religious ones, thus threatening the spiritual destiny of individuals. While Kierkegaard certainly had no clear conception of ideology in the Marxist sense of a class-based control of consciousness, he anticipated what the neo-Marxist Antonio Gramsci later termed "hegemony," that is, social control gained more through manipulated consent than force, a control more powerful because it is hidden and mystified, and penetrates the most intimate aspects of everyday life. This monopoly of speech allows the press—as it did *The Corsair*—to influence, standardize, and create "public opinion" and to slander and misrepresent with impunity. Regarding this latter point, Kierkegaard said: "The press enables one single rabble-rouser [*Corsair* writer P. L. Möller] to do incalculable harm" (1978: 57). Something is drastically wrong when such a "cowardly wretch can sit in hiding and write and print for the thousands!" (1982: 218). In a statement that shows insight not only into the power of the press over private individuals but into the relative autonomy of the press from the state, Kierkegaard writes: "*The Corsair* wheels and deals in people, holds their honor and serene private lives in its hands as if they were trifles. For all the rest of us are tied hand and foot, and the authorities have to put up with being made to look like fools lest they abuse their power, because in that case the voice of accusation would become strident" (1982: 169).

Kierkegaard attributes the power of the media in part to its anonymity as an abstract mediation between people, as compared to the situation in antiquity when the vilifier "had to appear in person in the square, and if he lost, he had to take the consequences" (1982: 178). A lurid, satirical paper such as *The Corsair* "caters to the passion and is out for profit—and benefits splendidly from the license of contemptibleness to be completely unconstrained" (167). The press appeals to our lowest instincts and so encourages us to remain within the aesthetic rather than the religious dimension, to revel in base desires rather than authentic passions and genuine ideas. A paper like *The Corsair* "does great harm because it seduces the unstable, the irresponsible, the sensate, those who are lost in earthly passions" (180). The press, then, panders to the lowest desires, cheapens the currency of discourse, and helps to foster the spread of inauthentic communication ("loquaciousness" and gossip). It is a tool of diversion that seduces people into the lure of superficial spectacle and triviality. Well before the age of film and television, Kierkegaard also observes that the press has led to a decline in book reading and so has contributed to the deterioration of the mind (221): "With the press as degenerate as it is, human beings will surely be transformed into clods" (220).

Crucially, the press for Kierkegaard creates an age of "publicity" that devalues action and manufactures an illusory pseudo-world, which Baudrillard would later call a "hyperreality" (see Chapter 3): "The public likes to transform all actuality into a theater" (Kierkegaard, 1978: 136), and the press assists this proclivity. It helps to further the abstraction process initiated by reflection and so contributes to a derealization of the world, encouraging individuals to live in unreal fantasy worlds and causing them to lose their authentic possibilities for thought and action in a mass society. Thus, for Kierkegaard modern society is an abstraction, a nonentity, a fiction, a mass of individuals without connection to one another or to the religious Idea. In language that could come straight from Baudrillard, who characterizes the public by terms such as "black hole" or "opaque nebula" (Baudrillard, 1983b: 3), Kierkegaard describes the social as "an evasion, a dissipation, an illusion" (1978: 6). He goes on to say: "The public may take a year and a day to assemble and when it is assembled it still does not exist" (91). The public is "an abstract void and vacuum that is all and nothing" (93), and so "the category 'public' is reflection's mirage ... the fairy tale of an age of prudence" (93).

Kierkegaard also attacks reflection and an abstract rationalism that he believed his age was nurturing, condemning reflection as a "danger" that ensnares people in logical traps. He constantly compares reflection to a prison, a captivity, a bondage that binds the individual and the whole age and "can only be broken by [passionate] religious inwardness" (1978: 81). Reflection seduces individuals into thinking its possibilities are "much more magnificent than a paltry decision" (82). It leads them to act "on principle," to dwell on the deliberation of the context of their actions and the calculation of their worth or outcome. Kierkegaard argues that this drives away feeling, inspiration, and spontaneity, all of which are crucial for true inner being and a vital relation to God. For Kierkegaard, as Nietzsche would later agree, genuine inner being is characterized by the tautness and tension of the soul, which for Kierkegaard characterizes passionate existence. But the "coiled springs of life relationships . . . lose their resilience" in reflection (78) and "everything becomes meaningless externality, devoid of [internal] character" (1978: 62).

Kierkegaard thus contributes to the development of an irrationalist tradition that has echoes in some later postmodern thought. Kierkegaard might well agree with his contemporary Fyodor Dostoyevsky, who wrote: "An intelligent [reflective] man cannot seriously become anything . . . excessive consciousness is a disease" (Dostoyevsky, 1974: 3, 5). In an age overtaken by rules and regulations, genuine action—which Kierkegaard assumes to be subjective and spontaneous—is frustrated at every turn. Kierkegaard complains we are too "sober and serious" (1978: 71) even at banquets, and he bemoans the fact that even suicides are premeditated (68)! "That a person stands or falls on his actions is becoming obsolete; instead, everybody sits around and does a brilliant job of bungling through with the aid of some reflection and also by declaring that they all know very well what has to be done" (73). Thus, it is passion, not reflection, that guarantees "a decent modesty between man and man [and] prevents crude aggressiveness" (62). "Take away the passion and the propriety also disappears" (64).

The ambiguity in the word "passion" may cause some confusion here. To say that the age and its individuals are "passionless" is not to say that there are no emotions whatsoever but, rather, that there is no true spiritual inwardness and depth, no intensively motivated action and commitment. It suggests that passion exists only in a simulated pseudo-form, "the rebirth of passion" through "talkativeness" (Kierkegaard, 1978: 64). "Chattering" for Kierkegaard gets in the way of "essential speaking" and merely "reflects" inconsequential events (89–99). Hence, in the present age, emotions—which in fact are all too pronounced—have been transformed into negative forces.[5] Anticipating Nietzsche's genealogy of the "slave revolt" in morality, Kierkegaard claims that the "enthusiasm" of the prior Age of Revolution, a "positively unifying principle," has become a vicious "envy," a "negatively unifying principle" (81), a leveling force in its own right insofar as those lacking in talent and resources want to tear down those who have them. Both Kierkegaard and Nietzsche, then, reduce egalitarian politics to herd envy of the strong or noble.

Yet Kierkegaard systematically champions passion over reason. For Kierkegaard, there are three stages of existence: the aesthetic, the ethical, and the religious. In each of these stages, passion and nonrational components are deemed superior to rationality. In the aesthetic stage, it is the sensual pleasures of culinary taste, art, and eroticism that provide the earthly delights of everyday life, and not the machinations of reason. In the ethical stage Kierkegaard valorizes (against Kant) the ethical passion of resolve, choice, and commitment over universal principles and the faculty of ethical judgment. And in the religious stage, the highest stage of existence for Kierkegaard, it is the infinite passion of the choice of Christian belief, the absurd faith in the Christian mysteries and paradoxes, and the subjective yearning for salvation and redemption that Kierkegaard defends as the heart and soul of the religious life.

Moreover, truth is subjectivity for Kierkegaard. Rejecting objectivist theories of truth, Kierkegaard acclaims the subjective passion and commitment whereby an ethical, or Christian, subject lives in the truth, makes the truth (of ethics, religion) the form and substance of everyday life. Such existential truths are of far more value for Kierkegaard than the truths of philosophy and science. In particular, Kierkegaard mocks Hegel with his pretensions of absolute and objective truth collected into a totalizing system of knowledge. Likewise, Kierkegaard ridicules the pretensions of Enlightenment reason and modern science to provide infallible methods of securing objective knowledge. Such "truths," for Kierkegaard, are of little existential import in contrast to the pleasures and insights of art, the imperatives of ethical commitment, and the infinite and inexpressible value of religious redemption.

For Kierkegaard, the modern subject is a solipsistic monad, yearning for salvation and infinite happiness, plagued by anxiety and guilt, obsessed with God and religious transcendence. The social bonds, community, and forms of association that modern social theory would valorize as the distinctive achievements of modernity, with modes of social integration, interaction, and social norms, are volatilized into the ghostly aura of the phantom public, leaving the individual in fear and trembling alone before God and the passion of religious choice. Hence, Kierkegaard carried out a critique of reason, reflection, objectivist knowledge, totalizing thought, and modern society that has influenced the postmodern turn in theory. His critiques have been supplemented by powerful attacks on modern theory and society by Marx and the Marxist tradition, by Nietzsche and his progeny, and by contemporary writers that together would produce the postmodern turn in theory.

MARX AND THE DAWN OF ABSTRACTION

> A commodity appears, at first sight, a very trivial thing, and easily understood. Its analysis shows that it is, in reality, a very queer thing, abounding in metaphysical subtleties and theological niceties.
>
> —KARL MARX

For Marx, capitalism represents a rupture in history, the overthrow of the medieval era by a radically secularized modern world organized around the production, distribution, and consumption of commodities. In capitalism, we find the dissolution of organic social and natural relations in the development of a fragmenting division of labor and an unprecedented organization of society where commodity production structures social life and maximization of profit is society's sole purpose. Before capitalist society, commodity production existed, but it was always marginal in relation to social and religious values. Capitalism is thus the triumph of the economy over its human producers, the first social organization in which market relations subsume and dominate the totality of social life.

Marx sketches three different phases of exchange, distinguishing among the Middle Ages, when only the excess of production, what was superfluous beyond simple need satisfaction, was exchanged, an early capitalist period in which all industrial products were subject to exchange, and a more advanced stage of commodification

> when everything that men had considered as inalienable became an object of exchange, of traffic and could be alienated [i.e., sold]. This is the time when . . . virtue, love, conviction, knowledge, conscience, etc.—when everything finally passed into commerce. It is the time of general corruption, of universal venality, or to speak in terms of political economy, the time when everything, moral or physical, having become a marketable value, is brought to the market to be assessed at its truest value. (in Marx and Engels, 1976: 113–114)

In this remarkably prescient passage, Marx anticipates the creation of the consumer society and our present moment when nothing escapes the juggernaut of commodification. This triumph of the market can be interpreted in terms of inversion and abstraction. Capitalist inversion takes place on several levels, beginning with the inversion of subject–object relations and the domination of the subject by the object. Ironically, the subject is posited by modern theory as the sovereign power that is to rule nature and the world of objects, but under capitalism objects come to rule human beings: "The increase in the quantity of objects is accompanied by an extension in the realm of the alien powers to which man is subjected" (Marx and Engels, 1978: 93). Under the capitalist mode of production, the forces of production come under private ownership, commodity production multiplies, a parcelizing division of labor spreads, and subjects are displaced from concrete life activity to a thralldom within a world of alien objects. Forced to sell his or her labor power to survive, "the worker sinks to the level of a commodity and becomes indeed the most wretched of commodities" (70).

Instead of securing and enhancing the existence of subjects, productive activity under capitalism weakens and degrades them. Marx supplied a vivid description of the reversal that occurs with the alienation of labor in capitalist society: "The more the worker produces, the less he has to consume; the more values he creates, the more valueless, the more unworthy he becomes; the better formed his product, the more deformed the worker; the more civilized his object, the more barbarous becomes the worker; the mightier labour becomes, the duller becomes the worker and the more he becomes nature's bondsman" (Marx and Engels, 1978: 73). Marx saw this alienation as a "loss of [human] reality" in which the worker "mortifies his body and ruins his mind" (74). The dehumanization of work reverses the normal relation between the human and animal functions of the worker: "Man . . . no longer feels himself to be freely active in any but his animal functions . . . and in his human functions he no longer feels himself to be anything but an animal" (74). The creative, imaginative, and transformative activity that defines the "genuinely human" atrophies and dies. Ultimately, the estrangement of the worker from the products and process of production leads to an estrangement from his or her own human potential and from other human beings. The effects of the alienation of labor carry over into all aspects of social and personal life.

This process could also be described as the inversion of exchange value over use value. Whereas the dominant goal of all premodern production was the simple satisfaction of needs, and money was a mediating element, the realization of surplus value (profit) is the goal of the capitalist system, and commodities are mere mediating figures in the accumulation of capital, no longer tied to human needs except in an accidental way. Before capitalism, exchange was driven by the need for the other's use value. Capitalism eliminates individual exchange and subordinates use value to exchange value, reversing the hierarchy of value and establishing exchange value as the dominant logic. The goal of commodity production is not the creation of use value, the satisfaction of needs, and the simple reproduction of traditional ways of life but, rather, exchange value, the creation of profit, and endless accumulation for its own sake.

This inversion is rooted in a process of social abstraction that dissolves the concrete and the particular. The logic of abstraction is based on exchange value and its law of equivalence which dissolves qualitative differences into quantitative values. Exchange value solves the problem of exchange: how to trade qualitatively unique goods in fair and equal quantitative ratios that estimate the amount of labor time necessary to produce them. This process, however, results in the quantification of qualitative relations and phenomena, their reduction to a single standard (money), and the commodification of what was "considered as inalienable." Intrinsic value—whereby something is valued for its own sake rather than for any advantages it may bring—is eclipsed by instrumental value, such that knowledge, friends, love, and so on are valued only as means to the single true end of financial gain.

An object becomes a commodity when, over and above its use value, it assumes an exchange value that allows its sale, putting it within the realm of the quantitative. The insertion of subjects and objects into an economic calculus transforms them into abstract entities; it strips away their unique characteristics and reduces them to a numerical expression, a quantitative sign. Thus depersonalized, human beings themselves become raw materials and mere commodities. As analyzed by Erich Fromm (1947), people adopt the characteristics of the commodity, creating themselves as "marketable personalities," in order to compete successfully in a market economy. With the spread of money, commodification, and quantifying logic, a general abstraction process thus envelops society: "Just as money reduces everything to its abstract form, so it reduces itself in the course of its own movement to something merely *quantitative*" (Marx and Engels, 1978: 93). This is not to equate abstraction with commodification alone, for science, mathematics, technology, and technological reason also are key factors (see Weber, 1958; Marcuse, 1964; Lukács, 1972; Adorno and Horkheimer, 1972), but the commodity form constitutes an important source of abstraction in its own right, one that permeates social relationships and everyday life.

The triumph of the commodity installs money as the dominant social power and value and produces an inverted world in which money allows one to simulate various human qualities: "That which exists for me through the medium of *money*, that which I can pay for, i.e., which money can buy, that *am I*, the possessor of money. The extent of the power of money is the extent of my power. Money's properties are my properties and essential powers—the properties and powers of its possessor" (Marx and Engels, 1978: 103). Under these conditions of abstraction, in which meaning and exchange are abstracted from everyday life (at the same time that reified exchange *is* everyday life), one is able to *buy* qualities—if only in a temporary and illusory way—and the qualitative becomes a function of the quantitative. "Thus, what I *am* and *am capable* of is by no means determined by my individuality. I am ugly, but I can buy for myself the most *beautiful* of women. Therefore, I am not *ugly*, for the effect of *ugliness*—its determinant power—is nullified by money" (103). Under the conditions of "commodity fetishism," human beings becomes things, and things (commodities and money) take on human powers: "All the things which you cannot do, your money can do" (96); and, conversely, what money cannot do seemingly cannot be done.

With the valorization of exchange value comes the abolition of society—social needs, values, and relations—as the aim and referent of production. The function of capitalism is to satisfy the avaricious drives of private individuals, not to secure the needs of the social body. The organic community, *Gemeinschaft*, is replaced by a fragmentary *Gesellschaft* ruled by exchange value, commodification, lust for profit, and abstract powers. Marx described this inverted world as "perverted, enchanted," and "topsy-turvy." Yet the object of Marx's criticism was not the commodity per se, as a simple object in trade necessary to sustain social existence, but, rather, its *fetishization*

in capitalist conditions of production and exchange, its magnification to the point that it subsumes and mystifies the underlying relations of production, transforming subjects into objects and objects into subjects. Here, *abstractions* take over, and with them comes the danger of obscuring the nature of social reality, increasing depersonalization, and so deepening domination.

As a result of the hegemony of the commodity form and money under capitalism, the abstraction inherent in exchange value becomes even greater. In W. F. Haug's words: "As soon as exchange value has become independent in the form of money, the whole exchange value standpoint is provided with the precondition for its independence too" (1986: 18). Money provides a means to facilitate and generalize exchange, serving as "a uniform language of valuation" (14). In the most extreme, but quite common, form of trading stocks, there is no earthly referent at all, especially in the case of "junk bonds," "futures," and "derivatives," in which one speculates on future prices and values and such that investors, corporations, universities, and entire regions can go bankrupt because of risky, speculative investments. Since the 1980s, speculation in finance capital has intensified, and in 1995 bad investments in this realm bankrupted both England's oldest bank and Orange County, California. In finance capitalism, money is made out of money; the creation of profit occurs through manipulation of abstract figures with no apparent connection to the commodity world, which is itself already abstracted from social relations and activity, and is governed by the frenzied anarchy of speculation and trade. The capitalist economy, therefore, is inherently crisis ridden and subject to potential collapse.

Once the circulation of capital has been abstracted from sensuous needs and qualities and from any social referent, social life becomes degraded, privatized, and fragmented among competing private interests. When the logic of capital extends beyond the factories to penetrate all cultural and interpersonal relations, it has a profound corruptive and distorting effect. The inversion that occurs in the economy, and that affects the whole of social life, is then directly transferred to the cultural and personal realms which become commercialized and saturated with commodity fantasies, eventually producing the commodity self, whose identity and happiness is realized in narcissistic consumption and the worship of celebrities.

Detached from a reality of quid pro quo in which human qualities depend on human powers, removed even from a specific commodity body, exchange value assumes hyperabstract powers in the form of money. As a disseminating, corrosive, abstract form, money determines the nature of reality itself and builds its empire on simulacra and illusions. "Being the external, common *medium* and *faculty* for turning an *image* [or imagination] into *reality* and *reality* into a mere *image* (a faculty not springing from man as man or from human society as society), *money* transforms the *real essential powers of man and nature* into what are merely abstract conceits" (Marx and Engels,

1978: 105). Thus, just as money transforms an unreality (the absence of a power or quality) into a reality, so it transforms a reality (the presence of a power or quality) into an unreality. Real human powers pale in comparison with the illusory powers one can buy; real needs go unheeded where people pursue the dominant pseudo-need, the need for money.

In the early Marx there is already a heightened sense of the consumer self described by Fromm, Marcuse, Debord, Baudrillard, and others, in addition to the desubstantialization process, thematized by postmodernists, in which individuals live in consumer and media fantasy worlds. Marx provides a vivid description of the beginnings of an abstract commodity phantasmagoria, a process we will term *the commodification of reality*, and he also anticipates Baudrillard, as we shall explore in the next chapter. The movement and generalization of money and the commodity form is simultaneously the fetishization, mystification, quantification, and, ultimately, dissipation of social reality. But however abstract this inversion process was, Marx understood that it emerged from concrete historical conditions, in the production and reproduction of social life in capitalism. And however unreal social reality was becoming, Marx saw that it was also becoming increasingly real, all too real, with the proletarianization and immiserization of the masses, with "the real subsumption of labor" that mortified the mind and body of the worker while also generating the reality of class struggle and the hope of socialism and revolution.

Marx astutely identified the first powerful forms of inversion and abstraction that began with an autonomous economy, a self-contained capitalist system of production organized around the imperatives of profit and accumulation. He saw not only incipient forms of imperialism and monopoly but also the first manifestations of an emerging consumer society and self. In what could be an accurate description of today's advertisers, for instance, Marx wrote that the industrial capitalist "puts himself at the service of the other's depraved fancies, plays the pimp between him and his need, excites in him morbid appetites, lies in wait for each of his weaknesses" (Marx and Engels, 1978: 94). But while Marx displayed far-reaching insights into the new tendencies and developments of the commodity form, he could not have developed any concrete theory of how commodification, inversion, and abstraction would mature in the 20th century—a topic that we shall take up in the next chapter.

Nonetheless, Marx is important for our narrative of the postmodern turn because he helped to inaugurate the tradition of modern social theory and a discourse of rupture that would be deployed by postmodern theorists, who claim that a divide between modern and postmodern societies had occurred that was as significant as the break theorized by Marx between modern and premodern societies. His account of commodification would be radicalized by later theorists, like Jameson and Harvey, who would argue that in contemporary (postmodern) capitalism commodification processes are generalized in a hypercommodified world where everything is privatized and subject to market logic. Marx's account of the subject–object inversion, whereby the object comes to dominate the subject, and of the decline of subjectivity and individuality would be taken up by Baudrillard and postmodern theory. Marx's account of the inversion of exchange value and use value, whereby exchange value becomes dominant over use value in the process of the fetishization of commodities, which articulates value in terms of a calculable exchange value, anticipates Baudrillard's analysis of how sign value comes to govern in contemporary capitalist society. And Marx's analysis of how an abstract logic of capital comes to control every aspect of life anticipates Baudrillard's explanation of how simulation models constitute new forms of postmodern society and culture (see Chapter 3). But it is Friedrich Nietzsche who is a major figure in the story of the postmodern turn in theory.

NIETZSCHE AND MODERNITY

> There is nothing I want more than to become enlightened about than the whole highly complicated system of antagonisms that constitute the "modern world."
>
> —FRIEDRICH NIETZSCHE

Like Kierkegaard, Nietzsche perceived the importance of new modes of communication and technologies in the development of modernity: "The press, the machine, the railway, the telegraph are premises whose thousand-year conclusion no one has yet dared to draw" (Nietzsche, 1986: 378). Moreover, like Kierkegaard, he saw the press and mass culture engendering a leveling process that was destroying individuality and community while producing homogenized, herd conformity. Nietzsche believed that modern society had become so chaotic, fragmented, and devoid of "creative force" that it had lost the resources to create a vital culture and that ultimately, modern society greatly advanced the decline of the human species that had already begun early in Western history. In Nietzsche's view, two trends were evident that were producing contradictory processes of massification and fragmentation—whose extreme consequences would be a central theme of postmodern theory. On the one hand, modern society was fragmenting into warring groups, factions, and individuals without any overriding purpose or shared goals. On the other hand, it was leveling individuals into a herd, bereft of individuality, spontaneity, passion, or creativity. Both trends were harmful to the development of the sort of free, creative, and strong individuality championed by Nietzsche, and he was sharply critical of each.

Nietzsche shared Kierkegaard's belief that contemporary morality and religion were contributing to the leveling process, but unlike Kierkegaard,

who had positive conceptions of morality and religion, Nietzsche tended to see all existing forms of morality and religion—and Christianity in particular—as repressive of vital life energies and inimical to individuality. Thus, Nietzsche radicalized the Enlightenment critique of ideology and like Marx advocated a relentlessly secular approach to values and theory. Nietzsche carved out two different roads into the postmodern, a sociological critique and a philosophical critique, which mutated first into modern existentialism and then into postmodern theory, making him a master theorist of both of these traditions and a link from existentialism to the postmodern turn. Thus, we will argue in this chapter that Nietzsche anticipated postmodern theory both in his sociological explorations of mass society and in his critique of the subject and reason, his deconstruction of modern notions of truth, representation, and objectivity, his perspectivism, his attacks on mechanism and determinism, and his highly aestheticized philosophy and mode of writing. Nietzsche is thus a key figure in the postmodern turn, one who is a crucial source for the postmodern turn in theory as well as a social critic who also anticipated core insights of postmodern social theory.

Art, Society, and the State

The early Nietzsche championed an aestheticization of life, which would be important for later postmodern theory. In *The Birth of Tragedy* (1967a), he argued that "art represents the highest task and the truly metaphysical activity of this life" and that existence could only be justified as an aesthetic phenomenon (1967a: 32–52). From his early work on *The Birth of Tragedy* (1967a [1872]) to one of his last published texts *Twilight of the Idols* (1968b [1889]), Nietzsche contrasted the vital Dionysian culture evident in pre-Socratic Greece and early Greek tragedy with the more rationalistic Apollinian strains evident in Socratic reason and later Greek tragedy. Dionysian culture was eminently life affirming, expressive of bodily energies and passions, and it bound people together in shared cultural experiences of ecstasy, intoxication, and festivals, which Nietzsche thought created strong and healthy individuals and a vigorous culture. For Nietzsche, the Apollinian represented the principle of form, order, and individuation traditionally associated with classical Greek culture, while the Dionysian represented those powers of intoxication, disorder, and the dissolution of the individual ego in collective ecstasy and sensual surrender.

Nietzsche's celebration of the Dionysian and his critiques of Socratic reason and later rationalist Greek tragedy represented an attack on figures of Enlightenment rationality and modern science. Nietzsche later made it clear that the Socratic, or "theoretic," man who was the target of his critique in *The Birth of Tragedy* stood for modern science and rationality, and in an

"Attempt at a Self-Criticism" of his earlier work, Nietzsche claimed that "it was *the problem of science itself*, science considered for the first time as problematic, as questionable," that distinguished his critique (1967a: 18). Indeed, Nietzsche (1968a) was one of the first to question the value of science for life, suggesting that the "will to truth" and scientific lust for objectivity are masks for a will to power and advancement of repressive ascetic ideals. Moreover, although it is often not noted, Nietzsche was one of the first to put in question the organization of modern society and to develop a critique of modernity.[6]

Some of Nietzsche's salient texts focus on contemporary society and the state and exhibit a sharp analysis and critique of the modern age. Although he was presented by Lukács (1980) as an apologist for an imperialistic capitalism, in fact Nietzsche loathed what he saw as capitalism's base concern for merely monetary and bourgeois values, its alienated labor, and its tendency to turn everyone into "industrious ants" (Nietzsche, 1982 [1881]: 126–127). Like Marx, Nietzsche assailed the modern tendency to reduce all value to mere utility (1974: 77), although he failed to grasp the specific economic dynamics behind this process, which Marx analyzed through the profit motive and commodity form. Whereas Marx advanced critiques of capitalist self-interest and greed and advocated cooperation and solidarity as important social values, Nietzsche asserted self-affirmation and attacked benevolence and altruism. But though Nietzsche defended some key bourgeois values, he also composed an all-out assault on the capitalist values of mendaciousness, flattery, sycophancy, and mediocrity (1954 [1883]: 163–166). He distinguished between "actors" in the marketplace and "inventors of values," using the metaphor of pesky flies to denote the purveyors of the goods, services, and values of the modern economy and enjoining his readers to flee from its temptations and seductions.

In a prescient anticipation of role theory, Nietzsche saw that modern societies were producing a differentiated set of occupational, professional, and gender roles, ones that imposed conventional and conformist behavior on individuals. Nietzsche maintained that modern role playing is governed by sterile models that destroy particularity, spontaneity, and individual self-expression. Modern role players act, instead, in a highly conventional and unreflective manner, adopting "many roles," which they play "badly and superficially," in the mode of a mechanically acted "puppet-play." Wearing these social masks reduces individuals to "shadows" and abstractions, turning them into simulacra. Anticipating Baudrillard's postmodern theory (1983a, 1993), Nietzsche contended that simulation is so pervasive that it is hard to distinguish between the role and the person, between the simulated behavior and the real individual. Modern role players are so docile, Nietzsche claimed, that they "prefer the copies to the originals" (1974: 84–86), as if, as Baudrillard might say, the copies were more "authentic" and "real" than the "originals."

Nietzsche also developed strong criticisms of the machine and the machine age, writing:

> The machine of itself teaches the mutual cooperation of hordes of men in operations where each man has to do only one thing: it provides the model for the party apparatus and the conduct of warfare. On the other hand, it does not teach individual autocracy: it makes of many one machine and of every individual an instrument to one end. Its most generalized effect is to teach the utility of centralization. (1986: 366)

In this passage, it is as if Nietzsche had read Marx's analyses in the *Grundrisse* (Marx, 1973: 690ff.) and *Capital* concerning the dialectic of technology in capitalist modernity. As Marx analyzes it, capital multiplies the productive powers of labor by introducing new modes of cooperation and centralization, while the consequent specialization robs individual labor of its creativity and autonomy, subordinating the worker to the machine. Describing the energies set forth by the machine as "lower, non-intellectual" (1986: 366), Nietzsche concluded: "It makes men *active* and *uniform*—but in the long run this engenders a counter-effect, a despairing boredom of soul, which teaches them to long for idleness in all its varieties" (367). He claimed that wage workers under capitalism are worse off than slaves because they are at the "mercy of brute need" and of employers who ruthlessly exploit them:

> the workers of Europe ought henceforth to declare themselves *as a class* a human impossibility and not, as usually happens, only a somewhat harsh and inappropriate social arrangement; they ought to inaugurate within the European beehive an age of a great swarming-out such as has never been seen before, and through this act of free emigration in the grand manner to protest against the machine, against capital, and against the choice now threatening them of being compelled to become either the slave of the state or of the party of disruption. (1982 [1881]: 125–127)

In this passage, Nietzsche advocates individualist revolt against the capitalist machine, though he rejects joining the opposition party ("of disruption") because he feels that all social reform movements are rooted in the herd psychology of resentment of "higher types." Like Kierkegaard, he shows occasional insight into the social forces degrading and massifying individuals while rejecting social movements out of hand as a means of destroying these forces, opting instead for individualist means of transcendence. In this vein, Nietzsche developed a vitriolic attack on the modern state, finding it to be a "new idol" that is "the coldest of all cold monsters," run by "annihilators" who continuously lie and manipulate: "Everything about it is false," Nietzsche claims, and it "devours . . . chews . . . and rechews" its citizens (1954: 160–163). Nietzsche consistently attacked as well German nationalism, writing: "If one spends oneself on power, grand politics, economic affairs, world commerce, parliamentary institutions, military interests—if one expends oneself in this direction the quantum of reason, seriousness, will; self-overcoming that one is, then, there will be a shortage in the other direction" (1968b: 62). Nietzsche is distinguishing here between culture and the state and is clearly championing the former over the latter, writing: "All great cultural epochs are epochs of political decline: that which is great in the cultural sense has been unpolitical, even *anti-political*" (63).[7]

Nietzsche thus anticipated the critiques of the modern state and politics, of mass society and culture, and of their normalizing and homogenizing tendencies that were developed by the Frankfurt School, Foucault, and various postmodern theorists. He was beloved by critical theorists (see Horkheimer and Adorno, 1972 [1946]) and by postmodern theorists like Foucault (1972a, 1975) precisely because of his powerful critiques of mass culture, society, the state, and their homogenizing and normalizing features, along with his assaults on idealist theories. For Nietzsche, the state and the marketplace were bitter antagonists of culture, and he saw both as leveling individuals, producing mediocrity and cultural backwardness, and generating such mass hysterias as nationalism and anti-Semitism.

Consequently, Nietzsche took on the key institutions of modernity in his critique. Yet unlike Marx and Engels, he did not think that a new modern economy, society, and state could be formed that would realize the promises of modernity. Thus, Nietzsche combines highly modern ideas and impulses with distinctly antimodern positions, consistently attacking those ideas and institutions that he saw as fettering individuals and inhibiting to a more intense existence, as dulling creativity, or as otherwise foreclosing the myriad experiences and possibilities for richer individuality and a fuller life opened by the modern era.

Nietzsche's Attack on Reason and the Advent of Nihilism

From his early writings on, Nietzsche, like Kierkegaard, railed against a life-denying rationalism and idealist philosophy that championed reason over the passions. Nietzsche saw the "subject" as a mere construct, an idealized sublimation of bodily drives, experiences, and a multiplicity of thoughts and impulses. This "little changeling," Nietzsche complained, this subject, "is believed in more firmly than anything else on earth" but is a simple illusion created out of modern desperation to have a well-grounded identity. Belief in the subject is promoted by the conventions of grammar, which utilizes a subject–predicate form, giving rise to the fallacy that the "I" is a subject or substance (1968b: 37–38). For Nietzsche, "the doer" is "merely a fiction added to the deed—the deed is everything" (45). "The subject," he concluded, was thus merely a shorthand expression for a multiplicity of drives, experiences, and ideas.

Nietzsche thus broke with unified notions of the subject, arguing: "My *hypothesis*: subject as multiplicity" (1968a: 270). Consequently, anticipating poststructuralism and postmodern theory, he saw the rational subject celebrated by philosophy as a maelstrom of drives and impulses, most of which are not even conscious. Moreover, the subject was a product of modern culture and society that produced docile, disciplined, and conformist herd individuals. Yet, unlike Kierkegaard, Nietzsche also advocated during some stages of his work a critical reason that affirmed the body, passions, and individual development.[8] Attacking the "highest ideals" of Western culture, he opposed an experimental philosophy and "little truths" to empty generalizations of Enlightenment rationalism: "It is the mark of a higher culture to value the little unpretentious truths which have been discovered by means of rigorous method more highly than the errors handed down by metaphysical and artistic ages and men, which blind us and make us happy" (1986: 13). He valued experimental science and critical thought as the highest ideals and assailed the pretensions of idealist philosophy and religion (see Nietzsche, 1982: 101, 185, 204, and 1974: 110, 253, 324 *passim*).

In the spirit of Enlightenment, Nietzsche also polemicizes against metaphysics, arguing that it illicitly generalizes from ideas in one historical epoch to the entirety of history. Against this form of philosophical universalism, Nietzsche argues that "there are *no eternal facts*, just as there are no absolute truths. Consequently, what is needed from now on *is historical philosophizing*, and with it the virtue of modesty" (1986: 13). Castigating traditional philosophy and values from a critical Enlightenment perspective, Nietzsche anticipated later postmodern critiques of metaphysics, assailing the concept of enduring knowledge and the notion of a metaphysical world and presenting metaphysical thought—in a modernizing vein—as a thoroughly obsolete mode of thinking. He attributes the "metaphysical need," at the heart of Schopenhauer's philosophy, to primitive yearnings for religious consolation for the sufferings of life and urges "free spirits" to liberate themselves and pursue thinking and living *experimentally* (8).

Against metaphysics, Nietzsche, like Marx, champions historical and physical explanations, condemning all obscurantist thinking and writing. His assault on metaphysics can, therefore, be read as a broadside against traditional thought that clears the way for a thoroughly modern mode of thought. On the other hand, Nietzsche's attack on foundationalism, universalizing thought, and metaphysics is often taken as beginning a postmodern turn in theory through a radical deconstruction of modern theory. Nietzsche's critique thus contributed to development of a form of thought that was this-worldly, historical, and focused on the present.

His conception of the "death of God" described a modernity that in its fundamental imperatives no longer believed in transcendent values and in which a materialistic skepticism reigned. But Nietzsche did not see this feature of modernity as emancipatory per se; he equated it with nihilism and decadence and saw it as a failure to create original values and as expressive of a depletion or exhaustion of creative life energies. Profoundly prescient at diagnosing nihilism as the worm eating at the heart of modernity, Nietzsche wrote: "What I relate is the history of the next two centuries. I describe what is coming, what can no longer come differently: the *advent of nihilism*" (1968a: 3).

For Nietzsche, nihilism results from the exaggerated, false, overly optimistic belief in humanity, rationality, and life itself. Both Christianity and modernity contributed decisively to this, with Christianity promising peace, harmony, and salvation, placing its faith in God, and modernity promising progress, prosperity, and happiness, placing its faith in reason, science, and the marketplace. Both devalue the present in favor of a future (of salvation or of worldly progress), and both value a fictive world more than the existing world of appearance and becoming. In terms of modernity, Nietzsche states: "The faith in the categories of reason is the cause of nihilism. We have measured the value of the world according to categories *that refer to a purely fictitious world*" (1968a: 13). When peace is promised but the world explodes in war, when prosperity is guaranteed but the majority languish in sickness and poverty, when freedom is championed but only domination appears, and when reason is proclaimed but only insanity rules, nihilism is the practical result.

Thus, nihilism occurs when the highest values become devalued. Nietzsche acutely notes a human psychological trait: the tendency to pass from one extreme to another, from the ecstasy of absolute faith to the agony of disenchantment. Nietzsche assails nihilism as the true sickness of modern times, but he also affirms it, provisionally, as a necessary stage in the development of new, secular values that promote higher types. "Now that the shabby origin of [modern values] is becoming clear, the universe seems to have lost value, seems 'meaningless'—but that is only a *transitional stage*" (1968a: 10–11).

From our present-day perspective, this "transitional stage" of skepticism, nihilism, and world-weariness appears to have become the current "postmodern condition." In the spirit of Nietzsche, we can usefully read postmodern skepticism and despair as the flip side of modern faith and optimism, as an extreme reaction to modern exaggeration of the powers of reason compared to premodern piety. Rather than merely deploring the postmodern turn à la Habermas, we can also see postmodern developments as a *transition to potentially better modes of thought and values*, whereby postmodern thought can contribute important critiques of modern theory, as a demystification of the demystifying mind, and as part of the permanent revolution of critique unleashed by modernity. Of course, for extreme postmodern theory, modernity is nothing but a "purely fictitious world," merely a series of illusions and errors. But if, as we posit, we are between the modern and the postmodern, then we can draw on both discourses to reconstruct theory and practice for the present age.

Of course, there are important differences between Nietzschean and postmodern theory. Where many postmodernists affirm nihilism for its own sake and indulge in despair, Nietzsche used negation only as a means to affirmation, to a celebration of the will to power, of everything that enhances human strength, health, creativity, independence, and well-being:

> It is my good fortune that after the whole millennia of error and confusion I have rediscovered the way that leads to a Yes and a No.
>
> I teach the No to all that makes weak—that exhausts.
>
> I teach the Yes to all that strengthens, that stores up strength, that justifies the feeling of strength. (1968a: 33)

Where deconstructionist philosophies typically terminate in the No, merely seeking to unravel a positive modern value system into a heap of disconnected fragments, Nietzsche starts and finishes with a big Yes, a life-affirming value, deconstructing only to reconstruct. Nietzsche always criticizes from a *normative standpoint*, from a coherent system of values and beliefs that he affirms, never confusing ontology with history, never mistaking for the totality of life one of its distorted aspects: "Modern pessimism is an expression of the uselessness of the *modern* world—not of the world of existence" (1968a: 23). Moving far away from Schopenhauerian pessimism, back toward a Greek view of tragedy, toward a Dionysian view of existence, Nietzsche seeks "a *justification of life*, even at its most terrible, ambiguous, and mendacious" (521), a justification found in art, creativity, independence, and the emergence of "higher types" of humanity.

Thus, Nietzsche provides us with both a historical contextualization of modern/postmodern nihilism and a critique of it. Against nihilism, Nietzsche called for the creation of values that would be life affirming and original, positioning superior individuals against society. His assault on Christianity and otherworldly idealism can be read as a culmination of Enlightenment modernity, as a modernizing demystification and secularization of the world. For Nietzsche, as for the Young Hegelians, modernity comes into its own with the death of God, though from this event Nietzsche drew more significant philosophical implications, arguing that there no longer exists a solid, stable, eternal foundation for values and ideas. In a radically transitory, finite, and unstable world, there can be no foundation for philosophy or values. Nietzsche overturned and revalued (*Umwertung*) the highest philosophical values, championing becoming over being, appearance over reality, this world over the "true world," and body over spirit (1968b: 35ff.). He thus anticipated Derrida's deconstructive task of taking apart traditional philosophical hierarchies, and postmodern theory's rejection of foundationalism.

Nietzsche and Multiperspectival Vision

In addition, Nietzsche's perspectivism denied the possibility of affirming any absolute or universal values: All ideas, values, positions, and so on are posits of the existing individual, constructs of a will to power, which are to be judged according to the extent to which they do or do not serve the values of life and strong individuality. For Nietzsche, there are no facts, only interpretations, and he argued that all interpretation was thus inevitably laden with presuppositions, biases, and limitations. For Nietzsche, a perspective was thus an optic, a way of seeing, and the more perspectives one had at one's disposal the more one could see, the better one could understand and grasp specific phenomena. To avoid one-sidedness and partial vision, one should learn "how to employ a *variety* of perspectives and interpretations in the service of knowledge" (1968a: 119).

On the one hand, the notion of perspectival seeing indicates the limitations whereby each individual's body, history, culture, and location restricts what he or she can see. On the other hand, the doctrine points to the necessity of learning to see from a variety of perspectives and of cultivating multidimensional seeing as part of superior individuality. Nietzsche claimed that "every elevation of man brings with it the overcoming of narrower interpretations; that every strengthening and increase of power opens up new perspectives and means believing in new horizons" (1968a: 330). Acquiring perspectival vision enables individuals to overcome one-sidedness and thus to grasp aspects of life previously overlooked. It also enables individuals to overcome myopic modes of thought:

> We usually endeavor to acquire a *single* deportment of feeling, a *single* attitude of mind toward all the events and situations of life—that above all is what is called being philosophically minded. But for the enrichment of knowledge it may be of more value not to reduce oneself to uniformity in this way, but to listen instead to the gentle voice of each of life's different situations; these will suggest the attitude of mind appropriate to them. Through thus ceasing to treat oneself as a *single* rigid and unchanging individuum one takes an intelligent interest in the life and being of many others. (1986: 195–196)

Nietzsche's assumption is that reality is too complex and many-sided to be grasped from a single perspective: "A *multiplicity* of hypotheses, for example as to the origin of the bad conscience, suffices still in our own time to lift from the soul that shadow that so easily arises from a laborious pondering over a single hypothesis which, being the only one visible, is a hundredfold overrated" (1986: 304). A single hypothesis can provide but a partial grasp of things and thus requires refinement and rethinking. In *On the Genealogy of Morals* (1967b), for example, Nietzsche provides an interpretation of ascetic ideals, insisting that such ideals are very different in artists, philosophers, priests, and scientists. There is thus no essence of the ascetic ideal, no single hypothesis that explains them, but a wealth of phenomena that require different interpretations and valuations.[9]

The concepts of perspectival seeing and interpretation provide Nietzsche with a critical counterconcept to essentialism: Objects do not have an inherent essence but will appear differently according to the perspective from which they are viewed and interpreted and according to the context in which they appear. He spoke of his own "search for knowledge being manifested in the dream of having the 'hands and eyes' of many others and of being 'reborn in a hundred beings'" (1974: 215). Cultivating this approach required *learning to see* and interpret: "habituating the eye to repose, to patience, to letting things come to it; learning to defer judgement, to investigate and comprehend the individual case in all its aspects" (Nietzsche, 1968b: 65).

This passage points to another virtue of a perspectival optic: it enables one to grasp the specificity and particularity of things. Nietzsche mistrusted the distorting function of language and concepts that were overly abstract and general, and he required perspectival seeing and interpretation to grasp the particularity and specificity of concrete individual phenomena. Perspectival seeing allowed access to "a complex form of specificity" (Nietzsche, 1968a: 340), which made possible a more concrete and complete grasp of the particularities of phenomena. Seeing from conflicting perspectives also opened people to an appreciation of otherness and difference, and to a grasp of the uncertain, provisional, hypothetical and "experimental" nature of all knowledge.

Thus, the multiperspectival approach allows the theorist to grasp the particularity and concreteness of individual phenomena; to perceive and interpret phenomena from different perspectives; and thus to avoid homogenizing and essentializing approaches. Nietzsche's multiperspectival approach undermines claims for absolute truth or for a single, infallible method that will guarantee truth and objectivity. Nietzsche was in favor of experimental science, of gaining knowledge through the senses, and of testing hypotheses and attaining cumulative knowledge, but he attacked the belief in objectivity, in an immaculate perception, in a completely nonbiased and noninterested mode of seeing. Perception and cognition were always perspectival, Nietzsche argued, and he scorned those who believed that science alone could attain truth or that the scientist has privileged access to reality (1968a: 146ff.).

Rejecting this positivist notion of science, Nietzsche also anticipated the postmodern critique of mechanism, objectivistic causality, and determinism (see Chapter 5). Indeed, his critique of modern science is as radical as his critique of philosophy, morality, and religion; in each case, he digs into the roots of the phenomenon under question, ferreting out basic assumptions and effects of different beliefs and practices. As he does for philosophy, Nietzsche claims that the central concepts of modern science—truth, objectivity, mechanism, atoms, laws, causality, and so on—are anthropomorphically rooted illusions, rooted in the human, all-too-human need for order, meaning, and clarity in a universe indifferent to human fancies: "Ultimately, man finds in things nothing but what he himself has imported into them: the finding is called science" (1968a: 327). We should not suppose that because "truths" like "unity," "causality," and "laws" are needed or useful in order to live and think, that they are in fact metaphysical realities instead of cognitive constructs and pragmatic tools to achieve certain goals.

Nietzsche sees the mechanistic worldview as little different from the animistic worldview that preceded it, with both representing anthropomorphic illusions, to be contrasted to his own "anti- metaphysical" conception of the world. Nietzsche blasts positivist notions of objective, presuppositionless truth, insisting that all concepts are posits of the will to power, constructs that we use to augment human life. Rejecting the static basis of the mechanistic worldview, as well as the second law of thermodynamics, Nietzsche posits a world of becoming, development, and creativity. Assuming a vitalist and open-ended conception of the world, Nietzsche writes: "If the world could in any way become rigid, dry, dead, *nothing*, or if it could reach a state of equilibrium . . . , then this state must have [already] been reached. But it has not been reached: from which it follows— . . . the mechanistic theory stands refuted" (1968a: 548–549). Anticipating the conceptions of postmodern science (see Chapter 5), Nietzsche replaces the static conception of the world with a dynamic conception of a world of incessant becoming: "There is nothing unchanging in chemistry: this is only appearance, a mere school prejudice. We have *slipped in* the unchanging, my physicist friends, deriving it from metaphysics as always" (1968a: 333).

Anticipating the probabilistic epistemology of contemporary chaos theory, Nietzsche suggests that scientific theories proceed on the basis of a static conception of the world, whereby a more dynamic analysis would be far less steady and certain: "A world in a state of becoming could not, in a strict sense, be 'comprehended' or 'known'" (1968a: 281). Moreover, Nietzsche deplores the reduction of the world to mere quantitative terms: "The calculability of the world, the expressibility of all events in formulas—is this really 'comprehension'? How much of a piece of music has been understood when that in which it is calculable and can be reduced to formulas has been reckoned up? . . . what has one achieved?" (334). Nietzsche is suggesting, in other words, that in both cosmic and aesthetic phenomena, there is "more than meets the eye," and that this "more" transcends quantitative thinking. A crudely reductive scientific interpretation of the world, Nietzsche contends, "would be one of the poorest in meaning" (1974: 335).

Nietzsche is at his most provocative in his rejection of the notions of "law," "rule," and "causality" as human projections. Interestingly, he finds that the notion of physical "law" stems from the human need to believe that the world is governed by *moral* laws, and is one obedient to a divine or human force. Such a conception of the world implies that there is a commanding subject behind the laws, an anthropomorphic fantasy: "Let us beware of saying that there are laws in nature. There are only necessities: there is nobody who commands. Nobody who obeys, nobody who transgresses" (1974: 168). He similarly disposes of standard concepts of causality and of cause and effect relations as illicit. Nietzsche begins with Hume's skeptical observation that we never see or experience a "cause," that a cause is only inferred as a force of habit developed through constant association of one phenomenon with another. A relation of contingency is replaced with one of necessity. Following Hume's attack on Kant, who claimed that causality was one of the major concepts given to the mind independently of any empirical experience, Nietzsche claims that "the law of causality" is one of "the most strongly believed a priori 'truths'" but that it is nothing but a provisional assumption convenient for practical life. Here again, Nietzsche anticipates the insights of pragmatism and postmodern science.

Belief in causality and a world of causes thus rested on attribution of occult connections where they could not be demonstrated (Nietzsche, 1968a: 295ff., and 1968b: 48ff.). Nietzsche goes beyond Hume's initial psychological insight to root causality in something even more fundamental than habit: the "belief in the living and thinking as the only effective force—in will, in intention—it is belief that every major event is a deed, that every deed presupposes a doer, it is belief in the 'subject'" (1968a: 295). Here we find the basis for Wittgenstein's claim that the source of philosophical pseudoproblems is the deception of language and that thus a potential cure for the illness of metaphysics is linguistic analysis. For Nietzsche, language presents us with the grammatical illusion that every predicate has a subject, that something created requires a creator, that there is a doer behind every deed, a (Cartesian-like) "subject" behind every action. We project from cause–effect relations in our own mental world, he argues, onto the world at large. But, "when one has grasped that the 'subject' is not something that creates effects, but only a fiction, much follows" (297).

Nietzsche does not deny that things in the world interact with one another; he is prepared to allow certain kinds of causal notions, but he prefers to adopt a different language, his own terminology of *power relations*, of superior and inferior forces. If we rid ourselves of grammatical illusions and replace the language of agency with the language of power, then, Nietzsche argues, we can arrive at the following insight:

> Two successive states, the one "cause," the other "effect": this is false. The first has nothing to effect, the latter has been effected by nothing.
>
> It is a question [not of "cause and effect" but] of a struggle between two elements of unequal power: a new arrangement of forces is achieved according to a measure of power of each of them. The second condition is something fundamentally different from the first (not its effect): the essential thing is that the factions in struggle emerge with different quanta of power. (1968a: 337)[10]

Ultimately, the world is not as determined and orderly as it appeared in the Newtonian view; the modern paradigm suppresses becoming and thereby chaos. For Nietzsche, the world does not have the qualities that human beings typically impose on it, whether those of a rational and moral order, of natural "laws," of a predetermined harmony unfolding according to a perceived purpose or plan, or of anything flattering to our needs and desires. But neither is the world totally random and devoid of pattern or order of any kind. Nietzsche's conception of chaos is much like what has emerged in postmodern science (see Chapter 5), which defines the world as ordered complexity. Nietzsche's broadest metaphysical vision is of a cosmos governed by incessant change, becoming, mutation, and flux, which is arranged in patterns of order, regularity, and similarity. His view is stated powerfully in the final, most dramatic passage of *The Will to Power*:

> And do you know what "the world" is to me? Shall I show it to you in my mirror? This world: a monster of energy, without beginning, without end . . . a play of forces and waves of forces, at the same time one and many . . . eternally changing, eternally flooding back, with . . . an ebb and flow of its forms; out of the simplest forms striving toward the most complex, out of the stillest, most rigid, coldest forms toward the hottest, most turbulent, most self-contradictory, and then again returning home to the simple out of this abundance, out of the play of contradictions back to the joy of concord . . . this, my *Dionysian* world of the eternally self-creating, the eternally self-destroying, this mystery world of the twofold voluptuous delight . . . without goal . . . without will. . . . (1968a: 549–550)

Nietzsche also strongly opposed the mechanistic and scientific view of the world that would rob it of its aesthetic qualities, scornfully saying: "That the only justifiable interpretation of the world should be one in which you are justified because one can continue to work and do research scientifically in your sense (you really mean, mechanistically?)—an interpretation that permits counting, calculating, weighing, seeing and touching, and nothing more—that is a crudity and naiveté, assuming that it is not a mental illness, an idiocy" (1974: 335).

Nietzsche's conception of a multiperspectival discipline led him to question the claims of science, philosophy, or any one discipline to be the sole road to truth. His own approach combines many perspectives, including philosophy, science, history, social analysis, psychology, aesthetics, and even myth. He himself at various stages appeared as a philologist, cultural historian and critic, philosopher, aestheticist and artist, psychologist, and, as we are suggesting, social theorist. From this multidisciplinary space, the notion of perspectival interpretation gives Nietzsche a powerful weapon with which to criticize the one-sidedness and reductionism of many forms of modern theory. He wrote and lived before disciplinary specialization and professionalization appeared in the academic division of labor and thus provides a model of a transdisciplinary approach that is able to experiment with different perspectives and modes of discourse to gain a deeper understanding of the world.

Yet while arguing for a perspectival way of seeing, Nietzsche is also aware that sometimes a single strong hypothesis is valuable, claiming that in a modern "democracy of concepts," "a single concept that *wanted* to be master would now be called an *idée fixe*. That is *our* way of disposing of tyrants—we direct them to the madhouse" (1986: 369). A single strong and original perspective will obviously illuminate features missed by those who restrict their focus to the specifics and particulars of objects. Indeed, Nietzsche had his own strong and privileged perspectives that he believed provided unique insights that were of utmost significance for human life, just as Marx privileged the economic dimension, providing a powerful interpretive perspective on modern societies.

Contemporary postmodernists therefore sometimes mistake Nietzsche's perspectivism for an "anything goes" type of relativism and irrationalism. But this is precisely the type of intellectual indolence that he despised. Perspectivism is, instead, a highly disciplined mode of thought. Practiced "modestly" and with a sensitivity to limits, science is itself a perspectivist resource that provides access to multiple realities, provides knowledge about specific conditions and particular situations, and challenges error and dogma. On these grounds, Nietzsche considered science as a sign of "a higher multiplicity of culture." After stripping away the positivist veneer, he praised its "disciplining of the intellect, clarity, and severity in matters of intellectual conscience, noble coolness and freedom of the intellect" (1968b: 122–126, 162–165, 171–175).

Thus, Nietzsche did not completely reject the concept of truth as a lie or an illusion. Rather, he inquired into its value for life and ruthlessly exposed many taken-for-granted truths as sublimations of a will to power, delineating the interconnectedness of truth and power. His aphorisms and essays provide different optics that are sometimes contradictory but that provide stronger ways of seeing and understanding than more myopic, disciplinary optics. Nietzsche himself is clearly a seeker of truth, but he dramatically redefined that search and the nature and value of what was being sought.

Truth would not be found in the authoritative voice of one privileged perspective, even his own, but in a dialogue of many perspectives that fully cultivated a perspectival concept of knowledge. As a philosopher of difference, Nietzsche was aware that individuals had different perspectives according to their class, gender, race, and ethnicity, and his multiperspectival approach suggests that *all* subject positions are necessarily biased and limited and that therefore dialogue and debate from many positions is more likely to illuminate a phenomenon than is a single perspective. Yet Nietzsche believed that stronger perspectives would prevail in sustained debate and that one's perspective could be strengthened in some cases by appropriating one's opponents' insights and taking them to a higher level.

NIETZSCHE'S PROGENY

Nietzsche's legacy is highly complex and contradictory, and in retrospect he is one of the most important and provocative figures in the transition from modern to postmodern thought. His assault on Western rationalism profoundly influenced Heidegger, Derrida, Deleuze, Foucault, and other postmodern theorists, leading many to break with modern theory and to seek alternative theories. Nietzsche himself, however, combines modern, premodern, and postmodern motifs. Some of Nietzsche's positions, such as the theory of perspectivism that we just elucidated, are arguably modern concepts that take its motifs to higher levels but which pave the way for later ruptures with modern theory. Some of his own ideals and positions, however, are anti-modern and even premodern, such as his celebration of preSocratic thought, his romanticizing the Greeks, and his flirtation with quasi-metaphysical notions such as the eternal recurrence, fatalism (*Amor Fati*), and some of the prophetic motifs in books like *Thus Spake Zarathustra* (1883–1886).

As Heidegger suggests, one could read Nietzsche's doctrine of the will to power as a reversion to metaphysics, to the old Greek notion that there is a single metaphysical key to the universe.[11] Yet one could also read it as a uniquely modern idea, as an expression of the growing power of individuals and social forces in modern societies, in which science, technology, industry, and the new forms of social organization are creating new forms of power—forms that would, as Nietzsche and later the Frankfurt School and Foucault remind us, provide society with new powers over individuals and individuals with new powers of self-expression and development. And the modern expansion of power would also lead to imperialist wars, which Nietzsche anticipated.

Nietzsche's concept of the *Übermensch*, the overman who constitutes a higher form of human being, can also be seen as a typically modern idea, as expressive of the modern will to development, growth, and innovation. An ancient premodern sage, Zarathustra, however, is Nietzsche's chosen figure for the teacher of the *Übermensch*, and his championing of the cultural warrior, the noble aristocrat of the spirit, the sage, and the artist philosopher can be read as a replication of ancient ideals. Moreover, it is hard to read Nietzsche's ideas of the eternal recurrence and of *Amor Fati* and his celebration of Dionysus as anything other than premodern ideals. Nietzsche's concept of *Amor Fati*, his affirmation that "every man is himself a piece of fate," and his arguments for fatalism (1986: 325) reveal the lasting influence of certain Greek ideas on Nietzsche. Moreover, and crucially, his celebration of Dionysus as a figure of liberation shows his proclivities to premodern thought.

Nietzsche anticipates the postmodern vision of a world of chance and contingency but turns from it to Greek myths of fatalism (*Amor Fati*) and the eternal recurrence. Nietzsche thus ultimately moves from a skeptical and critical vision of a world without fixed order, determinacy, or law to seek refuge in myth and a metaphysical philosophy of life. Nietzsche thus stood on the threshold of the postmodern turn in theory, but he turned back, fixated on premodern forms of society with their warriors, strong individuals, and unified cultures, for him an attractive alternative to the weak personalities of a life-negating modern society.

For Nietzsche, modernity is the twilight of the West, an era of degradation and degeneration rather than progress and advancement as in Enlightenment mythology. Hence, Nietzsche carries out a perverse historicizing whereby his current 19th-century modern era is seen as a weaker, life-negating, ascetic, and mediocre era, far inferior to the robust civilizations of classical Greece and Rome or even to Renaissance civilization.[12] Nietzsche, like Kierkegaard, has a one-sided, limited, and flawed interpretation of democracy, liberalism, socialism, and modern politics, failing to see their positive potentialities. Instead, the entire modern age for Nietzsche signifies the advent of nihilism, a period of world-weariness in which all has been said and done, and cynicism and skepticism rule.

In general, Nietzsche sees Greek and Roman civilization and the Renaissance as stronger eras when more life-affirming values reigned, when more unified cultures and traditions nourished individuals, when superior individuals were appreciated and celebrated. In Nietzsche's Dionysian philosophy of life, whatever affirms life is good whereas whatever negates it is devalued. From this perspective, Nietzsche attacks metaphysics, morality, and religion but also the progressive institutions and social movements of the modern world. And yet Nietzsche articulates some of the deepest ideas of modernity, advocating such key modern values as individualism, growth, development, innovation, and the destruction of the old and the development of the new. A pathos of the new permeates his work, one that seeks new beginnings, values, and even a new era via negation of the old. In his emphasis on self-overcoming and growth, Nietzsche embodies a Promethean–Faustian spirit that Marx also exemplified, evincing a modern rejection of the past and present along with the development of new values and modes of life. But unlike Marx, who advocated a fundamentally scientific discourse for social theory, Nietzsche in both the style and content of his texts, deploys the strategies of aesthetic modernism, exemplifying in form and content Rimbaud's exhortation to *changer la vie* and Pound's demand to "make it new." His aphoristic texts are experimental and philosophize in a new manner. Likewise, *Thus Spake Zarathustra* is a modernist text, finding a new dramatic, literary form to express Nietzsche's philosophy.

One could also argue that the Enlightenment form of critique is radicalized in Nietzsche and applied to areas that were neglected by the *philosophes*, and indeed the Enlightenment itself is subject to Nietzsche's withering scrutiny. Foreshadowing Horkheimer and Adorno (1972), Nietzsche sees a dialectic of Enlightenment in which reason turns into its opposite, becomes myth, but a life-negating one. Reason is supposed to free the individual from bondage to nature, irrational social authority, and one's body, but ends up, in Nietzsche's vision, imprisoning individuals in a life-denying culture and in new myths such as "objectivity" or "progress."

Thus, Nietzsche's thought contains a unique synthesis of modern and premodern elements that envisage a postmodern break with modernity.[13] Many of his core values are premodern; his critiques of religion, philosophy, and morality are thoroughly modern, as are his celebrations of individuality, freethinking, and self-overcoming, while he dreams of a new (postmodern) era that overcomes the ills and limitations of the present age. Nietzsche is his contradictions, and key antinomies are apparent between the modern and premodern elements in his thought. Thus, although his thought is, on the one hand, highly modern and even modernist in its formal qualities, Nietzsche only affirms some modern ideas while rejecting others. Moreover, he yearns for a higher civilization that builds on the ancient cultures, which he continues to valorize, along with the most vital elements of modern culture, envisaging a postmodern rupture and creation of a new society.

Martin Heidegger combines Nietzsche's radical critique of modernity with nostalgia for premodern social forms and a hatred of modern technology, which he sees as producing new forms of domination. In *Being and Time* (1962 [1927]), Heidegger developed Kierkegaard's and Nietzsche's critique of mass society through his concept of *das Man*, the impersonal One, or They–Self, which dominates "average everyday" being. The They–Self for Heidegger is a form of tyranny that imposes the thought, tastes, language, and habits of the mass onto each individual, creating a leveling process, such that "authentic" individuality demands radical self-differentiation from others (see Kellner, 1973). The process is facilitated by meditation on death and the contingency and finitude of human existence, which lends an urgency to creative endeavors.

For the later Heidegger, the critical focus shifts from the existential structures of individual existence and modern society to modern technology, which produces a *Gestell*, a conceptual framework that reduces nature, human beings, and objects to "a standing reserve," as resources for technical exploitation. Heidegger (1977) renounces modern and technological modes of thought and values in favor of premodern forms of contemplation and "letting Being be," thus rejecting modernity in its totality. Like Nietzsche, he ultimately harkens back to premodern values, and with Junger, Spengler, and others he furthers a German anti-rationalist tradition that ultimately helped produce fascism, an anti-modern culture that Heidegger affirmed and promoted.

Heidegger's assault on modernity was developed by Foucault and other postmodern theorists, while his attacks on metaphysics and modern thought became central to Derrida. Heidegger (1977) argued that modern subjectivity set itself up as a sovereign instrument of domination of the object and that its own forms of representations of the world were taken as the measure of the real. For Heidegger, the representational form of modern thought and subsequent subject–object metaphysics illicitly enthroned the subject as the Lord of Being and set individuals in an inauthentic relation with Being. Derrida radicalized Heidegger's critique of dualistic metaphysics, while Rorty (1979) developed Heidegger's critique of representation into a critique of philosophy as the mirror of nature. These ideas would eventually coalesce into a powerful assault on modern philosophy, leading many to call for a new form of postmodern thought (see the collection of essays in Baynes et al., 1987, for examples of contemporary reconstructions of modern philosophy).

While Heidegger appropriated themes from Nietzsche into an antimodern and irrationalist mode of thought, Max Weber combined ideas from Nietzsche and Marx into a critical theory of modernity. Weber agreed with Marx that capitalism was the decisive event of the modern world and that societal rationalization and domination were consequences of the triumph of the capitalist mode of production over previous modes of social organization. But Weber thought that Marx excessively privileged the economic dimension and underplayed the importance of religion, culture, and technical rationality. Weber saw the latter as congealing in the capitalist labor process, commodification dynamics, bureaucracy, law, technology, and the dominant modes of thought. These forms of technological rationality produced an "iron cage" of domination that threatened "freedom" and "meaning," leading to a soulless proliferation of "sensualists without heart and specialists without spirit" (Weber, 1946). For Weber, socialism contributed to the erection of this "iron cage," further undermining democracy and freedom. As did later postmodern theorists such as Foucault, Weber stoically accepted the impossibility of radical social transformation and the necessity of limiting politics to reforms within the existing system, and he thus affirmed a conservative liberalism as the only viable politics for the modern age.

Lukács (1971) was the first neo-Marxist to combine Nietzsche's and Weber's insights with those of Marx, analyzing the new modes of reification, whereby human beings are transformed into things, in the capitalist labor process, in the state and bureaucracy, and in cultural and intellectual realms. Drawing from Weber's work, while rejecting its fatalist conclusions, Lukács argued that rationalization, domination, and reification were all results of the capitalist production process, which led to a greater administration of human beings, as reification made social relations appear to be relations among things, thereby blunting workers' experience of class and the social relations of oppression. In a scenario that out-Hegeled Hegel, Lukács saw the proletariat, the object of capitalist exploitation, becoming the subject of history by attaining revolutionary class consciousness and overthrowing capitalism.

At yet a higher level of synthesis, the Frankfurt School combined Marx's theory of capitalism, Nietzsche's critique of mass society and culture, Lukács's theory of reification, and Weber's concept of the "iron cage," developing a critical theory of modernity in *Dialectic of Enlightenment* (Horkheimer and Adorno, 1972), *One-Dimensional Man* (Marcuse, 1964), and other writings (see Kellner, 1989a). Horkheimer, Adorno, Marcuse, Fromm, Benjamin, and others associated with the Frankfurt School traced the gradual bureaucratization, rationalization, and commodification of social life in the new stage of state capitalism. They described how the "culture industry" defused critical consciousness, providing a key means of distraction and stupefaction, and they developed the first neo-Marxist theories of the media and consumer society. The Frankfurt School analyzed how contemporary capitalism produced a homogenization of needs, a flattening out of class contradictions, and the integration of the working class, thus putting in question the classical Marxian concept of revolution. Despairing of a political revolution, its members tended to limit radical potential to the works of modernist art, although Marcuse also embraced the radical movements of the 1960s as possible instruments of social transformation (see Kellner, 1984a).

On Habermas's (1987) analysis, the pathway into postmodern theory through Nietzsche and Heidegger, shaped as well by Bataille and an aestheticist French tradition, spawned a French irrationalist progeny that wanted to jettison the project of the Enlightenment, rejecting reason for aesthetic and erotic experience. This narrative reduces a complex story to a single line of development (see Best, 1995), whereas we would argue that there are many pathways into the postmodern, each with different trajectories and effects.

For instance, one road moves through Kierkegaard, Marx, and Nietzsche, is mediated by Weber, Lukács, and the Frankfurt School, and leads into the French neo-Marxist tradition of Debord and the Situationists and, finally, into Baudrillard and other postmodern theorists who contribute to developing a critical theory of the present age. In the Chapter 3, we shall pursue this pathway, delineating the shift from 19th-century competitive capitalism, organized around production, to a later form of capitalism, organized around consumption, media, information, new technology, and new forms of domination and abstraction, providing a transition to what some claim to be a new postmodern era.

NOTES

1. See the documentation in Kierkegaard (1984) and Perkins (1990).

2. As Habermas (1989) argued, the press during the 18th century was conceived by democratic theory to be part of the public sphere of debate and consensus, though it also often served specific political and economic interests.

3. Big lies have been promoted by the press to legitimate war and military interventions, ranging from Hearst's yellow journalism in the Spanish–American War to the manufacture of the Gulf of Tonkin incident in the Vietnam War to the lies and distortions of the Persian Gulf TV war; see Kellner (1990, 1992).

4. Curiously, Marx, Kierkegaard, Engels, Bakunin, and others were all in Berlin during the same period, studying Hegel and attending philosophy lectures at the university. There is no evidence, however, that Marx and Kierkegaard ever met or were familiar with each other's works.

5. In his book *Two Ages: The Age of Revolution and the Present Age*, a commentary on a popular Danish novel entitled *The Present Age*, Kierkegaard (1978) distinguishes between antiquity and modern society and between the previous Age of Revolution and the present age (i.e., the 1840s), characterizing the latter as having undergone a precipitous decline in passion; see our detailed analysis of this text in Best and Kellner (1990).

6. On Nietzsche's critique of modernity, see Kellner (1991) and on the neglect of Nietzsche in classical social theory, see Antonio (1995). The following discussion is indebted to our work with Robert Antonio on Nietzsche and social theory over the past decade.

7. This passage leads Peter Bergman (1987) to label Nietzsche the "last anti-political German," and although Nietzsche carries out as sharp a critique of German nationalism as one could imagine, he affirms quite specific and often problematic political positions (documented in Kellner, 1991); thus, it is a mistake to see him as simply "anti-political."

8. While the early Nietzsche can be read as a staunchly anti-Enlightenment critic of reason, in the 1870s Nietzsche moved toward a position more sympathetic toward a critical Enlightenment reason beginning with *Human, All Too Human* (1986 [1878]). This and the succeeding aphoristic works mark an attempt to develop a

modern, experimental type of writing that would serve as a weapon of critique against ruling ideas—precisely the strategy of the Enlightenment. Henceforth, Nietzsche displayed a much more positive relation toward reason and science, coupled with sharper critiques of idealist philosophy, religion, morality, and contemporary society than was previously evident in his work.

9. Theodor Adorno and Max Horkheimer applied this multiperspectival method to a study of anti-Semitism in their *Dialectic of Enlightenment* (1972), approaching the phenomenon from economic, political, sociological, religious, anthropological, psychoanalytic, and other perspectives. For our own explications of a multiperspectival method, see Best and Kellner (1991) and Kellner (1995a).

10. Elsewhere, Nietzsche adds:

> My idea is that every specific body strives to become master over all space and to extend its force (—its will to power:) and to thrust back all that resists its extension. But it continually encounters similar efforts on the parts of other bodies and ends by coming to an arrangement ("union") with those of them that are sufficiently related to it: they then conspire [!] together for power. And the process goes on—. (1968a: 340)

Clearly, such language is as anthropomorphic as anything invented by premodern or modern worldviews, and it ultimately seems far more arbitrary than the language of causation, stemming from Nietzsche's cosmological monism and political biases against democracy and equality in a favor of an aristocratic scheme of masters and slaves. His approach completely obscures the role of cooperation and solidarity in nature and society.

11. Heidegger interprets Nietzsche as the last metaphysician, claiming that he continues the Western metaphysical practice of seeking a single key to the universe, though he destroys Platonist metaphysics and finds the essence of being in the sensory world (Heidegger, 1977: 53ff.).

12. In *Twilight of the Idols*, Nietzsche wrote:

> We modern men, very delicate, very vulnerable and paying and receiving consideration in a hundred ways, imagine in fact that this sensitive humanity which we represent, this *achieved* unanimity in forbearance, in readiness to help, in mutual trust, is a positive advance, that with this we have gone far beyond the men of the Renaissance. What is certain is that we would not dare to place ourselves in Renaissance circumstances, or even imagine ourselves in them: our nerves could not endure that reality, not to speak of our muscles. This incapacity, however, demonstrates, not an advance, but only a different, a more belated constitution, a weaker, more delicate, more vulnerable one, out of which is necessarily engendered a morality *which is full of consideration*. If we think away our delicacy and belatedness, our physiological ageing, then our morality of "humanization" too loses its value at once—no morality has any value in itself:—we would even despise it. (1968b [1889]: 89–90)

13. For other takes on the relationship between Nietzsche and postmodern theory, see the essays collected in Koelb (1990), McGowan (1991), and Oliver (1995).

CHAPTER THREE

From the Society of the Spectacle to the Realm of Simulation

Debord, Baudrillard, and Postmodernity

> But certainly for the present age, which prefers the sign to the thing signified, the copy to the original, fancy to reality, the appearance to the essence, . . . illusion only is sacred, truth profane. Nay, sacredness is held to be enhanced in proportion as truth decreases and illusion increases, so that the highest degree of illusion comes to be the highest degree of sacredness.
>
> —LUDWIG FEUERBACH

Karl Marx was the first major social theorist to conceptualize the break between modern and premodern societies and thus to develop comprehensive theoretical perspectives on modernity. For Marx, the emergence of the capitalist mode of production generated a new form of modern society whose motor and infrastructure was organized around the production of commodities and the accumulation of capital. In the Marxian vision, capital created a world in its own image, and the commodity form of its products became a constitutive principle of social organization. Subsequent Marxists (Lukács, Adorno, Marcuse, et al.) have shown how, in a neo-capitalist consumer economy, commodification processes have permeated new realms of experience and social life. In their theories, capitalism has become a reified and self-legitimating system in which the object world assumes command, human well-being is defined by the conspicuous consumption of goods, and critical thought is vanquished in a "one-dimensional" culture of mass conformity.

In the 1950s and 1960s, a largely French avant-garde movement, the Situationist International, emerged that further developed these themes.[1] Led by Guy Debord, they combined a theoretical critique of consumer society with a radical artistic and social politics. They identified consumer capitalism as a new mode of social control, as a "society of the spectacle," that pacifies its citizens by creating a world of mesmerizing images and stupefying forms of entertainment. The Situationists contrast passive spectators to active subjects who would carry out a total subversion and revolution of the established society.

Building still further on this neo-Marxist tradition, Baudrillard argued that the commodity form has developed to such an extent that use and exchange value have been superseded by "sign value" that redefines the commodity primarily as a symbol to be consumed and displayed.[2] For Baudrillard, political economy and the era of production are finished, and we live in a new, dematerialized society of signs, images, and codes. Baudrillard claims that the massification process analyzed by Kierkegaard, Marx, and Nietzsche has developed to the point where distinctions among classes, social groups, political ideologies, and other aspects of the social differentiation process that characterized modernity have collapsed. Baudrillard argues that the distinction between reality and unreality itself has been erased, that technology and media and consumer culture totally control our lives, and that we have irrevocably passed from a modern to a postmodern world.

In this chapter, we shall accordingly trace a theoretical trajectory from Marx through Debord and the Situationists to Baudrillard and postmodern theory. This is a movement from the society of the *commodity* to the society of the *spectacle* to the society of the *simulacrum*, paralleled by increasing commodification and massification to the point of implosion of the key phenomena described by modern theory. Where one might speak of the "commodification of reality," we wish to invert this formula to speak, equally, of the "reality of commodification" and of the capitalist world, which Baudrillard and his followers claim dissolves in the postmodern world of hyperreality and implosion.

Although we chart a shift from modern to postmodern theory, we argue that the categories of Marxian theory still help to conceptualize our present moment, which is between the modern and the postmodern. It is our argument that modern theory—such as was developed by Marx, Nietzsche, Weber, the Frankfurt School, and others—is not only far from being obsolete, but is of central importance in making sense of our contemporary situation. We also focus on the claims that modernity has been surpassed by postmodernity and intend to elucidate these terms and indicate why it is wrong to speak of postmodernity as an absolute break from capitalist modernity.

THE SITUATIONISTS: COMMODIFICATION, SPECTACLE, AND CAPITALISM

> The commodity can only be understood in its undistorted essence when it becomes the universal category of society as a whole.
>
> —GEORG LUKÁCS

> The spectacle is the moment when the commodity has attained the *total occupation* of social life. The relation to the commodity is not only visible, but one no longer sees anything but it: the world one sees is its world. Modern economic production extends its dictatorship extensively and intensively.
>
> —GUY DEBORD

We interpret the emergence of Debord and the Situationist International as an attempt to update the Marxian theory in the French post-World War II conjuncture—a project that was also deeply influenced by French modernist avant-garde groups—and we believe these currents ultimately helped generate the postmodern turn in France. Debord and his friends were themselves initially part of a French avant-garde artistic milieu that was shaped by dada, surrealism, lettrism, and other attempts to merge art and politics (see Marcus, 1989; Plant, 1992; Wollen, 1993). Unorthodox Marxists like Henri Lefebvre (himself at one time part of the surrealist movement and the creator of a critique of everyday life) influenced Debord, as did groups like Socialism or Barbarism and Arguments, both of which attempted to create an up-to-date and emancipatory Marxist theory and practice. Rapid modernization in France after World War II and the introduction of the consumer society in the 1950s provoked much debate and contributed to generating a variety of discourses on modern society, inspiring Debord and others to attempt to revitalize the Marxian project in response to new historical conditions and new aesthetic and theoretical impulses.[3]

Yet the Situationist revision of Marxism developed significant differences from the classical project, resulting in new motifs and emphases. Whereas classical Marxism focused on production, the Situationists highlighted the importance of social reproduction and the new modes of the consumer and media society that had developed since the death of Marx. Whereas Marxism focused on the factory, the Situationists focused on the city and everyday life, supplementing the Marxian emphasis on class struggle with a project of cultural revolution and the transformation of subjectivity and social relations. And whereas Marxian theory focused on time and history, the Situationists emphasized the production of space and the constitution of liberated zones of desire.

The work of Debord and the Situationists can thus be interpreted as contemporary French efforts to renew the Marxian project. Their program was to reinvigorate Marxian revolutionary practice and to supplement Marx's critique of capital and the commodity, attempting to trace the further development of the abstraction process inherent in commodity production. Influenced by Sartre's notion that human existence is always lived within a particular context or situation and that individuals can create their own situations—as well as by Lefebvre's concept of everyday life and his demand that it be radically transformed—Debord and his colleagues began devising strategies to construct new "situations" (see Debord in Knabb, 1981: 17ff.). This project would merge art and everyday life in the spirit of the radical avant-garde and would require a revolution of both art and life.

Interestingly, as we will see, some of the Situationist aesthetic projects anticipated postmodern culture—such as the emphasis on pastiche and quotation and the collapsing of boundaries between high and low art and between art and life—though Situationist practice was always geared toward a revolutionary transformation of the existing society, both bureaucratic communist and capitalist ones. From a more strictly theoretical perspective, Debord and his colleagues synthesized Marx, Hegel, Lefebvre, and Lukács (whose 1923 work *History and Class Consciousness* had been translated into French in 1960 by the Arguments group) into a critique of contemporary society published in Debord's *Society of the Spectacle* in 1967. Politically, Debord and the Situationists were deeply influenced by the council communism promoted by the early Lukács, Korsch, and Gramsci and by a tradition taken up in France by both the Socialism or Barbarism and the Arguments groups.[4] This tradition was radically democratic, emphasizing the need for workers and citizens to democratically control every realm of their life from the factory to the community, and it influenced Debord and the Situationists' positive ideal.

The Society of the Spectacle

> When the real world changes into simple images, simple images become real beings and effective motivations of a hypnotic behavior. The spectacle as a tendency *to make one see the world* by means of various specialized mediations (it can no longer be grasped directly), naturally finds vision to be the privileged human sense which the sense of touch was for other epochs; the most abstract, the most mystifiable sense corresponds to the generalized abstraction of present day society.
>
> —Guy Debord

Debord's analysis of contemporary capitalism developed Marx's theory of commodification to its latest stage, which he described as "the becoming-world of the commodity and the becoming-commodity of the world" (#66). For the Situationists, the current stage of social organization is a mutation in capitalist organization, but it is still fully accessible to a Marxist interpretation. Beneath the new forms of domination, there is "an undisturbed development of modern capitalism" (#65). Also influenced by Gramsci (1971), the Situationists saw the current forms of social control as based on consensus rather than force, as a cultural hegemony attained through the metamorphosis of the consumer and media society into the "society of the spectacle." In this society, individuals consume a world fabricated by others rather than producing one of their own.

Paraphrasing Marx's opening to *Capital*, Debord said: "In the modern conditions of production, life announces itself as an immense accumulation of spectacles" (#1). The society of the spectacle is still a commodity society, ultimately rooted in production but reorganized at a higher and more abstract level. "Spectacle" is a complex term that "unifies and explains a great diversity of apparent phenomena" (#10). In one sense, it refers to a media and consumer society, organized around the consumption of images, commodities, and spectacles. But the concept also refers to the vast institutional and technical apparatus of contemporary capitalism, to all the means and methods power employs, outside of direct force, which subject individuals to societal manipulation, while obscuring the nature and effects of capitalism's power and deprivations.

Under this broader definition, the educational system and the institutions of representative democracy, as well as the endless invention of consumer gadgets, sports, media culture, and urban and suburban architecture and design, are all integral components of the spectacle. Schooling, for example, involves sports, fraternity and sorority rituals, bands and parades, and various public assemblies that indoctrinate individuals into dominant ideologies and practices. The standard techniques of education—which involve rote learning and mechanical memorization of facts presented by droning teachers, to be regurgitated through multiple-choice exams—are very effective for killing creativity and choking the spirit and joy of learning. Currently, the use of video technologies in the classroom can reinforce this passivity and commodify education, with TV "news" punctuated with ads by corporate sponsors. For example, the Whittle Corporation's Channel One is widely used in thousands of schools across the United States, exposing students to commercial messages and reducing news to infotainment. Of course, contemporary politics is also saturated with spectacles, ranging from daily "photo opportunities" to highly orchestrated special events that dramatize state power to TV ads and image management for prepackaged candidates.

For Debord, the spectacle is a tool of pacification and depoliticization; it is a "permanent opium war" (#44) that stupefies social subjects and distracts them from the most urgent task of real life: recovering the full range of their human powers through creative practice. In Debord's formulation, the concept of the spectacle is integrally connected to the concept of separation, for in passively consuming spectacles, one is separated from other people and from actively producing one's life. Capitalist society separates workers from the product of their labor, art from life, and spheres of production from consumption, all of which involve spectators passively observing the products of social life. The Situationist project in turn involved an overcoming of all forms of separation by advocating that individuals directly produce their own life and modes of self-activity and collective practice.

The spectacle spreads its narcotics mainly through the cultural mechanisms of leisure and consumption, services and entertainment, as ruled by the dictates of advertising and a commercialized media culture. This structural shift involves a commodification of previously noncolonized sectors of social life and the extension of bureaucratic control to the realms of leisure, desire, and everyday life. Parallel to the Frankfurt School's conception that society is "totally administered" or "one dimensional" (Horkheimer and Adorno, 1972; Marcuse, 1964), Debord states that "the spectacle is the moment when the commodity has attained the *total occupation* of social life" (#42). Here exploitation is raised to a psychological level; basic physical privation is augmented by "enriched privation" of pseudo-needs; alienation is generalized and made comfortable, and alienated consumption becomes "a duty supplementary to alienated production" (#42).

The shift to a "bureaucratic society of controlled consumption" (Lefebvre, 1984) organized around the production of spectacles is a means of advancing profit and gaining ideological control over individuals. Unlike early capitalism, whose structural exigencies lay in the forceful exploitation of labor and nature and in defining the worker strictly as a producer, the society of the spectacle defines individuals as consumers and attempts to constitute their desires and needs, first creating and then exploiting them.

The spectacle not only expands the profits and power of the capitalist class but also helps to resolve a legitimation crisis of capitalism. Rather than venting anger against exploitation and injustice, the working class is distracted and mollified by new cultural productions, social services, and wage increases. In consumer capitalism, the working classes abandon the union hall for the shopping mall and celebrate the system that fuels the desires that it ultimately cannot satisfy. But the advanced abstraction of the spectacle brings in its wake a new stage of deprivation. Marx spoke of the degradation of *being into having*, in which creative praxis is reduced to the mere possession of an object, rather than its imaginative transformation, and in which need for the other is reduced to greed of the self. Debord speaks of a further reduction, the transformation of *having into appearing*, where the material object gives way to its semiotic representation and draws "its immediate prestige and ultimate function" (#17) as image—in which look, style, and appearance function as signs of social prestige. The production of objects *simpliciter* gives way to "a growing multitude of image-objects" (#15) whose immediate reality is their symbolic function as image. Within this abstract system, it is the *appearance* of the commodity that is more decisive than its actual use value, and the symbolic packaging of commodities—be they cars or presidents—generates an image industry and new commodity aesthetics (see Haug, 1986).

While spectacles, such as Roman bread and circuses, have long distracted the masses and celebrated state power, the society of the spectacle has more immediate origins in 19th-century capitalism organized around commodity spectacles and consumption. As Walter Benjamin (1973) argued (discussed in Buck-Morss, 1989), the commodity-phantasmagoria of the spectacle began in the 19th century in the Paris Arcades, which put on display all the radiant commodities of the day. Department stores soon appeared in Paris and elsewhere that exhibited commodities as a spectacle and soon became coveted temples of consumption. Sears catalogues offered customers entrance to commodity paradise, and companies began using images and advertising to market their wares, creating a society in which images offered fantasies of happiness, luxury, and transcendence (see Ewen and Ewen, 1983).

By the 1920s, advertising had become a major social force, and films were celebrating affluence and consumer lifestyles, but the depression of the 1930s and World War II prevented the consumer society from developing. After the war, however, the consumer society took off as soldiers returned with money in their pockets to start families and to buy the new products offered and promoted on radio and television. Life in the suburbs was centered on consumption, and new shopping malls gathered together a diversity of department stores and specialty shops in an environment now scientifically designed—right down to subliminal messages in the Muzak—to promote consumption.

The consumer society is currently so highly developed that even alternative grocery stores and bookstores are organized around the principle of spectacle, dazzling the customer with their displays of wares. In Austin, Texas, for example, there is the Whole Foods shop, which provides a mesmerizing array of health and gourmet foods from the entire world. Next door there is a Book People, exhibiting three resplendent stories of books of all types, focusing on the alternative and countercultural. In the midst of this consumer paradise, the Buddhism section has a rock garden, a meditation space, and a giant statue of the Buddha, presented as a commodity icon, a god of mass-marketed spirituality. Even Christian bookstores are now feeling the pressure to diversify by resurrecting themselves as book-and-coffee emporiums in order to compete with the secular marketplace of spectacle and consumption, dramatically furthering the implosion between the sacred and secular that began with the initial commodification of religious texts and televangelism.

With cable and satellite TV, the spectacle is now so ubiquitous and accessible that one need not even rise from the reclining chair to shop; only a telephone and credit card is required to purchase a vast array of products from TV home-shopping networks. And advertisers are already creating new malls in cyberspace that will provide virtual shopping environments of the most exotic kind to parade a surfeit of products. Indeed, corporations are currently establishing Web sites on the Internet, offering all sorts of visual spectacles to entice customers to buy their goods and to provide consumer profile information for future advertising and commercial ventures. Like the industrial commodity markets that preceded it, the spectacle has gone global, with the proliferation of satellite dishes beaming Western sex and violence to all corners of the globe. Elections from Israel to Russia reduce politics to a battle of image and media spectacle, with Hollywood-style media campaigns for candidates intent on selling personalities more than political platforms.

Entertainment is a dominant mode of the spectacle; its codes permeate news and information, politics, education, and everyday life. Newspapers like *USA Today* fragment news into small stories, illustrated by graphs, charts, and color pictures, while both local and national TV news is saturated by happy talk and human interest stories. Cable TV promises to deliver over 500 channels by the year 2000, and Internet Web sites may offer even more "infotainment" spectacles, as multimedia technologies develop and the computer absorbs more and more cultural forms.

The infotainment society reduces all genres, from news to religion to sports, to the logic of commodity spectacle. It appears that professional sports, a paradigm of the spectacle, can no longer be played without the accompaniment of cheerleaders, giant mascots who clown with players and spectators, and raffles, promotions, and contests that hawk the products of various sponsors. Instant replays turn the action into high-tech spectacles, and stadiums themselves show electronic reproductions of the action, as well as displaying giant advertisements for various products, which often rotate for maximum saturation—previewing forthcoming environmental advertising in which entire urban sites will become scenes to promote commodity spectacles. Sports stadiums, like the new United Center in Chicago, Illinois, or the America West Arena in Phoenix, Arizona, are now named after corporate sponsors. The Texas Rangers stadium in Arlington, Texas, is as much a shopping mall and commercial area as it is a sports arena, with its office buildings, stores, and restaurant in which, for a hefty price, one enjoys a view of the athletic events, while consuming food and drink.

Moreover, sports is increasingly subject to market logic and commodification, with professional athletes making millions of dollars in salaries and in fees for product endorsements and advertising. Media presentation of events like basketball games are increasingly commodified, accompanied as they are by the "Bud player of the game," "Miller Lite genuine moments," the "Reebok half-time report," the "AT&T time out," and "Dutch Boy in the paint," along with ads featuring star players hawking merchandise. TV networks bid astronomical sums for the rights to broadcast live professional sports events, and superevents like the Superbowl and NBA championship games command some of the highest advertising rates in TV. It probably will not be too long before the uniforms of professional sports players are as littered with advertisements as racing cars. In a way, this has already happened with superstars like Michael Jordan who, commodified from head to foot, has sold his various body parts to the highest bidders and in winter 1996 added film to his image repertoire with *Space Jam*. Indeed, the top athletes augment their salaries, sometimes spectacularly, by endorsing products and appearing in TV shows and films, thus collapsing sports, commerce, entertainment, and advertising into dazzling spectacles that celebrate the products and values of corporate culture.

It appears in the society of the spectacle that a life of luxury and happiness is open to all, that anyone can buy the sparkling objects on display and consume the spectacles of entertainment and information. But in reality only those with sufficient wealth can fully enjoy the benefits of this society, whose opulence is extracted out of the lives and dreams of the exploited. The poor souls who can't afford to live out their commodity fantasies in full are motivated to work harder and harder, until they are trapped in the squirrel cage of working and spending, spending and working—and increasingly borrowing money at high interest rates. Indeed, consumer credit card debt has skyrocketed 47 percent in recent years, as credit cards are easier to get and interest payments rise; the average debt per household is now over $3,000, up from barely more than $1,000 per household in 1985 (*The New York Times*, December 28, 1995: C1); by the mid-1990s, over one in 90 credit card holders was declaring bankruptcy, and by the end of 1996 consumer debt had skyrocketed to another all- time high (*NBC Nightly News*, December 8, 1996).

When images determine and overtake reality, life is no longer lived directly and actively. The spectacle involves a form of social relations in which individuals passively consume commodity spectacles and services without active and creative involvement. The popular MTV animated series *Beavis and Butt-Head* provides contemporary examples of such passivity, as the two characters sit in front of the television watching music videos and are usually only incited to action by something they watch. Their entire vocabulary and mapping of the world derives from the media and they describe media bites as something that's "cool" or that "sucks" according to whether the images do or do not conform to dominant forms of sex and violence (see Kellner, 1995a; Best and Kellner, forthcoming).

Media spectacles are financed by advertisers, who in turn pass along costs to the consumers, who are doubly exploited in work and consumption. Consumers pay for the spectacles of entertainment, subsidized by advertising, in the form of higher costs for products. Moreover, the entertainment and information offered is a function of what the culture industries think will sell and advance their own interests, producing more desires for their goods and way of life. In an attempt to further control the benighted couch potatoes of consumer capitalism, the entertainment industry has invented "interactive TV"—an oxymoron if there ever was one—which allows the viewers to be their own director, to call their own shots, to edit their own videos, or even to project their own image onto the screen (especially enticing with porn videos) to "interact" with the programmed dialogue. Thus, we can now go into the TV, becoming a part of it as it has become a part of us. With every passing day, people become more and more like characters in David Cronenberg's film *Videodrome*, or like the "Television Man" satirized by the Talking Heads:

> I'm looking and I'm dreaming for the first time
> I'm inside and I'm outside at the same time
> And everything is real
> Do I like the way I feel? . . .
> Television made me what I am . . .
> (I'm a) television man.

Virtual reality devices promise to take individuals into an even higher and more powerful realm of social abstraction such that one may forget that he or she really is interacting with the environment projected by the device, be it a war game or a pornographic fantasy. Of course, "virtual" and "interactive" technologies merely seduce the viewer into an even deeper tie to the spectacle, and there is no media substitute for getting off one's ass, for interactive citizenship and democracy, for actually living one's life in the real world. Indeed, advocates of the superiority of cyberworlds denigrate the body as mere "meat" and "real life" ("RL") as an annoying intrusion into the pleasures of the media and computer worlds of cyberspace.

The correlative to the spectacle is thus the spectator, the passive viewer and consumer of a social system predicated on submission and conformity. In contrast to the stupor of passive consumption, Debord and the Situationists champion active, creative, and imaginative practice in which individuals create their own "situations," their own passionate existential events, fully participating in the production of everyday life, their own individuality, and, ultimately, a new society. Thus, to the passivity of the spectator they counterpose the activity of the radical subject who constructs his or her own everyday life against the demands of the spectacle (to buy, to consume, to conform, etc.). The concept of the spectacle therefore involves a distinction between passivity and activity and between consumption and production, condemning passive consumption of spectacle as an alienation from human potentiality.

The concept also involves distinctions between the artificial and the real and between the abstract and the concrete. Unlike real human needs for creativity and community, commodity needs and spectacles are artificial, with capitalism endlessly multiplying needs for the latest gadget or product line while creating a fantasy world of imagined self-realization and happiness. In place of concrete events and relations with others, the spectacle substitutes abstract images, commodity fantasies, and relations with technology. The spectacle escalates abstraction to the point where one no longer lives in the world per se—"inhaling and exhaling all the powers of nature" (Marx, in Marx and Engels, 1978: 115)—but in an abstract *image* of the world. "Everything that was directly lived has moved away into a representation" (Debord,

#1), which Debord describes as the "philosophization of reality": "The spectacle does not realize philosophy, it philosophizes reality" (#19). By this he means that spectacle and image constitute an ersatz reality, an ideal world of meanings and values to be consumed by the commodity self. The realization of philosophy, as conceived by Marx, entailed the abolition of "philosophy"—that is, of an abstract ideology constituted above and against the concrete conditions of social existence—and the synthesis of theory and practice. For Marx, revolutionary struggle seeks to realize the ideals of the Enlightenment, creating equality, freedom, individuality, and democracy as the form of social life, thus actualizing Western culture's highest philosophical ideals.

The philosophization of reality, on the other hand, separates thought from action as it idealizes and hypostatizes the world of the spectacle. It converts direct experience into a specular and glittering universe of images and signs, where instead of constituting their own lives, individuals contemplate the glossy surfaces of the commodity world and adopt the psychology of a commodity self that defines itself through consumption and image, look, and style, as derived from the the spectacle. Spectators also project themselves into a phantasmagoric fantasy world of stars, celebrities, and stories, a world in which individuals compensate for unlived lives by identifying with sports heros and events, movie and television celebrities, and the lifestyles and scandals of the rich and infamous.

Individuals in the society of spectacle constitute themselves in terms of celebrity image, look, and style. Media celebrities are the icons and role models, the stuff of dreams whom the dreamers of the spectacle emulate and adulate. But these are precisely the ideals of a consumer society whose models promote the accumulation of capital by defining personality in terms of image, forcing one into the clutches and clichés of the fashion, cosmetic, and style industries. Mesmerized by the spectacle, subjects move farther from their immediate emotional reality and desires and closer to the domination of bureaucratically controlled consumption: "The more [one] contemplates the less he lives; the more he accepts recognizing himself in the dominant images of need, the less he understands his own existence and his own desires . . . his own gestures are no longer his but those of another who represents them to him" (Debord, #30). The world of the spectacle thus becomes the "real" world of excitement, pleasure, and meaning, whereas everyday life is devalued and insignificant by contrast. Within the abstract society of the spectacle, the image thus becomes the highest form of commodity reification: "The spectacle is *capital* to such a degree of accumulation that it becomes an image" (#34).

But, as inverted and abstract, the spectacle is simply "the other side of money" (#49) and the social relations of capitalism. Earlier, in Marx's day, money directly dominated society as the representation of general equivalence, allowing the exchange of incompatible use values. In contemporary capitalism, however, the hegemony of money is indirect, mediated through the production of images that allow a more generalized equivalence. "The spectacle is the developed modern complement of money where the totality of the commodity world appears as a whole, as a general equivalence for what the entire society can be and can do. The spectacle is the money which one *only looks at*, because in the spectacle the totality of use is already exchanged for the totality of abstract representation" (#49). Where qualitative differences previously were erased in the serial production of objects, now they evaporate in the stratosphere of images and signs. In TV land, all images, whether of war in Bosnia or of Kathie Lee Gifford soaking up sun on a luxury cruise ship, attain a fundamental equality, if not for spellbound spectators then at least for industry executives, for whom news, sports, and sitcoms ultimately are nothing but interchangeable frames for generating advertising revenue.[5]

The world of the spectacle is thus the world of capital, of commodities and consumer and media fantasies. Debord emphasizes the super-reification of image objects as a massive unreality, an inversion of reality and illusion. The spectacle is "the autonomous movement of the non-living" (#2). The actual class divisions of society, for example, are abolished in the spectacle and replaced with signs of unified consumption that address everyone equally as consumers. But, like Kierkegaard, Feuerbach, and Marx, Debord saw not simply the blurring of illusion and reality but the authentication of illusion as more real than the real itself. "Considered in its own terms, the spectacle is *affirmation* of appearance and affirmation of all human life, namely social life, as mere appearance" (#10). The universalization of the commodity form is then interpreted by Debord as the reduction of reality to appearance, its subsumption to the commodity form, and its subsequent commodification.

Along these lines, as we will show, Debord's thinking is remarkably congruent with Baudrillard's key themes, specifically his notions of simulation, implosion, hyperreality, and the proliferation of signs and images in postmodern culture. But Debord was more a good Hegelian–Marxist than a proto-Baudrillardian. Like Marx, as much as Debord emphasized the commodification of reality, he also emphasized the *reality of commodification* and the ability of individuals to see through capitalist illusions and fantasies. Despite the pronounced emphasis on the artificiality of the spectacle, Debord refused to abandon the attempt to interpret and change social reality. Peering into the shadows of a reified unreality, Debord drew back to report and critique what he had seen; there is an implosion of opposites, but the separate poles retain their contradictory identity; illusion overtakes reality, but reality can be recuperated for Debord through a *critical hermeneutics* that sees through appearances, illusions, and fantasies to the realities being masked and covered over. In addition, Debord urged radical practice, the construction of situations, to overcome the passivity of the spectator.

Situationist Practice

> Today the revolutionary project stands accused before the tribunal of history—accused of having failed, of having engendered a new alienation. This amounts to recognizing that the ruling society has proved capable of defending itself, on all levels of reality, much better than revolutionaries expected. Not that it has become more tolerable. Revolution has to be reinvented, that's all.
>
> —INTERNATIONALE SITUATIONNISTE

Like Marx and unlike Kierkegaard and Nietzsche, Debord and the Situationists believe that working people can see through illusion and can organize and struggle for their own interests. The fundamental goal of Situationist praxis was to reconstruct society and everyday life to overcome the apathy, deception, passivity, and fragmentation induced by the spectacle. The recovery of active existence was possible only by destroying spectacular relations and by overcoming passivity through the active creation of "situations" and the use of technology to enhance human life. Situationist praxis was rooted in a technique of deconstructive contrast that foregrounds the contradiction between the emancipatory potential of technology and its actual application toward the perfection of domination. Through a cultural revolution aimed at the creation of a revolutionary subjectivity, the Situationists hoped to extend the cultural struggle to a general social revolution that would overthrow the society of the spectacle, both capitalist and communist.

The aesthetic strategies of the Situationists included the *détournement*, a means of deconstructing the images of bourgeois society by exposing the hidden manipulation or repressive logic (e.g., by changing the wording of a billboard); the *dérive*, an imaginative, hallucinatory "drift" through the city (an urban variation on the surrealist stroll through the countryside); and the *constructed situation*, designed to unfetter, create, and experiment with desires (see the texts in Knabb, 1981: 5–13, 43–47, 50–59).

These strategies were embodied in a wealth of imaginative projects. The concept of *détournement* involves a turning around and turning upside down that inverts the inversion. That is, just as capitalism inverts value by putting its values and forms over life and human self-activity, so does Situationist practice invert the anti-life and anti-creativity perspectives of capitalism by intensifying life and creative practice. There is a certain Nietzschean vitalism in Situationist practice, which attempts to enhance, intensify, and increase life energies against the banal death culture of the existing society. This emphasis on passion, spontaneity, and realizing desires would be taken up by later postmodern theorists such as the early Lyotard, Foucault, Deleuze, Guattari, and others as well (Best and Kellner, 1991).

Situationist practice involves the disarticulation of conventional forms of culture and their rearticulation into forms of oppositional culture. Situationist *détournement* used the form of comic strips, for example, in their publications to promote revolutionary slogans; they changed advertising slogans into political ones by defacing billboards; they created urban graffiti to disseminate their ideas; and they used every media from film to painting to transmit their politics, attempting in the process to destroy the bourgeois forms of the medium in question, which would be restructured à la Bertolt Brecht into organs of revolutionary transformation. For instance, Debord suggested a *détournement* of D. W. Griffith's racist film *The Birth of a Nation* "by adding a soundtrack that made a powerful denunciation of the horrors of imperialist war and of the activities of the Ku Klux Klan, which are continuing in the United States even now" (in Knabb, 1981: 12). And Debord made his own films employing such techniques, while Situationist artist Asger Jorn painted over existing paintings to attack conventional art forms, to implode the distinction between high and low art, and to create new forms of aesthetic practice.

Interestingly, the Situationist advocacy of pastiche and plagiarism anticipated one of the major strategies of postmodern art, which utilized quotation and outright stealing of images and ideas of the past to make aesthetic statements (see Chapter 4). Like other groups in the aesthetic avant-garde, the Situationists attacked the distinction between high and low art, a gesture that would be definitive for postmodernism in the arts. But for the Situationists, quotation and appropriation were geared to produce radical political effects, to see behind the illusions of contemporary media culture, to promote critical consciousness, and to produce new forms of culture and everyday life, and not merely to make ironic commentary on the institution of art or the dissolution of subjectivity, or to merely play with the pieces of a moribund culture.

After a period of predominantly aesthetic activity in the 1950s, Debord and the Situationists turned toward politics and political practice as the focus of their endeavors in the 1960s. Debord involved himself with the Socialism or Barbarism group, which emphasized the primacy of workers' self-activity and participated in Lefebvre's seminars on everyday life. Dissatisfied with both approaches, however, the Situationists turned to developing strategies that would directly revolutionize everyday life. First published in 1967, the same year as Debord's *The Society of the Spectacle*, Raoul Vaneigem's *The Revolution of Everyday Life* (1983) contained utopian proposals for revolutionary practice that supplemented Debord's rather abstract analysis of contemporary forms of social domination. Its French title, *Traité de savoir-vivre à l'usage des jeunes générations*, expresses the practical orientation of the text, which manifests the Situationist appeal for realizing desire, living creatively, and liberating oneself from the ideologies and banalities of the existing system.[6]

Following the lead of "Western Marxists" such as Lukács, Gramsci,

Korsch, and the Frankfurt School, the Situationists combated the mechanistic assumption that a worldwide socialist revolution would result automatically from the entropic breakdown of the capitalist economy: "The growth of the productive forces cannot guarantee such power" (#188). Instead, the subjective conditions of the revolution must be extensively developed through a cultural politics. But where Lukács sought a "class consciousness" in the narrow sense of political awareness, the Situationists, following the lead of Wilhelm Reich (1972), extended this consciousness to include an entire new subjectivity, new desiring forms, and a revitalized creativity and imagination. "The consciousness for desire and the desire for consciousness are identically the project which, in its necessary form, seeks the abolition of classes, the workers' direct possession of every aspect of their activity" (#53). With Marcuse (1955, 1969, 1972, 1978), Debord believed that a general social liberation required an immediate self-emancipation and self-creation. And as Marx sought the "abolition" of philosophy in its merger with praxis, so Debord, in a broader conception, sought the supersession of art in the aestheticizing of everyday life (#191).

Updating the sociological analyses of culture and communication begun by Kierkegaard, Nietzsche, the Frankfurt School, and others, Debord and the Situationists thematized the loss of genuine communication and social relations in the age of electronic media and information. Moreover, they included bureaucratic forms of "radical" political organization (Leninism) within the sphere of spectacular relations, seeing that nondemocratic leftist structures also pacified individuals and subjected them to the rule of alien powers. In opposition to Lenin's concept of a "vanguard party," the Situationists argued for councils as the proper mode of organization: "Only there is the spectacular negation of life negated in its own turn" (#117).

Consequently, Debord and the Situationists attempt à la Marx to merge theory and practice and to continue the revolutionary project in contemporary conditions. Moreover, Debord followed Marx in utilizing the discourse of the modern to describe contemporary societies. In post-World War II France, rapid modernization gave rise to a plethora of critical analyses of society, and Debord, Lefebvre, the early Baudrillard, and others associated with the Left utilized the term "modern" to describe contemporary capitalist societies. It was the task of Baudrillard, however, to declare Debord's neo-Marxist–modernist framework a fiction, to push the analysis of abstraction and inversion to its ultimate consequences, to obliterate the subject and embrace the object, to cross that threshold where opposites lose their identity, where truth ceases to exist, where politics dies with the collapse of the social, where history loses its redemptive meaning, and where reality disappears altogether. This threshold point for Baudrillard is the doorway from modernity to postmodernity, leading to the turn from modern to postmodern theory.

BAUDRILLARD: DEATH OF THE REAL

> Abstraction today is no longer the map, the double, the mirror or the concept. Simulation is no longer that of a territory, a referential being, or a substance. It is the generation by models of a real without origin or reality: a hyperreal. The territory no longer precedes the map. Henceforth, it is the map that precedes the territory. . . . it is the map that engenders the territory.
>
> —JEAN BAUDRILLARD

Jean Baudrillard was deeply influenced by Debord and the Situationists. Both Baudrillard and Debord theorized the abstraction involved in the development of the consumer and media society. For both, the electronic media were a new stage in abstraction in which interpersonal relations become technologically mediated. Both saw the media as one-way modes of transmission that reduced audiences to passive spectators;[7] both were concerned with authentic communication and a more vivid and immediate social reality apart from the functional requirements of a rationalized society. For Baudrillard, this entailed a destruction of all media, for their function is precisely to *mediate*, to prevent genuine communication, which, in a strangely Rousseauean metaphysics of presence, he conceived to be direct and nonmediated. Debord's conception of media as "unilateral communication" is similar (see #24, #28), though he attempted to devise media practices that would transform the media, and thus, unlike Baudrillard, he championed the development of alternative media against existing society and culture.

And yet despite similarities to his predecessors, Baudrillard takes us into a whole new era of social development: beyond Marx, beyond neo-Marxism, beyond the Situationists, beyond modernity, and beyond theory itself. We leave behind the society of the commodity and its stable supports; we transcend the society of the spectacle and its dissembling masks; and we bid farewell to modernity and enter the postmodern society of the simulacrum, an abstract nonsociety devoid of cohesive relations, shared meaning, political struggle, or significant change. For Baudrillard, postmodernity marks the horizon where modern dynamics of growth and explosion reach their limits and begin to turn inward, resulting in an implosive process devouring all relational poles, structural differences, conflicts, and contradictions, as well as "truth," "reality," and even "power."[8]

Yet in his early works, *Le système des objets* (1968), *La société de consommation* (1970), and *For a Critique of the Political Economy of the Sign* (1981 [1972]), Baudrillard pursued an analysis of commodities and consumer society. Until *The Mirror of Production* (1975 [1973]), Baudrillard could be described, like Debord, as a neo-Marxist whose project was to retain the basic theoretical framework of Marxism, organized around class and production, while supplementing it to account for the changes in the nature of domination effected by the shift to a society based on mass media, consumption, and what Baudrillard called a "political economy of the sign."

Debord and Baudrillard were doing sociological studies of the new consumer society and everyday life in France simultaneously in the 1960s; both worked with Lefebvre and were part of a similar political and intellectual milieu. Just as Baudrillard was aware of the work of the Situationists, there is evidence they were aware of his work, since in one text they denounced him as a "decrepit modernist-institutionalist" (in Knabb, 1981: 211). But it seems the Situationists were more an influence on Baudrillard than vice versa. For Baudrillard, the Situationists were "without doubt the only ones to attempt to extract this new radicality of political economy in their 'society of the spectacle'" (1975: 120). At one time, in fact, Baudrillard considered himself a Situationist: "Pataphysician at twenty—Situationist at thirty—utopian at forty—transversal at fifty—viral and metaleptic at sixty—that's my history" (1990: 131). Yet he rejected the Situationist analysis as itself bound to an obsolete modernist framework based on notions like history, reality, and interpretation, and he jumped into a postmodern orbit that declared the death of all modern values and referents under conditions of simulation, implosion, and hyperreality.

Thus, the genesis of the postmodern turn in France was a passage through neo-Marxism and the political and artistic avant-garde, including Debord and the Situationists. Baudrillard, Lyotard, Foucault, Derrida, Deleuze, Guattari, and other French postmodern theorists lived through this situation, combining impulses from Marxism, surrealism, and other avant-garde aesthetic and political movements. From this perspective, Baudrillard's work can be seen as an attempt to assess the catastrophic fallout of the commodification and abstraction process traced by Marx and Debord, the processes of massification described by Kierkegaard and Nietzsche, and the hegemony of ideology and illusion assessed by theorists as disparate as Kierkegaard, Marx, Nietzsche, Gramsci, the Frankfurt School, Debord, and others. Whereas Kierkegaard and Nietzsche held out promises of greater individuality and self-determination for superior persons, and Marx and Debord believed in the potential freedom of people through revolution, Baudrillard abandons the project of individual transformation as well as of social transformation, declaring both subjectivity and sociality to be illusions. And whereas Marx analyzed the reduction of materiality into quantitative commodities and Debord depicted the absorption of the commodity world into a specular empire of images, Baudrillard describes an even more advanced state of abstraction in which the object is absorbed altogether into the image and dematerializes in closed cycles of semiotic exchange.

Baudrillard theorizes a cybernetic, self-reproducing society based on consumption, media, information, and high technology, a society in which exchange occurs at the level of signs, images, and information, thereby dis-

solving Marx's distinction between "superstructure" and "base," as well as Debord's distinction between appearance and reality. Emphasizing contemporary capitalism as a rupture in the old mode of organization, Baudrillard's work was well distanced from classical Marxists but much akin to the Situationists, whom he credited for having grasped consumption as a new form of domination. But the early Baudrillard broke with the Situationists on both theoretical and political grounds. He understood contemporary society not in terms of spectacle but of "sign value," rooting the development of the commodity in the structural logic of the sign, rather than vice versa (1981). Baudrillard sometimes spoke of the "spectacle," but only provisionally. He rejected the term for two reasons: because it implies a subject–object distinction that he feels implodes in a hyperreality and because the Situationists theorize the spectacle as an extension of the commodity form rather than as an instantiation of a much more radical and abstract order, the political economy of the sign, or as the proliferation of signs and simulation models.

In his work of the early 1970s, Baudrillard claims that the "political economy" theorized by Marx is first and foremost a *semiological revolution*: a massive restructuring and reduction of complex (precapitalist) symbolic formations to the rationalized formulas of industrialized society. Political economy, for Baudrillard, represents an "immense transmutation of all values . . . into economic-exchange value" (1981: 113). The semiological project involves the encoding of economic values and logic into a system of commodity signs that signify wealth, affluence, power, success. In Baudrillard's analysis, in addition to use value and exchange value, the commodity sign radiates *sign value*, whereby a system of meaning differentially organizes images, objects, and practices into a hierarchy of prestige and value. The process of commodification for Baudrillard thus contains a logic of status and distinction in which value is defined by the position of the commodity sign in a differential system that valorizes some products over others in a code of social prestige.

Political economy for Baudrillard thus reduces the richness of symbolic ambivalence and exchange (premodern religious and cultural practices that have polyvalent, nonutilitarian meanings) to rationalized equivalence, in which the codes of the commodity world rule and determine all value as instrumental value. In this world, the fundamental fact is not the exploitation of labor but, more generally, the rationalization and functionalizaton of the entire world. In his work of the early 1970s, Baudrillard could be assimilated, à la Debord, to a neo-Marxist problematic that pursued a hybrid analysis of commodity and sign production and consumption, class, and prestige. But already—although relegated to skeptical footnotes and repressed intuitions—he showed strong signs of his later deconstructive radicalism, which would lead him to reject Marxism (and modern theory) altogether.[9]

We would suggest, however, that even in his early neo-Marxian work,

Baudrillard frequently followed Weber more than Marx. In *For a Critique of the Political Economy of the Sign* and *The Mirror of Production*, he interprets political economy as a gigantic system of social administration, of which capitalism and industrial organization are only moments of a larger rationalization process. Baudrillard radicalized this broader emphasis in his later mid- to late-1970s work, in which the focus shifts from consumer society to an even more intense form of control in the mode of cybernetic codes and simulation models. And just as Weber saw Protestantism and Calvinism as causal influences on capitalism rather than as mere effects of the market, Baudrillard insists on the causal primacy of semiotic over material exchange.

Baudrillard's argument against Debord is that during the phase of political economy theorized by the Situationists in terms of the society of media, consumption, and spectacle, a generalization and complexification of the sign form extended throughout the entire culture and environment leading to a hegemony of sign value in which commodities are produced, distributed, and consumed for their conspicuous social meaning. The object is converted into a mere sign of its use, now abstract and divorced from physical needs. The whole cycle of production, distribution, and consumption, Baudrillard claims, is transformed into a semiotic system of abstract signifiers with no relation to an objective world. In the imaginary world of sign value, one consumes the signs of power or prestige through driving a certain type of car or wearing designer clothes.[10] This is a new stage of abstraction, a dematerialization of the world through semiological (re)processing in which images and signs take on a life of their own and provide new principles of social organization.

Rule by sign value involves an autonomization of the signifier in which the relative unity and stability of the industrial world and its system of signs breaks apart. No longer constrained by an objective reality or by pre-existing needs or use value, the signifier is free to float and establish its own meanings through its manipulation in coded differences and associative chains (as occurs in advertising, fashion, and environmental design). Freed from any stable relationship with a signified, where the sign structure points to a distinct referent in the world, the signifier becomes its own referent, and this autonomization becomes the basis of semiological domination. The commodity form is eclipsed by the "sign form" and subsequently "bears no relation to any reality whatever: it is its own pure simulacrum" (Baudrillard, 1983a: 11). Signification is now radically relativized, and anything can pass as "meaning" or "reality."

The hegemony of the sign which appears in contemporary capitalism is not an extension of the commodity, as Debord saw it, but, rather, the unfolding of a more dominant logic, of which the commodity was only an initial form and alibi. Lacking an adequate semiotic perspective, bound with Marx to the mirror of production, Debord, in Baudrillard's view, failed to grasp

> the passage from the form-commodity to the form-sign, from the abstraction of the exchange of material products under the law of general equivalence to the operationalization of all exchanges under the law of the code [i.e., the semiological structure that governs all meaning, reducing value to merely utilitarian form]. With this passage to *the political economy of the sign*, it is not a matter of a simple "commercial prostitution" of all values [as Marx says]. . . . It is a matter of the passage of all values to exchange-sign value, under that hegemony of the code. That is, of a structure of control and of power much more subtle and more totalitarian than that of exploitation. *For the sign is much more than a connotation of the commodity*, than a semiological supplement to exchange value (1975: 121–122).

In Baudrillard's scheme, the referential world of the commodity—needs, use value, and labor—was only a historical passageway for a radical semiurgy that aims at the liquidation of society and the real, their displacement through structural codes and signs. Radical semiurgy—that is, the proliferation and dissemination of signs—constitutes a new source of abstract power that lies not in the commodity nor in the organization of the economy but in the autonomous development of the sign, whose genealogy Baudrillard traced in his studies of simulations (1983a, 1993, 1994). Under conditions of radical semiurgy and implosion, Baudrillard claims not only that political economy and the era of production are dead but also that more recent theorizations of the disciplinary society and the spectacle itself are obsolete:

> We are witnessing the end of perspective and panoptic [disciplinary] space (which remains a moral hypothesis bound up with every classical analysis of the "objective" analysis of power) and hence of the *very abolition of the spectacle*. . . . We are no longer in the society of the spectacle which the Situationists talked about, nor in the specific types of alienation and repression which this implied" (1983a: 56).

Baudrillard's point is that the distinctions and assumptions that Foucault's and Debord's analyses depended on—the subject–object distinction, a subject of alienation or coercion confronting objective reality—have been obliterated. The inversion of illusion and reality that Marx and Debord described is now radicalized, finalized, and pushed to its highest degree in Baudrillard's analysis. His postmodern world could be described well in Debordian terms—"the affirmation of all human life . . . as mere appearance" (#10)—with the key proviso that for Baudrillard there is no longer a real to be recovered behind the illusion (and so there is no illusion either). Drawing a crucial distinction between dissimulation and simulation, Baudrillard implies that both terms involve a feigning and a faking, but whereas dissimulation masks reality, and so ultimately reaffirms it, simulation devours the real and, like a grinning Cheshire cat, leaves behind nothing but commutating signs, self-referring simulacra that feign a relation to an obsolete real.

Simulation for Baudrillard thus describes a process of replacing "real" with "virtual" or simulated events, as when electronic or digitized images, signs, or spectacles replace "real life" and objects in the real world. Simulation models generate simulacra, representations of the real, that are so omnipresent that it is henceforth impossible to distinguish the real from simulacra. The world of simulacra for Baudrillard is precisely a postmodern world of signs without depth, origins, or referent. As he put it in his travelogue *America*: "Why is L.A., why are the deserts so fascinating? It is because you are delivered from all depth there—a brilliant, mobile, superficial neutrality, a challenge to meanings and profundity, a challenge to nature and culture, an outer hyperspace, with no origin, no reference points" (1988a: 123–124).

Simulacra are mere signs and images of the real that come to constitute a new realm of experience, the "hyperreal." Baudrillard's hyperreal is the end result of a historical simulation process in which the natural world and all its referents have been gradually replaced with technology and self-referential signs. This is not to say that "representation" has simply become more indirect or oblique, as Debord would have it, but that in a world where the subject–object distance is erased, where language no longer coheres in stable meanings, where originals are endlessly reproduced in copies, and where signs no longer refer beyond themselves to an existing, knowable world, representation has been surpassed. The real, for all intents and purposes, is vanquished when an independent object world is assimilated to and defined by artificial codes and simulation models, as when the events of the social world attain significance through the entertainment codes of mass media or when men and women judge themselves according to conformity to the dominant ideals of masculinity and femininity, ideals that are largely presented by fashion and advertising (the most extreme example being Cindy Jackson, the "Barbie Doll Woman," who had over 30 different surgical alterations in an attempt to look just like the figure she had worshipped since childhood).

Thus, hyperreality signifies a rupture in the notion of the real brought on by techniques of mass reproduction. "Reality" implies something singular, sui generis, a touchstone by which to measure everything else. But in the conditions of reproduction, Baudrillard claims, all this is lost: Reality becomes what can be infinitely extended and multiplied in a series, through a reproductive medium. No longer sui generis, it infinitely resembles itself in identical copies. No longer the touchstone of everything, it is confused with its copies or even devalued in light of them. Once, perhaps sacred, it becomes strictly operational in reproduction, no more unique or definable than any one of the images of Campbell's Soup cans or Marilyn Monroe in Andy Warhol's paintings.

Thus, for Baudrillard, hyperreality is the transmogrification of reality within the conditions of simulation and reproduction. The Greek prefix "hyper, meaning "over, above, more than normal, excessive," is appropriate. For many, the world of media fantasies is more real than everyday life; video or computer games are more fascinating and alluring than school, work, or politics (often understandably so); porno videos stimulate libido in abstraction from the problems of real relations with others and reduce complex gender identities to mere sex puppets; and hyperreal theme parks like Disney World and simulated environments for many are more attractive than actual geographic sites. The hyperreal is thus the death of the real, but it is a theological death: The real dies only to be reborn, artificially resurrected within a system of signs, "a more ductile material than [representational] meaning in that it lends itself to all systems of equivalence, all binary oppositions and a combinatory algebra" (Baudrillard, 1983a: 4).

When the "real" is no longer directly represented and is artificially (re)produced, it becomes not unreal or surreal, a myth or a fantasy, but hyperreal, "a hallucinatory resemblance of the real with itself" (Baudrillard, 1983a: 3), a "reality" replicated from a model, doubled or multiplied within reproductive processes, volatilized from medium to medium, open to infinite multiplication. "The real is produced from miniaturized units, from matrices, memory banks and command models—and with these it can be reproduced an infinite number of times" (3). In a hyperreal culture, things are conceived from the point of view of reproducibility, as we come to think something is real only insofar as it exists as a serialized commodity, as able to be bought and sold, as able to be made into a novel or a movie. "The very definition of the real becomes: *that of which it is possible to give an equivalent reproduction . . . the real is not only what can be reproduced, but that which is always already reproduced.* The hyperreal" (146).

Baudrillard's analysis of hyperreality can be seen as an extension and radicalization of Walter Benjamin's theory of mechanical reproduction (1969: 217–251). Benjamin analyzed the social and ontological consequences of this advance in technique whereby everything can be technically reproduced. He showed how in the reproduction of a work of art, its aura—its singularity, uniqueness, and specific historical context—was erased, ontologically eroded; for example, what power could there be in viewing the original *Mona Lisa* after such frequent exposure to its reproductions and parodies? What Benjamin analyzed at the level of the work of art can be extended to representational reality in general. For Baudrillard, reality is "volatilized" in its assimilation to the various reproductive media of postmodern society. In a hyperreal postmodern world, reality is dissipated and depleted; it loses its power and force through its cultural processing, through mechanical reproduction and the proliferation of illusions and pseudo-forms.

But ontology abhors a void, and so illusion is taken as "real" and accorded full authenticity. Coca-Cola is the "real thing"; Memorex videos are more real than their referents; Ronald Reagan is a "real" president and Bruce Willis is a "real man." In this (hyperreal) world, Madonna is a fashion arbitrator, successively and successfully promoting different fashions, ranging from her early "boy-toy" attire to a later, more glamorous and upscale couture (see Kellner, 1995a). As we write in late 1996, the clothing and fashions from her movie *Evita* are setting the pace for contemporary fashion—even before the release of the movie (*The New York Times,* December 11, 1996: A21)! A complex Argentine political constellation and the problematic figure of Eva Perón are thus abstracted from the complexity of history into fashion icons for a new look and style.

Under such conditions, Baudrillard argues, an objective world is not able to be extracted from its entanglement within systems of signs and images. Reality and unreality are not mixed like oil and water; rather, they are dissolved like two acids. This is what we call "strong" hyperreality, a situation in which someone is unable to separate reality from illusion, truth from falsehood, original from copy. There is another form of hyperreality, what we call the "weak" version, in which, as both Debord and Baudrillard have observed, someone can make such discriminations but *prefers* the false, the illusionary, or the artificial. In these cases, people understand the differences all too well and consciously choose the illusion, the hyped, the fake, or the copy as somehow better, sexier, more exciting—more real.

This condition is innocent enough when some prefer electronic drums or digitally sampled violins to the real thing, but it is more problematic when students demand that classroom teaching entertain like TV and that books read as fast as a music video, when a significant percentage of a country embraces an actor who simulates presidential qualities rather than an informed, active leader, or when the virtual communities of the Internet *replace* face-to-face contact altogether. Lost in the funhouses of the hyperreal, the postmodern self dissolves in the realm of ersatz experience, becoming itself a mutating set of signs whereby identity is defined in terms of look, style, and image.

THE REALITY OF COMMODIFICATION: POLITICAL ECONOMY AND THE POSTMODERN

> The end of labor. The end of production. The end of political economy. The end of the signifier/signified dialectic which facilitates the accumulation of knowledge and meaning. . . . And at the same time, the end of the exchange value/use value dialectic which is the only thing that makes accumulation and social production possible. The end of the linear dimension of discourse. The end of the linear dimension of the commodity. The end of the classic era of the sign. The end of the era of production.
>
> —JEAN BAUDRILLARD

For Baudrillard, the era of modernity was defined by production and governed by the industrial bourgeoisie. In his view, modernity has now been superseded by a postmodern era of simulations characterized by information and signs and governed by models, codes, and cybernetics. Cumulatively, for Baudrillard, the emergence of simulation and hyperreality as organizing principles of society has effected a break with modernity and the emergence of a new postmodern historical era. But after having embarked upon a postmodern turn during the late 1970s and early 1980s, Baudrillard moved away from theorizing the society of simulation into metaphysics and literary experimentation, trying to develop a new fragmentary style of writing appropriate to the fragmentation of the postmodern world (see the discussion in Kellner, 1989b). He also proclaimed the triumph of the object and defeat of the subject, advising us to renounce the subject's project to dominate the object or change the world. This shift is articulated in relation to Debord and Situationism in a 1990 interview where, after indicating why he refuses to offer positive solutions and why he sees himself primarily as an intellectual terrorist, he mused:

> Of course, today, the real terrorists are not so much us, as the events around us. Situationist modes of radicalism have passed into things and into situations. Indeed, there's no need now for Situationism, Debord, and so on. In a sense, all that is out of date. The hyper-critical, radical, individual sensibility no longer exists. Events are the most radical things today. Everything which happens today is radical. There's a great wealth of radical events, and all one needs to do is to enter into its interplay. Nowadays, reality is radical. Reality is Situationist, not us! (1993: 170)

Baudrillard suggests here that the radicality that Situationists invested in practice, in world-transforming activity, has now become invested in the object world itself, which has been radically transforming itself, seemingly without the intervention of human subjects. In effect, the later Baudrillard moves toward a form of extreme technological determinism in which an autonomous technology produces a new postmodern world. Yet from another angle precisely such verbal provocation is consistent with Situationist practice, which always attempted to shock and be extreme. In an interview in 1991, Baudrillard admitted that "I have always been a bit of a Situationist" (1993: 181), and precisely his extremism, exaggeration, and provocations can be linked to Situationist strategy, despite his attacks on their project.

But there are important differences. The encounter between Debord and Baudrillard is a confrontation between modern and postmodern theories. At stake is nothing less than the possibility of interpreting and changing social reality. Against Baudrillard's decontextualized analysis of hyperreality, Debord situates the spectacle within the framework of advanced capitalism and its structural imperatives of accumulation, growth, and profit. As we have seen, while Baudrillard initially interrelated the political economies of signs and commodities, he later forfeited this dialectical vision to analyze signs, codes, images, and culture in general apart from capitalist social institutions, falling prey to technological determinism and semiological idealism. And where Baudrillard collapses into a nihilistic acceptance of the triumph of the object, Debord and the Situationists advocate struggle and resistance—though, as we shall see, ultimately Debord succumbs to Baudrillard's pessimism.

Though Baudrillard has important insights into contemporary society, we think our present social organization is better interpreted as an *intensification* of (capitalist) modernity rather than as a wholly new "postmodernity," and we believe that political economy and semiology should be synthesized rather than opposed. The contemporary form of society Baudrillard tries to describe is best understood as a generalized extension of capitalism—a more abstract and imagistic mode of commodity production (Debord), a higher realization of its fundamental structuring principles (Mandel, 1975), a capitalism in which postmodernism is "the cultural logic" that comes to dominate ever more spheres of life (Jameson, 1984, 1991), or as part of a new stage of globalized technocapitalism (Kellner, 1989a).

Thus, the dissolution of the social and the real has its roots in the capitalist economy, where abstraction finds its initial and continued support and nourishment. One can interpret postmodernity as a specific phase in an inversion and abstraction process that proceeds through the commodity form and capitalist social relations, assimilating first the physical object, then the entire sphere of culture and everyday life, and extending to the commodified exchange of images, signs, and events in a postmodern media culture and society. And here our current transitional era between the modern and the postmodern *can* be fruitfully understood through a *qualified* use of Baudrillard's key categories: hyperreality, simulation, and implosion, *if* they are synthesized with the categories of political economy. The productive use of Baudrillard therefore need not entail renouncing political economy and the whole modern lexicon; we can acknowledge the conceptual advances of Baudrillard's theorizations of media and semiotics while rejecting their extreme phrasing, their linkage to an apocalyptic concept of rupture, and their pessimistic and quietistic political conclusions.

In particular, we argue that capitalism is currently undergoing a global reorganization based on new technologies and a transnational corporate restructuring and that consequently our contemporary moment is between a disorganized and reorganized capitalism, a situation that requires intense focus both on political economy and on technology and culture. Part of this transformation involves new functions for the state, as well as the loss of national sovereignty to transnational forces, and the creation of new identities and experiences. But we believe that Baudrillard is wrong when he says that we are no longer within a disciplinary society, or a society of the spectacle, but are completely within a fully processed simulatory, cybernetic, postmodern society where distinctions such as those between subject and object, appearance and reality, surface and depth, and so on, are obliterated. Just as a feudal mode of production can coexist with capitalism, so too can disciplinary or "spectacular" forms of power coexist with the society of simulation. What we see today is not discipline *or* simulation, the society of the spectacle *or* the world of the panopticon, hyperreality *or* political economy, but a complex interplay of various mechanisms of social control that include discipline, spectacle, surveillance, simulation, the classic overt violence of the state, and the global restructuring of capitalism, along with good old sexism, racism, and class domination.

The reorganization of capital involves increasing commodification and its penetration into realms of life previously part of a public sphere, colonizing personal life previously resistant to market logic, and entering into new sectors of the world never before defined by the market. During the past decade, we have seen increased privatization of health, education, libraries, hospitals, prisons, and other institutions, accompanied by the systematic dismantling of the welfare state. Even fetal tissue, human body parts, human fertility, and children (through adoption agencies) are becoming increasingly commodified. DNA, the genetic basis of life, is now something genetic engineers and biotech industries can manipulate, patent, sell, and own, as was confirmed in March 1997 when British scientists successfully cloned a sheep. And if powerful interests have their way, communication and information on the Internet will be thoroughly commodified and controlled by giant corporations, as will television programs and other current forms of "free entertainment." Indeed, as we write, mammoth communications and entertainment corporations are merging, restructuring, and organizing themselves to control the entertainment and information systems of the future.

The rise to prominence of the Court TV channel represents a commodification of the judiciary system in which trials and legal processes are offered to audiences and growing audience numbers are sold to advertisers. The O. J. Simpson trial demonstrated that "justice" itself has become commodified and that millions of dollars are needed to buy "justice" in the United States—or, depending on one's opinion of the event, that anyone can purchase an acquittal on the commodity law market. As the old saying goes, capital punishment is punishment for those without capital. The Simpson spectacle also gave birth to a profusion of products and artifacts, leading the Simpson camp to trademark O. J. Simpson's name and image and to sue for a percentage of the profits. Simpson himself attempted to commodify his notoriety into a wealth of products and services that would net him (and his lawyers) millions, though widespread revulsion at his acquittal and his subsequent behavior may have limited the profitability of his enterprises.

Simpson was, of course, a superstar professional athlete, and we have already discussed how sports is a major pillar of the spectacle. Since the rise of televangelism in the 1980s, religion has been relentlessly commodified as well, with TV evangelists raking in millions of dollars from gullible contributors. Yet now even the pope has become a commodity machine, a global superstar whose image the Roman Catholic church recently licensed to sell official papal souvenirs, ranging from books and posters to watches, sweatshirts, and bottled (holy?) water—with a papal Web page to promote the Vatican's image and to sell their merchandise. Always a major site of the spectacle and a source of capital, religion itself has become commodified with TV religion, religion Web sites, and a dramatic increase in religious artifacts ranging from bibles on CD-ROM to Christian rock music videos and CDs.

In a global market capitalism, commodity markets are opening with great fanfare in China and Russia as capitalism exports its markets, products, McCulture, and status consciousness around the globe, bringing with the new goodies its seamy side in the form of crime (both organized and street thug), drugs, social decay, and pathological acquisitive individualism. It is therefore impossible to make sense of contemporary society without using the concepts of political economy, and it appears that Marx's nightmare of a totally commodified society is becoming a reality. Moreover, Baudrillard to the contrary, capitalism continues to depend on the disciplining and normalization of identities to maintain its "moral" order, as is evident by the continued attacks on sex and violence in the media, the furor over gays in the military in the first year of the Clinton administration, a later uproar over gay and lesbian marriages, calls for increased roles for the FBI and state surveillance after the 1995 bombing of the Federal Building in Oklahoma, the proliferation of prisons and calls for tougher law enforcement, the use of mounted video cameras on downtown streets that allow police to monitor activity and conversations 24 hours a day, increased employment of the death penalty, the reintroduction of chain gangs, and growing state regulation of pornography and political discussion on the Internet. Indeed, as vividly dramatized in the 1995 movie *The Net*, it has become all too easy to imagine the possibility of an all-knowing panopticon society as more and more personal information about our backgrounds, medical histories, consumer preferences, bank accounts, and credit histories are stored in computers. Finally, the imposition of work as a mode of discipline has advanced, rather than decreased: We now, on the average, work 163 hours more per year than in 1969 (see Schor, 1992).

In short, we are currently undergoing a great transformation that is redefining and sometimes strengthening many modern institutions while simultaneously generating new postmodern forms of economy, politics, society, and culture. Looking back to the Renaissance, we see that breaks and transitions can often occur over long periods of time, with the emergence of new modern forms coexisting with older premodern forms. Indeed, one could argue that the present moment consists of a contradictory matrix of modern, postmodern, and in many parts of the world the continuation of premodern forms. To use the terminology of Raymond Williams, there are emergent, hegemonic, and residual forms in a given social formation that are overdetermined by older and newer forms of life. Whereas one can interestingly debate whether modern or postmodern forms are currently hegemonic, we find it most prudent to theorize the postmodern as an emergent form of society and culture, albeit a constellation that is growing in power and influence.

Certainly, the phenomena of simulation, hyperreality, and implosion are evident today—yet we would argue that they are mixed with Debordian spectacles, Foucaultian forms of surveillance, and Marxian modes of commodification. During the past decades all the major political battles in the United States have taken place in the media, in a hyperreal world of media image and spectacle. The proliferation of advertising, entertainment, and tabloids produce new fantasy worlds that simulate the real, and indeed "reality programming" that feigns crimes, accidents, or domestic or legal conflicts has become a new popular mode of television for audiences hooked on simulation. The O. J. Simpson trials of 1994–1997 have been an unparalleled legal spectacle in which spectators lived in the hyperreal world of a courtroom drama organized around a struggle over images and with the participants gaining instant celebrity status. And the 1991 Gulf War demonstrated the incredible power mass media has in constructing "reality." Although the devastating effects of the war on the Iraqi people and the environment were all too real, its true enormity was buried in the barrage of media images that coded it as the struggle of Good against Evil and helped to mobilize the public in support of it. The fact that "smart" bombs, so often shown hitting their target with "surgical" precision, in reality usually missed their target (see Kellner, 1992) was merely one overt example of the conversion of falsehood into "truth" and the triumph of simulation and hyperreality.

In a world in which actors are constantly confused with their characters, so that soap opera villains or villainesses have to hire body guards, in which an actor playing O. J. Simpson in reenactments of his 1996–1997 civil trial is accosted in public and cheered on or attacked as if he were O. J. himself, in which vice-presidents attack fictional TV characters for corroding family values, the speaker of the house appears as himself on a situation comedy, the president plays himself on a TV movie, and politicians blame social problems on the media, it is clear that the codes of the real and the unreal have become confused. The triumph of the hyperreal is sure to intensify with the proliferation of computer culture, in which individuals live in the "virtual communities" of the Internet, have cybersex, shop in virtual malls, explore virtual environments, and construct their own genders and identities in new forms of cyperspace interactions. Media and computer culture are producing new areas of experience and interaction, and postmodern theory helps to illuminate the novelty and strangeness of these worlds (see, for example, the studies in Bogard, 1996).

Where the TV previously was a black hole of culture that absorbed all other cultural forms—radio, film, sports, advertising, and so on—now the computer has become the new epicenter of implosion, sucking in the forms and content of previous media. Digitization provides a common substance for text, image, sound, and video, creating a new technoculture. Through the computer, one can now receive phone calls, faxes, radio, TV, films, music, and the texts, sites, and sounds of the Internet, and can, in turn, post one's multimedia texts. Numerous corporations have developed units that attach to the TV to allow for full Internet access and electronic mail, but it's not clear whether this new technological mutation is a TV-becoming-computer or a computer-becoming-TV; at any rate, the lines between the two have definitely, perhaps unrecognizably, blurred.

But against Baudrillard, we argue that simulation and hyperreality ultimately cannot be divorced from the larger analysis of capitalism and political economy. Media and computer culture are vanguards for the commodity culture of global transnational capitalism (Kellner, 1995a, 1995b). Debord himself saw this trend arriving:

> Culture turned completely into commodity must also turn into the star commodity of the spectacular society. Colin Kerr, one of the most advanced ideologues of this tendency, has calculated that the complex process of production, distribution and consumption of knowledge already gets 29% of the yearly national product in the United States; and he predicts that in the second half of this century culture will hold the key role in the development of the economy, a role played by the automobile in the first half, and by railroads in the second half of the previous century. (1970: #193)

Indeed, computers and other new technologies are dramatically transforming the nature of work and production, community and social life, and individual identity, but they are part of a restructuring of capitalism and not independent technological forces. Moreover, a postmodern implosive event such as the erasure of boundaries between news and entertainment in television infotainment should be seen not as the fated effect of some autonomous "code," as Baudrillard sees it. Against postmodern formalism, one should stress the effect intense rating competition among the networks has on the rise of infotainment, pressures that compel them to package the news in a more visual, dramatic, interesting, and consumable way (see Best and Kellner, 1988; Kellner, 1990). Moreover, the current implosion between tabloid and hard news is driven by economics. Similarly, one could see the implosion of identities, gender and otherwise, as a result, at least in part, of the fashion industry and its economic and ideological function within capitalism, serving both to engulf the individual within a continuous cycle of commodity consumption (where the self must continually renew and recycle its own images) and to legitimate the society based on commodity production.

TOWARD A CRITICAL HERMENEUTICS

> We are condemned to interpretation.
> —PAUL RICOEUR

Unlike Baudrillard, who rejects all philosophical frameworks (save his own), proclaiming the death of hermeneutics in the implosion of depth into surface, message into medium, and dialectics in a cyberblitzed erasure of antagonism and contradiction, Debord maintains a critical hermeneutics that seeks to decipher and critique the underlying basis of a frozen history and social order. This means that Debord (1) holds onto a "dissimulation" thesis and so thinks that referential reason and its reality principle (a knowable world existing independently of consciousness) remain intact and recoverable; (2) maintains that there is a link between cultural production and the mode of social organization determined by the imperatives of capitalism; and (3) attempts to apply theory toward the revolutionary transformation of individual subjects and capitalist society. Thus, unlike Baudrillard, who moves from neo-Marxism into an extreme postmodern framework, Debord retains the key categories and assumptions of modern radical social theory, which attempts to map the complex interconnections of society, to discover the sources and mechanisms of oppression, to undertake interpretation and ideological critique, and to engage theory toward the cause of human emancipation and social transformation.

Baudrillard, as we have shown, came to reject all of these tenets of modern theory and politics. He pushed the Debordian framework, in which reification and abstraction are so strongly emphasized, to the point where dissimulation becomes simulation; complex, multidimensional reality becomes no more than a stage set; and social change is an impossibility or strategic illusion. For Baudrillard, the reified images of postmodernity are no longer simply the instruments of a hegemonic abstract capital; they move beyond capital, beyond representation, into the realm of pure signs, dead power, and hyperreality. Just as Aristotle thought the pre-Socratics were groping toward his conception of causality, so Baudrillard might think Debord was striving toward his conception of hyperreality, unable to make the decisive moves. Baudrillard, by contrast, advances a version of postmodern politics, one informed by pessimism and despair that rejects the norm of human emancipation and revolutionary social transformation.

On the other hand, to a certain extent Baudrillard's (1983c) concept of "the ecstasy of communication" is a radicalization of Debord's concept of the spectacle. For Baudrillard, communications technology has rendered everything transparent and visible, overt and obscene (in the sense that all is revealed). There are no more secrets, hidden realities behind appearances, lies to be exposed. Rather, everything is translucent when "exposed to the harsh and inexorable light of information and communication" (Baudrillard, 1983c: 130). Perhaps the "obscene" emerged as the ordinary when the media began going after political figures' private lives in a postWatergate culture that exposed every detail of public figures' personal lives to public scrutiny. Whereas Franklin Roosevelt managed to hide his physical disability from the public and the affairs of presidents never used to be mentioned in the media, every detail of Bill Clinton's sexual exploits and business/political misadventures have been exposed to the harsh glare of media scrutiny.[11] And everyday on TV talk shows, ordinary people expose their darkest, innermost secrets and confess their every transgression before the judgment of a TV nation, hoping to gaining absolution for their sins or at least to attain the 15 minutes of celebrity status that Andy Warhol deemed everyone's birthright in a media culture.

Although Baudrillard presents his concept of the "ecstasy of communication" as an antithesis to notions of the scene and spectacle, dismissed as obsolete modern concepts, surpassed in the dynamics of the postmodern, one can read Baudrillard's notion as a further stage of development of Debord's spectacle, in which it inhabits and colonizes ever more spheres of everyday life, exposing more and more spaces to the "ecstasy of communication." But against Baudrillard, Debord posits a world behind the spectacle, where the forces of oppression and resistance struggle and live.

Indeed, in August 1995, the Mark Fuhrman tapes, released in the O. J. Simpson trial, revealed to the public the vile racist attitudes and criminal behavior of members of the Los Angeles police force—as was dramatized earlier in the Rodney King videotapes. Until media exposure, such noxious attitudes and activities by the police were relatively hidden from view. So Baudrillard is right that a media society exposes ever-more hidden and often scandalous phenomena, but one suspects that such revelations are only the tip of the iceberg, that many, many more secrets and crimes in the sphere of business, politics, and everyday life are hidden from view, and that society is not as transparent as he claims. In fall 1996, for instance, there were many media revelations about CIA operations in the 1980s that were responsible for an influx of drugs into U.S. inner cities—a CIA "drugs for guns" operation used to finance the Nicaraguan contras.[12] Occasionally, such scandalous vities are exposed, but this still is not the normal state of affairs, and thus we believe that we are not yet in a Baudrillardian "ecstasy of communication" where all is clear and transparent.

Hence, we subscribe to Debord's Hegelian–Marxian distinction between appearance and reality and the need for critical thought to continue to probe behind appearances. Whereas Baudrillard surrenders to the surface of the simulacrum, Debord maintains a commitment to critical hermeneutics, attempting to get at the roots of human suffering and oppression through analysis of the capitalist mode of production and reproduction. Just as Marx discovered congealed human value in the commodity form, and from there identified the whole social basis of exploitation, Debord critically deciphers the congealed image object, the spectacle; penetrates its reified surface; and situates it within its context of social and historical relations. Although he maps out an advanced stage of reification, Debord argues that no object is fully opaque or inscrutable, standing outside of a social context that it cannot ultimately refer to, betray, and be interpreted against.

Where Debord's strategy reminds us of the ultimately antagonistic, conflictual, and contradictory aspects of a social reality open to critique and transformation, Baudrillard's radical rejection of referentiality is premised upon a one-dimensional, No-Exit world of self-referring simulacra. Baudrillard is right: "Reality" has become increasingly difficult to identify in "the empire of signs." But however reified and self-referential postmodern semiotics is, signs do not simply move in their own signifying orbit. They are historically produced and circulated, and though they may not translucently refer to some originating world, they nonetheless can be sociohistorically contextualized, interpreted, and criticized. Thus, even Warhol's "Diamond Dust Shoes" and other paradigms of postmodern flatness, seemingly purely self-referential (see Jameson, 1991), ultimately refer to a whole world, one, in this case, of advanced commodification and of the assimilation of art to media culture and the market (see Chapter 4).

Through Debord's work, we can grasp a point of singular importance: *Self-referentiality does not entail hyperreality*. Signs, images, and objects are not inscrutable and hermetic simply because they no longer stand within a classical space of representation. It is not that one signifier brings us a "real" world and another doesn't but that one occludes a larger social context more than does another, that contextualization may be more difficult in one case than in another. However self-referential and abstract the signifiers, a critical hermeneutics can uncover their repressed or mystified social content and social relations.[13]

Moreover, in a sense, Baudrillard's hyperreality is itself an illusion, the projection of a realer-than-real, constructed by the powers that be to obscure the deprivations, ugliness, and oppressiveness of social reality. Critical hermeneutics can always uncover the constructedness of the hyperreal, which is, after all, nothing more than a construct and model of the real; the hyperreal can always be contextualized, deconstructed, and unmasked. Indeed, the hyperreal is a challenge to critical hermeneutics to see precisely what realities and interests lie behind it, how it is constructed, and what it is concealing and mystifying and why.

Debord is right then to insist that "the spectacle is not [just] a collection of images, but a social relation among people, mediated by images" (#4). While Debord did not provide as powerful an account of postmodern media and signification as did Baudrillard, he correctly insists that the spectacle is "the other side of money" (#34) and so of capitalist social relations. For Baudrillard, however, the sign develops according to its own autonomous logic, and his one-sided analysis is decontextualizing and depoliticizing, serving to exonerate the "captains of consciousness" and media moguls of the present.

While critical hermeneutics does not posit a *Ding-an-sich* discoverable beyond a historical horizon and unmediated by ideology or language, it rightly tries to recover the distinction between reality and illusion as the preliminary basis for a sociopolitical criticism. It is the work of the culture industry to erase this distinction, and it should be the task of radical criticism to recover it, requiring a reconstructed notion of "representation." Though Baudrillard cogently problematizes certain realist views of representation (e.g., those developed in the 17th century, which see language as the "mirror of nature"; see Foucault, 1973; Rorty, 1979) or the realist fictions of our present-day media, he wrongly rejects the illuminating potential of other forms of "representation," such as Jameson (1988, 1991) has attempted to develop in his notion of "cognitive mapping." And while Baudrillard illuminates recent mutations in the sign brought on by media and advertising (which Foucault has failed to consider; see Baudrillard, 1986), he mystifies media culture by severing its dynamics from political economy and current political struggles.

To pass from the collapse of the classical *episteme* to the thesis of radical simulation and implosion, from the fragmentation of meaning to the "end of meaning," is far too hasty a move and obscures the ways in which we still can and must configure our world, not in an act of pictured reflection but, rather, in a theoretical and critical analysis that attempts to grasp the constitutive relations of society and to decode their ideological operations. In a critical hermeneutics, the surface appearance of things is unmasked to reveal not the "real" itself, which remains a dialectically mediated category, but the social forces behind these appearances: the actors, groups, policy makers, spin doctors, and institutions still identifiable and subject to a critically informed resistance. This suggests that "simulation" can be critically deconstructed and resolved into "dissimulation," an activity that reveals simulation to be wholly constructed, serving the interests of specific social groups and hiding certain alternative realities.

Moreover, there are different realities for different social actors: What may be "hyperreal" for one will be patently ideological for another. Thus, not Baudrillard's fatalistic formula—("this impossibility of isolating the process of simulation"; 1983a: 40) but Jean-Luc Godard's empowering suggestion ("trace the images back to their sources") should be the impulse behind a critical mapping of the postmodern terrain. Baudrillard's exaggerated articulation of potentially useful concepts demands that we speak not only of the commodification of reality, of the dissolution of the real through the movement of the commodity form, but also of the reality of commodification and the social forces behind it.

This critical reversal, implicit in Debord's radical hermeneutics, foregrounds what Baudrillardian postmodern theory consistently obscures: the continued existence of the capitalist mode of production, of consumer society, of the culture industries, of the state, and of coercive violence in the repression and determination of social being. Debord insists upon the reality of the spectacle as (1) an institutional apparatus governed by dominant social classes; and (2) an ideology, derived from real social conditions, which "has become actual, materially translated" (#5) and which possesses a real motivating force of "hypnotic behavior" (#18). Subsequently, from this perspective, Baudrillard obscures the suffering and misery of the present age. To employ Debord's useful distinction, "the reality of commodification" refers both to *privation* as the actuality of exploitation, suffering, hunger, and homelessness throughout the world, and to *enriched privation*: the impoverishment of everyday life under the power of hypercommodified leisure and culture.

The critique of the postmodern thesis is necessary insofar as its extreme versions conjoin with capitalism to obscure the most vicious and banal aspects of a violence no less real for being "media-tized." Debord's strategies, by contrast, can be likened to those of Marcuse, who, in Jameson's words, reminds us "that salvation is by no means historically inevitable, that we do not even find ourselves in a prerevolutionary, let alone a revolutionary, situation, and that the total system may yet ultimately succeed in effacing the very moment of the negative, and with it of freedom, from the face of the earth" (1971: 115).

But Debord's critique becomes a call to action, not an excuse for despair. What can give impetus to an oppositional politics is the fact that while negativity is largely contained, the negation of life is still experienced, and this experience can be strategically heightened: "The critique which reaches the truth of the spectacle exposes it as the visible *negation* of life, as a negation of life which *has become* visible" (#10). And though postmodern culture and new technologies portend new modes of domination, they also provide new political possibilities and structural openings (e.g., a radical appropriation of new communication technologies or the implosive erosion of traditional identities and hierarchical relations; see Kellner, 1995b).

Indeed, dissatisfaction with everyday life is dramatically evident in the products of rap music artists like Public Enemy, Ice-T, Ice Cube, Tupac Shakur, and others. Rap is an authentic discourse of the anger of black inner-city ghettos, but its popularity with white audiences as well points to general dissatisfaction, alienation, and anger over the general oppressiveness of everyday life. Showing no deficit of cynicism, many blacks described the Mark Fuhrman tapes as the soundtrack to the Rodney King beatings. Rap music, in fact, anticipated the ghetto explosions that erupted in 1992 after an all-white jury found the L.A. police innocent in the beating of African American Rodney King, despite evidence of the brutal beating captured on videotape. Rap focuses intensely on inner-city ghetto life and on the anger, ready to explode, of African American youth (see Kellner, 1995a), illustrating Debord's thesis that just as the spectacle occludes history, *history haunts the spectacle*—insofar as it can be recovered in a radical critique that exposes the deprivations of the contemporary world. The spectacle degrades social existence more than ever, but it cannot fully eclipse the experience of this degradation and so remains vulnerable to a politics of countermemory (Foucault, 1977) and utopian projection (Bloch, 1986). The recovery of a time before and after the hegemony of contemporary capitalism is an important part of a renewed critical awareness, and many radicals, women, people of color, and marginalized groups have been recapturing their history and using it to draw lessons for the present.

In their cultural politics, Debord and the Situationists maintained a link to "real" or "authentic" needs. Baudrillard, of course, denied *all* needs, and saw "primary" ("true") needs as only the "alibi" of "secondary" ("false") needs and of the consumer system produced and reproduced through the ideology of needs (1981: 63–87). Though Baudrillard provides a powerful critique of certain metaphysical discourses of need, such as we find in traditional political economy, and showed how use value too can become fetishized (see Baudrillard, 1975, 1981), thereby supplementing Marx's and Debord's theorization of the commodity in important ways, his rejection of needs in toto is an idealist move that denies the physical body and the ethical and psychological requirements of a healthy human being. His error was a failure to distinguish between the social *mediation* of needs and the social *construction* of needs (all needs are socially mediated, but not all needs are socially constructed) and to see how historically created needs could be positive (as Marx well saw) or employed as "surplus consciousness" to subvert the commodity system (Bahro, 1978).

Clearly, we need not articulate a metaphysics of nature or human essence to reject the extreme postmodernist position and construct some basis for political struggle. The alienation model can receive different articulations, not all of which are metaphysical or "humanist." Insofar as we can speak of unrealized historical possibilities, for example, we can retain the critical concept of alienation without positing a human essence, and so retain a normative impulse that provides the power behind any vision of an alternative future. With Debord, we must assert and reassert "the *absolute wrong* of being relegated to the margin of life" (#114) when human beings are capable of democratic self-organization and cultural production without the mediation of corporations or the state.

We must continue to see that even under conditions of relative affluence life could be better and that it is artificially impoverished under its current organization. Where capitalism has shown itself capable of buying off radical consciousness, through consumer seductions and a slight redistribution of the economic pie, radical critique must continue to focus on the full extent of human potential, the degree to which this has been falsified under conditions of relative leisure and comfort, and how its development requires an altogether different social order than is possible under capitalism. It must also expose the inherent imperatives of capitalism to degrade and destroy the natural world, and thus it must make the case that an ecological society is necessarily a noncapitalist one. And in conditions of growing impoverishment and diminishing possibilities for improvement, such as the majority of people face at present, radical critique should serve to indicate the source of human suffering and to offer possible solutions.

For Debord, in sum, the contemporary capitalist world remains accessible to interpretation and vulnerable to active transformation. The power of postmodern society and culture is by no means assured. Whereas Baudrillard would have us believe in the "fantastic perfection" of the schemes of control (1983a: 40), Debord reminds us that we need to remain vigilant to the "new signs of negation multiplying in the economically developed countries" that "already enable us to draw the conclusion that a new epoch has begun: now, after the workers' first attempt at subversion, *it is capital abundance which has failed*" (#115). Indeed, the consumer society itself might produce opposition and demands for radical change if it cannot fulfill the very needs that it produces and disseminates (Marcuse, 1972).

A dialectical analysis, for example, could seek the contradictions within the consumer society itself and locate the moments when its strategies of absorption fail and meet subject resistance (see de Certeau, 1984). The aesthetic promise of the commodity constantly gives way to a rude reality, an unsurpassable lifeworld where money can*not* buy everything and where many people do not have the money to fulfill their most basic needs for food, shelter, and health care. Indeed, mystified signifiers are irrelevant to actual human needs, and the promises of happiness through consumption are often false. Within postmodern consumer culture, subjects wear designer jeans or cruise the Internet hour after hour yet remain lonely and unhappy; the spectacle is ubiquitous, but people are still bored; everyday life is shit, and people know it.

There are, however, problems with Situationist theory and practice. Although Debord rejected the party form of vanguardism in orthodox Marxism, he uncritically replicated old-fashioned workerism, which privileged the working class as the revolutionary subject of history, an anachronism that New French Theorists such as Foucault (1972b) and Deleuze and Guattari (1987) rejected in favor of a more decentralized struggle of multiple groups, now termed "new social movements." In *The Mirror of Production* (1975), Baudrillard also embraced a concept of micropolitics organized around women, blacks, and other groups he thought to lie outside of the code of political economy. However, he very soon abandoned any hope of political change and found solace in nihilism.

For post-1975 Baudrillard, there is no one to turn to, nowhere to go, and nothing to be done. Taking postmodern theory to its ultimate conclusions, his thought leaves us paralyzed and without any ground from which to articulate opposition. Indeed, we cannot even draw the most elementary distinctions, such as that between left and right. And any imaginable opposition to the cybernetic order is a priori rejected as alibi, as simulated negativity, as just another route of social reproduction. We are left, if anything at all, with the antithesis of a Gramscian politics of counterhegemony, which Baudrillard describes as "fatal strategies" that assert that the "system's own logic" is "the best weapon against it" (1993: 4). But Baudrillard's work describes less the actual nature of our contemporary world than his own entombment within an increasingly bizarre theoretical orbit: the ideological perspective of a French, urban, middle-class intellectual enmeshed in a very specific mode of experience divorced from the complex realities of diverse groups of people and their daily struggles. Precisely this withdrawal from radical politics is a symptom of the post-'68 disillusionment with the radical project, a French replay of the postwar conservativism of American ex-radicals.

If reification is the transformation of the world into an occulted phantasmagoria of objects and signs without social relations, then this is precisely the task that Baudrillard's work helps to accomplish. His analyses erase the institutional forms of advanced capitalism, which seem to have determined his discourse to such an extent as to (mis)lead him into concluding that capitalism no longer exists. Baudrillard's insistence on the hermetic nature of postmodern simulacra is better seen as the introjection and projection of the capitalist imaginary—the dream of seamless closure, complete mystification, and perfected hegemony—than as an accurate description of contemporary social reality. Thus, Baudrillard ultimately provides a superimplosive postmodern theory that confuses *tendencies* of contemporary society with a *finalized state of affairs*.

Baudrillard is ultimately a cynical survivalist who defines his project as an attempt "to reach a point where one can live with what is left. It is more a survival among the ruins than anything else" (1975: 25). Debord and the Situationists, however, champion the construction of situations "on the ruins of the modern 'spectacle'" (in Knabb, 1981: 25), advocating cultural activism and struggle to transform everyday life and to create new desires, pleasures, and forms of culture and society. Interestingly, Debord and Situationist concepts are experiencing a curious afterlife in the proliferation of 'zines and Web sites that embody Situationist practice. The past decade has been marked by a profusion of cultural activism that uses inexpensive new communications technology to spread radical social critique and cultural activism. Many of these 'zines pay homage to Debord and the Situationists, as do a profusion of Web sites that contain their texts and diverse commentary. Situationist ideas are thus still an important part of contemporary cultural theory and activism.

EPILOGUE: DEBORD'S COLLAPSE

Unfortunately, however, Debord later renounced his activism and fell into a pessimism as deep as Baudrillard's. In the first decade after the publication of *The Society of the Spectacle*, Debord maintained his belief in "the inevitable fall of all cities of illusion" (1979: 23) and in the possibility of revolution through the creation of autonomous proletariat assemblies. He congratulated himself that every thesis in his book remained true and that nothing was contradicted by the development of historical events (11). In his *Comments on the Society of the Spectacle* (1990), however, written more than 20 years after his original reflections, Debord's confidence cracked and his theses, he implies, have been falsified by historical developments.

His main conclusion is that "the spectacle has . . . continued to gather strength" and that it has "learnt new defensive techniques" (1990: 2, 3), largely absorbing whatever critical opposition it may once have received from the quarters of everyday life. Having had two more decades to penetrate into the mass psyche, the spectacle has not only been extended quantitatively, it has also changed qualitatively, developing a new hybrid form. Earlier Debord (1979) distinguished between two forms of the power of the spectacle: the "concentrated," dictatorial form and the "diffuse" form. The former found its model in Soviet Russia, or perhaps in George Orwell's *1984*, and operates through overt oppression. The latter was perfected in the United States, is aptly represented in Aldous Huxley's *Brave New World*, and involves a more subtle form of power based on libidinal and psychic control and seduction.

Debord came to feel that there is a third form of spectacular power, the "integrated spectacle" pioneered by France and Italy, which combines basic aspects of the concentrated and diffuse forms and therefore represents a more formidable power than either form by itself. No more talk of the inevitability of the fall of capitalism; no more references to an autonomous and empowered proletariat; and no more strategies for revolution. Debord now complains more than he criticizes, joining hands with Baudrillard in the solidarity of despair and resignation. Indeed, Debord's pessimism was taken to its ultimate conclusion in 1994 when he committed suicide.

Debord's succumbing to the fatalism of the postmodern scene must be understood in part as a result of his identification of social change with the struggle of the proletariat, which for a long time has been neither a class for itself nor even a class in itself, and of his inability to find new sources of resistance in other political groups, such as feminist groups, gay and lesbian groups, peace and environmental activists, and media and computer activists. His despair can also be traced, as with Baudrillard, to an initially false hope for total revolution, an unattainable ideal whose inevitable failure will naturally lead to pessimism. Political defeatism is all too often the flipside of ultra-revolutionary utopianism

But Debord's collapse can perhaps also be related to an excessively unitary and monolithic concept of the spectacle. The concept ultimately operates as a fetish and cultural monolith for Debord, an all-encompassing and all-powerful force of cultural hegemony that functions as the totalizing concept of capital in some versions of Marxism. But just as there are contradictions and competing sectors within capital, so too are there competing forces within the spectacle that use media culture to advance different aims and agendas. Indeed, it is preferable to see a variety of types of spectacles, rather than a single spectacle, and to see media culture as a contested terrain rather than a monolithic concept of domination à la Debord. Such a pluralist and more open concept of culture opens the door for projects of radical cultural intervention and alternative media that Baudrillard eschews and that are problematic and paradoxical given Debord's more monolithic concept of the spectacle—which undermines his earlier emphasis on the importance of cultural activism.

Ultimately, however, the pessimism of Baudrillard and Debord is a response to the "God that failed," to the collapse of Marxism, of naive revolutionary ideals, and of the struggles for dramatic political change. There is indeed a good deal to be pessimistic about, as the techniques of power become increasingly stronger as ecological and economic crises intensify, and as the Right continues to define the terrain of culture and everyday life while the Left disintegrates into special interest groups preoccupied with identity politics and internecine squabbles. But the later Debord and Baudrillard both ignore the crisis tendencies that continue to threaten global capitalism and the deep dissatisfactions of everyday life that continue to erupt on occasions, such as the events in the aftermath of the Rodney King trials in Los Angeles, the epicenter of the spectacle. The powers of reason and argumentation are indeed ebbing (as perhaps up to one-third of the U.S. population is functionally illiterate), but the fuses of the powder kegs of anger and desire continue to burn. Indeed, a new "radical middle" is incubating in the United States, made up of angry voters who are irreconcilably alienated from both Republicans and Democrats, from the politics-as-usual corrupted by power, money, and special interests (see *Newsweek*, September 25, 1995). This growing disaffiliation with the status quo could be mobilized into progressive viewpoints and movements rather than rightist racist populism. As Debord himself once said, these are the flashpoints of collective action and social transformation.

But there are other problems with Debord's perspectives as well. He only theorized class alienation and targeted capitalism as the primary source of human misery, failing to address gender or race oppression. He sometimes formulated his political goals in excessively Hegelian and utopian forms as the overcoming of all alienation and separation, as the creation of higher and greater social unity, thus replicating uncritically some of the now-questionable Enlightenment goals of modernity. Debord often saw things in too starkly dichotomous terms, contrasting the passivity of the spectacle to the activity of self-creation. But consumption too can be active, as evidenced by some clever Situationist practice, and passivity can be a part of invigorating leisure and enjoyment and not just a mode of thralldom to the spectacle. And yet we do not accept the facile critique of Debord that he was an "elitist" whose work is a "sham piece of agitprop theory."[14] Indeed, we believe that it is precisely the radicalism and activism that characterize Debord's early politics which provide a refreshing antidote to the cynicism and pessimism of what would come to characterize much postmodern politics (see Chapter 6).

In sum, Debord and the Situationists worked within modern paradigms of theory, art, and politics, theorizing in the transitional zone between the modern and the postmodern. Acutely aware of the changes wrought by an advanced capitalist, consumerist, and media society dominated by spectacles and images, they sought to modify the basic imperatives of Marxian theory and politics in a way that broke down barriers between theory and art, politics and everyday life, and theory and practice while refusing to relinquish modern norms of interpretation, critique, revolution, and hope for a better world. They acknowledged the infiltration of power and ideology into the most intimate aspects of existence, but they also ferreted out the contradictions of capitalism and searched for new forms of opposition.

In theory, they exemplified the modern paradigm through Enlightenment norms of critique designed to bring about social transformation and human emancipation, through the humanist value of fully developed individuals, and through their belief in the liberating potential of science and technology and a vision of progress realized in a postcapitalist society. In politics, they urged a revolution of everyday life in order to create a global politics of human solidarity against mechanisms of alienation and exploitation. In art, they sought new artistic forms and experiences, they wrote scores of revolutionary manifestos, and they valorized hermeneutics and semiotic complexity. Although they developed an extended notion of art that challenged conventional boundaries, broke down the opposition between high and low art by drawing on popular culture, and called for an experimental play within the *dérive* and the constructed situation, they refused to indulge in the ludic and retained a serious conviction that art mattered and that the artist had moral obligations to humanity. Accordingly, in Chapter 4, we will see how Debord and the Situationists anticipated in fascinating ways the postmodern turn in the arts, and how their avant-garde approach in the 1960s was being challenged and superseded by emerging postmodern aesthetic forms and practices.

NOTES

1. On the history of the Situationists, see Marcus (1989), Plant (1992), and Wollen (1993). In addition to the works of Debord, which we discuss in this chapter, the main Situationist texts are collected in Knabb (1981) and include works by Vaneigem (1983a, 1983b). Debord's *The Society of the Spectacle*, first published, in French, in 1967, was published in translation in a pirate edition by Black and Red (Detroit) in 1970 and reprinted many times; another edition appeared in 1983 and a new translation in 1994; thus, in this discussion we cite references to the numbered paragraphs of Debord's text to make it easier for those with different editions to follow our reading. The key texts of the Situationists and many interesting commentaries are found on various Web sites, producing a curious afterlife for Situationist ideas and practices. See, for example, http://www.nothingness.org/Situationist International/journal.html and http://ccwf.cc.utexas/~panicbuy/HaTeMaiL/situationist.html.

2. On Baudrillard, see Kellner (1989b), Best and Kellner (1991), and Kellner (1994).

3. See the discussion in Poster (1975) of the new forms of Marxian theory in postwar France. Many discussions of Debord and Situationism downplay the Marxian and Hegelian roots of their project; for example, Marcus (1989) and Plant (1992) exaggerate the avant-gardist aesthetic roots of the Situationist project and underemphasize the Marxian elements.

4. Council communism rooted itself in the tradition of soviets, or workers' councils (in German, *Räte*), rather than parties. They opposed the bureaucratization of the Soviet Union and all Marxist–Leninist parties, which they thought were hopelessly hierarchical and bureaucratic. In opposition to bureaucratic communism, they championed workers' self-activity and self-organization; see the texts of Karl Korsch collected in Kellner (1977) and the discussion in Boggs (1976).

5. Kathie Lee Gifford suffered an image problem in the summer of 1996 when it was revealed that one of the products in her line of clothing was produced by a sweatshop subminimum-wage-labor company that had often even failed to pay workers their wages on time. This leads to a furious scramble to reframe her positive image, as it also exposed to a mass public the deplorable conditions of sweatshop labor. In the spring of 1997, the halo hovering above Kathie Lee's fabled and idealized

marriage with Frank Gifford was rudely knocked into the gutter when it was revealed that he had had an affair with a femme fatale, who apprently set him up with a hidden video camera, allowing her to sell the seamy and steamy story to a tabloid. Yet *TV Guide* helped resurrect Kathie Lee's angelic image by putting her on the cover of their July 19–25, 1997, issue, complete with a halo!

6. Vaneigem later resigned from the Situationist International after disputes with Debord; see his letter of resignation and Debord's scathing condemnation of his one-time comrade in Debord and Sanguinetti (1974: 123–138). Vaneigem later published *The Book of Pleasures* (1985) and a series of texts on revolutionary practice.

7. Debord's criticism that media communication "is essentially *unilateral*" (#24) was taken over directly by Baudrillard (1981: 169ff.); Baudrillard's stress on image and semiurgy—that is, the proliferation of signs and images—comes from Debord (#18, #34) and his notions of "map" and "territory" also derive from Debord, who wrote: "The spectacle is the map of this new world, a map which covers precisely its territory" (#31).

8. There are, of course, other accounts of postmodernity; see Best and Kellner (1991) and our following studies. Baudrillard's account, however, has been extremely influential and is symptomatic of the excesses of an extreme postmodern theory; thus we will engage his analysis in some detail.

9. This is especially evident in his essay "Requiem for the Media" (in Baudrillard, 1975: 164–184), where he says things such as "perhaps the Marxist theory of production is irredeemably partial, and cannot be generalized" (165) and theorizes the media in terms of their form, à la Marshall McLuhan, rather than their political economy, their ideological functions, and the ways that they reproduce capitalism. On Baudrillard's relations to Marxism, see Kellner, 1989b.

10. *Wired*, the publication of choice for the digerati who write about the information and computer culture and those who consume it, has a monthly feature that under the rubric "Fetish" presents the latest products to satisfy its consumers' techno-lust. According to *Newsweek* (January 8, 1996: 54–55), the latest lifestyle fetish is designer paint, such as that from Stewart, which costs up to $110 a gallon and comes in hundreds of different shades.

11. The hyperscrutiny and revelations of the most intimate details of political candidates' private lives began with media exposure of the sexscapades of Democratic party presidential candidate Gary Hart in 1980 and have hit their low point with exposés of Bill Clinton's allegedly sleazy relations with Genifer Flowers and Paula Jones, along with reports that the Arkansas state police allegedly procured women for Clinton. The question that received the most attention from Clinton's 1992 MTV interview with a group of young adults was: "Which is it Mr. President? Boxers or shorts?" The August 1995 issue of *Vanity Fair*, however, also exposed the private life of Clinton's major Republican opponent, Speaker of the House Newt Gingrich, to the sort of sordid tabloid treatment usually reserved for those in the entertainment industry.

12. This operation was widely discussed in the alternative press, in scholarly publications, and in books during the 1980s, but it rarely surfaced into mainstream media discussions; see the analysis in Kellner, 1990.

13. Some outstanding examples of this practice would include the deconstruc-

tion of advertisements carried out by Williamson (1978), Goldman (1992), and Goldman-Papson (1996), as well as Gottdiener's (1995) materialist semiotics. This has also been an intention of our work on media and consumer culture over the past two decades.

14. See the review by John Palattella of a 1996 conference on fashion, where a participant attacked Debord as elitist and declared that he "must be wrong about fashion since he 'obviously didn't like to shop'" (*In These Times*, April 15, 1996: 47). Against such inane dismissals of Debord, we argue that Situationist critique, despite its faults, is an indispensable part of a critical theory of contemporary society and radical cultural politics in the present age. The "populist" critique of Debord also misfires, since Debord more than anyone stressed alternative cultural practices in everyday life and the possibility of cultural resistance, which are the crux of a populist cultural studies that in the case cited wrongly distances itself from Debordian critique.

CHAPTER FOUR

Postmodernism in the Arts
Pastiche, Implosion, and the Popular

> Abstract expressionism in painting, existentialism in philosophy, the final forms of representation in the novel, the films of the great *auteurs*, or the modernist school of poetry (as institutionalized and canonized in the works of Wallace Stevens): all these are now seen as the final, extraordinary flowering of a high modernist impulse which is spent and exhausted with them. . . . the younger generation of the 1960s will confront the formerly oppositional modern movement as a set of dead classics, which "weigh like a nightmare on the brains of the living," as Marx once said in a different context.
>
> —FREDRIC JAMESON

As Debord and Baudrillard developed their critical analyses of consumer culture in the 1960s, as capitalism was becoming a full-blown society of the spectacle, and as oppositional political movements were contesting existing societies, new tendencies emerged in the arts in the form of new postmodern methods, styles, and consciousness. At this time, a "new sensibility" appeared in criticism and the arts that expressed dissatisfaction with prevailing modernist forms and ideologies. Seen as stale, boring, pretentious, and elitist, European and American high modernism were rejected. The new attitude pronounced the death of modernism and the arrival of "postmodernism," of a new ideology and new aesthetic forms exemplified in the novels of William Burroughs, the music of John Cage and the dance of Merce Cunningham, the paintings of Andy Warhol and pop art, and the architecture of Robert Venturi and Philip Johnson. Postmodernism not only brought dramatic changes in existing fields, such as architecture, literature, painting, film, music, and dance, it also involved a creation of new art forms, such as happenings, performance art, multimedia installations, and computer art, suggesting that we were indeed living in a new culture of the simulacrum (Baudrillard, 1983a, 1993, 1994; see Chapter 3).

In the following section, we will describe the lines of the historical shift from modernism to postmodernism in the arts and the emergence of a new postmodern culture. We will illuminate this transformation with discussion of the postmodern turn in a variety of aesthetic fields, ranging from architecture, painting, and literature to multimedia art and media culture. We argue that though there are a number of diverse postmodern expressions in the

ERECT NO SIGNS
POST
NO
SIGNIFIERS

arts, they share core stylistic features and, with postmodern interventions in social theory and science, are part of a shift to a new postmodern paradigm.

FROM MODERNISM TO POSTMODERNISM

> The category of the new has been central to art since the middle of the last century. . . . there has not been a single accomplished work of art in the last hundred years or so that was able to dodge the concept of modernism. . . . The more art tried to get away from the problematic of modernism, the sooner it perished.
>
> —T. W. Adorno

To elucidate the postmodern turn in the arts, we must begin with some reflections on the forms of modernism that postmodernists parody or repudiate. Beginning in the 19th century, "modernism" took shape as a tendency in the arts that articulated new artistic styles and techniques and new ideologies about the function of art and the role of the artist in society.[1] In the 1850s, Parisian poet Charles Baudelaire called for a form of modern poetry that would be able to capture the uniqueness of modern experience, especially the shocks of urban life. His successor Arthur Rimbaud demanded, "*il faut être absolument moderne*," that art be "absolutely modern," and poet Ezra Pound insisted that artists "make it new." Modernist art sought innovation, novelty, and contemporary thematic relevance, rejecting tradition by negating old aesthetic forms and creating new ones. In this sense, modernism in the arts followed the basic processes of modernity, which involved negation of the old and creation of the new, producing continual originality and "creative destruction" in all spheres of life (see Berman, 1982).

In response to the romantic failure to preserve a progressive role for art in bourgeois society and to the increasing encroachments of the market and mass society on the artistic world, modernist artists sought autonomy in the arts, aspiring to free art from religion, morality, and politics, thus allowing the artist to pursue purely aesthetic goals. Indeed, a primary characteristic of modernism is its belief in the autonomy of art, involving an active attempt by the artist to abstract art from social ideology in order to focus exclusively on the aesthetic medium itself. Belief in art for art's sake and the autonomy of art ultimately decentered the aesthetic project from representation and the imitation of reality to a concern with the formal aspects of art. Beginning with the French impressionists in painting, modernist art breaks with realist modes of representation and the concept of art as mimesis, an imitation of reality, in order to explore alternative visions and to experiment with the aesthetic possibilities of a given artistic medium. This modernist project echoed through the arts, generating experiments with new forms, styles, and modes of aesthetic creativity.

Consequently, modernist artists undertook a series of formalist experiments in an intensive search for new languages that would liberate them from traditional notions of arts and reality, often anticipating later concerns of science, as Cézanne's multiperspectival vision prefigured that of Einstein (see Chapter 5). In some cases, this preoccupation with form, technique, and mode of vision rendered modern art highly self-referential, more about itself, its own artistic form, than about the social world or even the artist's experience of the world. Schoenberg and his followers experimented, for instance, in the production of a radically new system of atonal music that

> gave each note in the chromatic (twelve-note) scale equal weight by the device of requiring all twelve notes to be used once before any were repeated. The seemingly arbitrary ordering of notes was called a "tone row" and functioned much like a melody in traditional music. When the same idea was applied to other elements of musical language (rhythm, dynamics, timbre), the all- inclusive name for the results was "serial" music. (Glass, 1987: 13)

This practice produced a very abstract and modern sounding music governed by an inventive technique and rigorous formalism. In each particular art, modernist artists sought to discover what was specific to painting, writing, music, and other arts, to eliminate extraneous elements derived from other spheres: modernist painters, for example, sought to exclude the literary or didactic from painting. Artists like Cézanne and Picasso experimented with abstract and geometric forms that broke with naturalistic representation; composers like Schoenberg and Webern created new atonal and formal systems of music; writers like Pound and Joyce used language in innovative ways and produced new modes of writing; modern architects devised novel modes of housing and forms of urban design, eliminating aesthetic decoration in favor of function and utility; and groups of modern artists in every aesthetic field created dramatically innovative works and techniques.

The movement toward innovation and purity in modernist art replicates the logic of cultural modernity not only in its drive for constant originality and novelty, thus producing a "tradition of the new," but also in its pursuit of the modern logic of cultural differentiation. On Habermas's (1983) account, modernity involves the differentiation of spheres of value and judgment into the domains of science, morality, and art, with each sphere following its own logic. Thus, the modernist celebration of the autonomy of art, the specialized development and refinement of specific artistic spheres, and the quest for formal invention follows the broader trends of modernity.

In a sense, the modernist imperative toward ceaseless change and development involves an embrace of the ethos of capitalism, in which variation of product means new markets, shifting tastes, and more profits. During the modernist century (approximately the 1850s to the 1950s), the artist was forced to sell his or her wares on the market, independent of the patronage systems that formerly supported art. This led to an internal contradiction within modernism between the need to produce novel and attractive products for the market and the urge to purify art of anything external or extraneous to the art object. Thus, conflicts erupted between the logic of aesthetic autonomy and its religion of "art for art's sake"—driving modernist artists to avoid contaminating their art with mass society and mass culture—and the imperative to sell their products for the highest price.

Modernist art became ever more complex and demanding as innovations proliferated, and by the 20th century modernism defined itself as "high art" distinct from the "low art" of the masses (Huyssen, 1986). Elitism became the corresponding attitude of high modernism and the modernist artist, whose genius and purity of vision was incomprehensible to the layperson. Leading modernists sought to develop their own private language, their own unique vision and style that expressed their singular self. Hence, the works of Eliot, Pound, Klee, and Kandinsky articulated an ideolect that proved incomprehensible to the uninitiated but was readily perceivable to the initiate. The search for a private code inscrutable to the masses, or to most critics for that matter, reached its height in Joyce's *Finnegans Wake*, a novel accessible only to patient polyglots. Walking on the clouds of genius, the modernist often feels intense alienation from the masses, such as evident in many of Baudelaire's poems.[2]

The modernist artist was thus driven to create the great work, the masterpiece, and his or her own unique individual style. Genius, monumentalism, and distinctive style and vision were thus intrinsic features of modernist aesthetics. One could easily recognize the paintings of Monet or van Gogh; the prose style of Kafka or Hemingway; the music of Schoenberg or Stravinsky; the theater of Pirandello or Brecht; or the buildings of the International Style. Modernist works also expressed the personal vision of the artist, his or her own unique view of the world, and the modernist masterwork attempted to generate new modes of art and new ways of seeing and thinking.

During the early decades of the 20th century, however, modernism split into different, often warring, camps. While a formalist modernism sought primarily to pursue pure aesthetic concerns, avant-garde modernist movements emerged that aspired to revolutionize society, culture, and everyday life by assaulting the institution of art, allegedly corrupted by the bourgeois market society, and redefining the relation between art and life.[3] Whereas modernism tried to transform (romantic) alienation into individual autonomy and creativity, the political avant-garde exploded the boundaries isolating the artist from society in order to use the unique gifts of the artist as a means of advancing radical social change. Paradoxically, the extreme individualists of avant-garde art worked in artistic *movements* that sought to align themselves with whatever social forces—scientific, technological, or political—that they believed augured emancipatory change.

For the most part, modernism was a male affair, although women were participating in the avant-garde movements by the 20th century. Women painters like Mary Cassatt and Berthe Morisot were active in the impressionist movement, Paula Modersohn-Becker and Käthe Kollwitz were major figures in the expressionist movement, and women became active participants in groups like the Bloomsbury circle, dada and surrealist movements, and the Paris arts community.[4] Thus modernist subcultures gave women an opportunity to participate in cultural creation in a wide variety of arts from which women had previously been excluded, although they continued to suffer prejudice and exclusion in many cases. Likewise, people of color and non-Western artists would eventually appropriate the techniques and practices of modernism, though this development would only gradually mature.

Although avant-garde movements like expressionism, futurism, dada, and surrealism built on the formal experimentalism of modernism, continuing the attack on realism and mimesis, they assaulted its ideology of aesthetic autonomy and assailed the bourgeois "institution of art" whereby art was produced, distributed, and received as a commodity and tool of political legitimation (see Burger, 1984). Against modernism, the avant-garde movements saw art as a means of social transformation and sought to integrate art into everyday life. Where high modernism was becoming largely conservative in its function, a bastion of elite taste and an "affirmative culture" (Marcuse) that ultimately legitimated bourgeois social and political domination, the avant-garde strove to subvert dominant aesthetic ideologies and to effect revolutionary social change.[5] Nevertheless, the avant-garde remained bound, with modernism, to the romantic notion of the artist as privileged social figure or visionary (as in Shelley's claim that artists are the "unacknowledged legislators of the world"). Moreover, much modern art continued to assume the idealist notion that language was an autonomous bearer of meanings and force of change, two key assumptions to be opposed by postmodern art and criticism.

Needless to say, the avant-garde failed to deliver on its promises to abolish oppressive ideologies and institutions or to merge art and life in a progressive social transformation. The militant rhetoric and manifestos of the avant-garde rang loudly for little more than a decade before being silenced by fascism, bureaucratic socialism, capitalism, and war. In Germany, the avant-garde tradition was stopped in its tracks in 1933 when Hitler came to power and banished all forms of modern art as decadent. In Soviet Russia, the last vestiges of a flourishing avant-garde tradition were exterminated by 1934 with the declaration of socialist realism as the official style under the cultural czardom of Zhdanov. In the United States, the avant-garde was defanged during the 1940s and 1950s, less harshly but no less decisively, with the canonization of modernism in the universities and museums and the commodification of art in a dramatically expanding art market. Modernist art lost its sharp critical and oppositional edge, becoming an adornment to the consumer society, while its techniques were absorbed into advertising, packaging, and design, as well as the aestheticization of everyday life.

From the Shock of the New to Postmodern Historicism

The postmodern turn in the arts maintains some links to earlier aesthetic traditions while also breaking sharply from bourgeois elitism, high modernism, and the avant-garde alike. With modernism and the avant-garde, postmodernists reject realism, mimesis, and linear forms of narrative. But while high modernists defended the autonomy of art and excoriated mass culture, postmodernists spurned elitism and combined "high" and "low" cultural forms in an aesthetic pluralism and populism. Against the drive toward militant innovation and originality, postmodernists embraced tradition and techniques of quotation and pastiche. While the modernist artist aspired to create monumental works and a unique style and the avant-garde movements wanted to revolutionize art and society, postmodernists were more ironic and playful, eschewing concepts like "genius," "creativity," and even "author." While modernist art works were signification machines that produced a wealth of meanings and interpretations, postmodern art was more surface-oriented, renouncing depth and grand philosophical or moral visions (Jameson, 1991).[6]

Yet a more activist wing of postmodernism advanced the anarchist spirit of the avant-garde through a deconstruction and demystification of meaning, but while breaking with its notions of agency, its idealist definition of language, and its utopian vision of political revolution (Foster, 1983, 1985). Postmodernists abandon the idea that any language—scientific, political, or aesthetic—has a privileged vantage point on reality; instead, they insist on the intertextual nature and social construction of all meaning. For postmodernists, the belief of the avant-garde in the integrity of the individual as an activist agent, in language as revelatory of objective truth, and in faith in historical progress remain wedded to the mythic structure of modern rationalism. As we will see, while some versions of postmodernism leave ample room for social criticism and political change, the postmodern turn in criticism and the arts abandons modern notions of the subject, the work of art, and political change.

On the whole, the postmodern turn in the arts reacted against what was seen as both the decay of an institutionalized high modernism and a failed avant-garde. Some critics, however, mourned the passing of modernism while others celebrated its demise. In 1959, Irving Howe lamented the end of the modern in his "Mass Society and Postmodern Fiction" (1970). A sad eulogy was given by Harry Levin in his 1960 article "What Was Modernism?" (1967). For both Howe and Levin, postmodernism was a symptom of decline; it represented the appearance of a new nihilism, an "anti-intellectual undercurrent" (Levin) that threatened modern humanism and the values of the Enlightenment.

In 1964, by contrast, Leslie Fiedler wrote two key articles, "The New Mutants" and "The Death of Avant-Garde Literature" (collected in Fiedler,

1971), which celebrated the flowering of a popular culture that was more playful, exuberant, and democratic, challenging the opposition between high and low art and the elitism of academic modernism. In the same vein, Susan Sontag published in 1967 an influential collection of essays entitled *Against Interpretation*, which attacked the elitism and pretentiousness of modernism and promoted camp, popular culture, new artistic forms, and a new sensibility over the allegedly stale, boring forms of entrenched modernism. Whereas modernism denigrated "kitsch" and "mass culture," those who took the postmodern turn valorized the objects of everyday life and of commercial culture. Moreover, against what Sontag considered an abstract hermeneutics practiced by modernist critics, she affirmed the immediate, visceral experience of art and form over content and interpretation. In 1968, Fiedler made an explicit appeal to "Cross the Border—Close the Gap" (Fiedler, 1971), and this exhortation to break down the boundaries between high art and popular culture became a rallying cry of the new postmodern attitude.

Generally speaking, the postmodern turn in literature was carried out against the canonized forms of high modernism that had emerged as dominant in the United States in the 1950s. Modernist writing sought the innovative, the distinctive, and the monumental. Modernist writers like Kafka, Hemingway, Joyce, Eliot, and Pound sought their own distinctive styles to articulate their unique visions. For postmodernists, the aesthetic of high modernism had run its course and depleted its possibilities; the notion of the artistic work as a hieroglyph understood only by experts was rejected for a more accessible, populist writing style; and the concept of the author as an expressive unitary consciousness was dismantled to place the writing subject within a dense, socially constructed, intertextual discursive field.

Apocalyptic references to the "literature of exhaustion" and the "death of literature" proliferated, along with corollary references to the "death of the novel" and the "death of the author." These moods elicited conflicting responses ranging from calls for a new "literature of replenishment" (Barth, 1988), which would revitalize traditional modes of writing, to calls for altogether new forms of writing and culture. A wide range of writers who were developing new experimental modes and styles of "surfiction" or "metafiction" were labeled "postmodern," including John Barth, Vladimir Nabokov, Thomas Pynchon, Donald Barthelme, John Hawkes, and Robert Coover. These styles employed self-reflexive and nonlinear writing that broke with realist theories of mimesis, depth psychology and character development, and views of the author as a sovereign subject in full command of the process of creation. Its "characters" are typically empty, depthless, and aimless, embodying "the waning of affect." Moreover, as in Alain Robbe-Grillet's work, moral, symbolic, or allegorical schemes are often abandoned in favor of surface meaning, or the depiction of the sheer "meaninglessness" of random events and fractured "narration." Where modernist novels still assumed some order and coherence in the world and, despite moral uncertainties, aspired to project schemes of redemptive vision, postmodern fiction took a more nihilistic stance in portraying the random indeterminacy of events and meaningless actions, projecting an epistemological skepticism later articulated in postmodern theory.

Crucially, postmodern writers implode oppositions between high and low art, fantasy and reality, fiction and fact. Spurning "originality," postmodern writers draw on past forms, which are ironically quoted and eclectically combined. Instead of deep content, grand themes, and moral lessons, ludic postmodernists like Barth, Barthelme, and Nabokov are primarily concerned with the form and play of language and adopt sportive, ironic, self-reflexive, "metafictional" techniques that flaunt artifice and emphasize the act of writing over the written word. Of course, some of the stylistic techniques of postmodern literature were defining features of modernism itself, motivated by its revolt against bourgeois realism—leading many critics to see postmodern literature as continuous with modernism rather than as constituting a radical break or rupture.

But one could twist the argument that almost all conceivable stylistic inventions were made by modern and avant-garde artists to lend credibility to the postmodern sense that there is nothing new for a writer to accomplish. All the postmodernists could do, then, would be to push these modernist moves further, to be more radically anti-realist and anti-narrative. Postmodernists, consequently, deployed language to turn in on itself with a new energy. The intense self-reflexivity of postmodern literature thus leads to a constant interruption of narrative, an untiring reminder to the reader that he or she is reading a text, language, a fiction, and not viewing a world without mediation. This is, of course, a technique that Brecht developed in the form of the "alienation effect" of his modern drama in order to break emotional identification between the spectator and the play and to awaken critical reflection instead.

Having arisen in the late 1950s and the 1960s, the trends of postmodernism in the arts quickly spread through literature, painting, architecture, dance, theater, film, and music, spilling over into philosophy, social theory, and science by the 1970s and 1980s. In contrast to the differentiating impetus of modernism, postmodernism adopts a dedifferentiating approach that willfully subverts boundaries between high and low art, artist and spectator, and among different artistic forms and genres. To return to Habermas's (1983) scheme, the dedifferentiation process that characterizes postmodernism began the moment the autonomous distance between art, science, and morality started to collapse, once science and money, as vehicles of social power, increasingly encroached on the autonomy of other social spheres in a process that Habermas terms the "colonization of the lifeworld." Once commodification dynamics had advanced to the point that modernism itself was assimilated to the market, the "shock of the new" had been defused, and avant-garde art became a sound investment. In the 1913 Armory Show in New York, European avant-garde art made its shocking debut in the United States; five decades later, it was hanging on the walls of bank lobbies, it provided background for advertisements, and it adorned the clothes and bedsheets of bachelor pads and middle-class homes.

At the same time that high art proved itself safely cornered and sanitized, popular art forms based on radio, film, television, advertising, and comics thoroughly saturated U.S. culture. Rather than snobbishly dismissing these "low culture" forms, artists of the 1960s embraced them as a refreshing alternative to high culture and assimilated their forms into their work. Where modern artists were typically insular, obscure, and idiosyncratic in their work, postmodern artists began to speak in the most available, public and commodified languages, such as exemplified in Warhol's use of media images or Venturi's desire to "learn from Las Vegas."

By the 1960s, there was a widespread feeling that novelty and innovation in the arts had been exhausted and that all that could be done had been done. The painting *White on White* by the Russian suprematist painter Kasimir Malevich, which features two barely discernible white squares superimposed on one another, was a clear indicator that in this field, for example, certain formal limits had been reached, and there was talk of "the end of painting." The search for new beginnings that characterized modernism gave way to a "sense of ending" (Kermode, 1967). All that was left for the postmodern artist to do, it was thought, was to play with the pieces of the past and to reassemble them in different forms.

Hence, rather than bold innovation and originality, postmodernists deployed eclecticism, pastiche, and parody. As the postmodern architect Venturi put it: "Creating the new means choosing from the old." The modern quest for the new was informed by a belief in the artist as a unique, expressive self. Modern art was one with modern philosophy in its belief in a transcendental self outside of space and time. Contrariwise, postmodern artists, articulating the same anti-humanist themes of poststructuralist and postmodern theory, abandoned the belief in a self, author, and creative genius. The artist is no longer the originary and unique self who produces the new in an authentic vision but, rather, a *bricoleur* who just rearranges the debris of the cultural past.

Rather than expanding on the themes of selfhood, authenticity, originality, and liberation, postmodern artists parody them. Rather than inventing new materials, postmodernists quote what's already around and combine fragments in a pastiche—as Robert Rauschenberg pastes texts from newspapers and images from classical paintings onto his canvases, or as rap artists "sample" riffs from past songs. The postmodern turn is well exemplified in the work of Andy Warhol, who boasted he could produce as many works of art in a day through mechanical reproduction as Picasso could in a lifetime.

The attack on the modern ideology of creativity and authenticity is also present in the work of the New York "artist" Mark Kostabi, who signs his name to works conceived and created by a crew of struggling painters who are paid paltry wages for works that fetch thousands of dollars, or in the work of Jeff Koons, who pays groups of artisans to produce "his" environmental sculptures and other works of "art."

Hence, the modern concern for monumentality, for great and original works, gives way to the postmodern theme of irony, and modernist seriousness passes over to postmodern play. The postmodern sensibility thus carries out the death of the author and the end of the great work. As Benjamin (1969) analyzed it, the "aura" of the artwork as something singular, sui generis, is corrupted and destroyed through the technology of mass reproduction, particularly in electronic form. Once the "original" is endlessly reproduced, a Baudrillardian state of hyperreality takes effect such that the original becomes indistinguishable from the copies and no more real than its reproductions (see Chapter 3).

The lack of aura of the mass-reproduced object and the belief in the end of the expressive self leads to another important contrast between modern and postmodern art: their respective emphases on depth and surface. The modernist focus on psychological and affective depth is related to the modernization processes. Where the capitalist experiences modern subjectivity as freedom from economic constraints, the artist experiences it as a distance from social conventions and the objective world and as a focusing inward. Just as rationalists like Descartes and empiricists like Hume grounded their new epistemologies in the isolated, experiencing self, so modern artists draw their creative resources from an exploration of their interior emotional worlds. The goal of exploring and processing experience was given ever-fresh stimulation through the dramatic changes overtaking modern society in the 19th and 20th centuries, leading to intense individual experiences and passionate, frenzied expressions of subjectivity and anxiety, such as are paradigmatically represented by Edvard Munch's *The Scream*.

In contrast to the expressive power of modern art, some forms of postmodern art display a "waning of affect." The term, coined by J. G. Ballard and popularized by Jameson (1991), suggests that the neurasthenia of the modern condition has given way to a widespread feeling of emptiness and blankness, as though the modern mind, addicted to cocaine, had taken massive doses of lithium to come down and cool out. Coolness, blankness, and apathy become new moods for the decelerating, recessionary postmodern condition in an age of downsizing and diminishing expectations. According to Jameson, the alienation of the subject in the modern era, which required depth of feeling and a critical distance between the subject and the objective conditions of its life, has been absorbed, as expressive subjectivities mutate into fragmented selves devoid of psychological depth and autonomy.

Although advocates of the postmodern like to champion it as a break from the modern, there are very few "postmodern" elements that are completely new or innovative. While postmodern discourse renounces originality and the celebration of novelty and innovation characteristic of modernism, it also continues the experimentalism of modernism and the avant-garde. Like these movements, it is committed to formalism, self-reflexivity, ambiguity, and a critique of realism. Against modernism and the avant-garde, however, postmodernism declares both the death of the author and of the work, replacing the former with the decentered self or *bricoleur* and the latter with the "text." In the poststructuralist lexicon, "text" refers to any artistic or social creation that signifies and can be conceptually interpreted. Thus, not only are artworks like novels and paintings "texts" but so too are buildings, landscapes, and cities. The shift from "work" to "text" is meant both to broaden the category of objects for critical interpretation and decoding and to suggest that the meanings of the text are usually multiple and conflicting, requiring new methods of interpretation that are multiperspectival and that decenter the "authorial voice."

Despite the heterogeneity of the various postmodern turns in the arts, they share key concerns and family resemblances. We believe this is the case because there are broad and sweeping changes occurring throughout the culture in general, and these same "epistemic" elements are being articulated in similar ways by various artists and theorists in different fields—an argument that we develop in Chapter 6. In fact, it should be no surprise that postmodern developments appeared first in the arts and only later in criticism and social theory as an explicit movement, since cultural changes are typically explored first phenomenologically, as experiences and moods, and only later reflexively, as theories. Yet, as we will see, the postmodern turn in the arts involves unusual intimacy between theory and culture, with developments in the arts influencing theory and movements in the arts illustrating trends of postmodern theory.

One key general characteristic that unites the various postmodern movements in the arts is that they are implosive and dedifferentiating. This is to say that they renounce, implode, deconstruct, subvert, and parody conventionally defined boundaries such as those between high and low art, reality and unreality, artist and spectator, and among the various artistic media themselves. These implosive tendencies are reactions against forms of modernist purism that seek unified or pure aesthetic styles defined according to a strict set of genre rules, as well as responses to the sociological conditions of media culture forms that are saturating culture and society. The implosion of well-defined boundaries means that postmodern cultural forms are typically eclectic and combine a host of different forms, often in playful and ironic ways.

Such play with different styles suggests yet another crucial characteristic of postmodern cultural forms: the rejection of structure, order, continuity, and cause–effect relations in favor of disorder, chaos, chance, discontinuity, indeterminacy, and forces of random or aleatory play. These motifs can take the form of an attack on narrative structure in literature, music, dance, or the willful combination of conflicting styles in architecture. John Cage (1961), for instance, made the postmodern paradigm of indeterminacy the organizing principle of his music, using what he called "chance operations" to organize sound in an aural collage that often picked up and delighted in accidental sounds from the environment or from electronic devices. Cage also undid the boundary between music and all other forms of sound, imploding music into sound, making all sound—including silence—a form of music.

Cage's collaborator, Merce Cunningham, likewise broke with theme and variation in dance, similar to the way Cage renounced theme and melody in music. In their collaborations, Cage and Cunningham produced their music and dance sequences separately, working within an agreed-upon time and structure, and thus were among the first to exhibit a dissociation of music and dance and the effects of an indeterminate chance juxtaposition. Just as Cage absorbed a wealth of sounds in his music, Cunningham absorbed a large repertoire of movements into his dance collages, including clumsy gestures, stillness, and habitual actions of everyday life previously excluded from dance. Like Cage, Cunningham spurned expressive dance and signification, rejecting the notion that were was an underlying idea or meaning in his work. Cunningham claimed that each of his dances produces a unique atmosphere and invites the spectator to interpret the significance, as he or she likes, just as Marcel Duchamp and Cage insist that the spectator produces the meaning of the artwork (see the discussion in Tomkins, 1965).

Because so many different elements work together in the postmodern text, postmodernists are typically multiperspectival in their sensibility, believing that no single perspective, theory, or aesthetic frame can illuminate the richness and complexity of the world of experience, or the "text." Postmodernists are also acutely self-conscious about their implosive play with formal elements, and they typically foreground the formal, semiotic, or linguistic nature of their art in a way that calls attention to the process of aesthetic creation as fictive, constructive, and artificial in nature. Just as language came to the foreground of theory in ways that saw words as constituting rather than reflecting reality, so in the arts postmodernists abandon realist principles, which allegedly reflect the world as it is without mediation, and emphasize the medium of language or form itself, a technique also employed by modernists like Brecht. Such an approach characterizes the "metafictional" status of postmodern literature as well as the "double coding" of architecture, and both are symptoms of the "linguistic turn" that informs all postmodern forms and practices (see Chapter 6).

As we shall see, some forms of postmodernism play with language, forms, images, and structures, appropriating material and forms from the past, finding aesthetic pleasure in appropriation, quotation, and the play of language, reveling in linguistic and formal invention, puns, parody, and pastiche. Such types of postmodern art are described by Hal Foster as a "postmodernism of reaction" (1983: xii) that is highly historicist, playing with forms of the past and generally affirmative toward the status quo, renouncing the modernist project of critique and opposition. Because of the element of play in this strain of postmodernism, which has its analogue in theoretical discourse, Teresa Ebert (1996) suggests the term "ludic postmodernism," which she and Foster distinguish from a "postmodernism of resistance" (Foster, 1983: xii; Ebert, 1996: ix, *passim*) that questions and deconstructs rather than exploits cultural codes and "explore[s] rather than conceal[s] social and political affiliations" (Foster, 1983: xii).

Ludic postmodernism describes the aesthetic advocated by Sontag, who affirmed the surface, forms, and erotics of art rather than content, meaning, and interpretation. Such an aesthetic finds its precursor in Nietzsche, who anticipates key aspects of the ludic postmodern turn in art, with his emphasis on joy in appearance and on the aesthetic of form and with his renunciation of a depth hermeneutic, writing: "Oh, those Greeks! They knew how to live. What is required for that is to stop courageously, at the surface, the fold, the skin, to adore appearance. Those Greeks were superficial—*out of profundity*" (1974: 38 and 1954: 683).

Consequently, we accept Foster's (1983) and Ebert's (1996) distinction between a ludic postmodernism that indulges in aesthetic play for its own sake while distancing itself from a politically troubled world, or even lending tacit or explicit support of the status quo, contrasted to a "postmodernism of resistance" that acknowledges its self-referential status but also seeks to engage political issues and to change the existing society.[7] The less-oppositional postmodernism often plays with contemporary culture, exulting in the profusion of existing culture and society while rejecting modernist tenets and returning to tradition and such things as ornamentation, decoration, and pastiche of past cultural forms. Such a ludic postmodernism thus abandons modernist pretensions to novelty, originality, purity, innovation, and the like and seeks pleasure in playing with the pieces of the past rather than in criticizing the present while envisioing at a new future.

Of course, as avant-gardists insisted, challenges to conventional modes of perception in themselves can assume a positive political character, but they can also have depoliticizing effects by limiting themselves to a merely formal level dominated by abstract or ludic functions. Oppositional postmodernists, by contrast, combine artistic and political levels of aesthetic production and employ the formalist inventions of postmodern culture toward political ends, keeping a critical distance and thereby breaking out of the self-referential funhouse of language characteristic of many forms of both high modernism and postmodernism—an argument that we will illustrate later in this chapter through discussion of some oppositional postmodern political artists.

On the other hand, such categorical distinctions between modes of postmodernism are only ideal types, and much postmodern culture partakes of both poles, often ambiguously. Thus it is frequently undecidable whether specific forms of postmodern culture are primarily oppositional or conservative, promoting desires for change or a pleasure in the existing order, and are thus "resistant" or "complicit." Indeed, such dichotomies are destabilized by audience reception in which differing audiences receive, use, and deploy various works in highly contradictory and unpredictable ways according to their own gender, class, race, ideological, and other "subject positions." Artworks coded as critical and oppositional may well have conservative effects, while reactionary works can be read against the grain and decoded to generate socially critical insights. Consequently, the categories one uses must be deployed in specific situations and appropriate qualifications made with attention paid to both encoding and aesthetic practice and form, as well as audience decoding and use of the artifact.

THE POSTMODERN MOMENT IN ARCHITECTURE

Each generation writes its biography in the buildings it creates.
—LEWIS MUMFORD

Architecture is *the* public art.
—CHARLES JENCKS

Postmodernism began appearing in a variety of artistic fields in the 1960s and 1970s, although it was most dramatically visible in the field of architecture, where it was adopted to describe the new forms of contemporary buildings, which returned to ornamentation, quotation of tradition, and the resurrection of past styles that a more purist modernist architecture had rejected. The rapid dissemination of postmodern discourse and forms in architecture helped to promote it in other aesthetic fields, thus providing concrete substance to the postmodern turn. Indeed, people live in houses, neighborhoods, and social environments, and so architecture is nothing less than the mode of construction of everyday life. Thus, shifts in how architecture is conceived and constructed inevitably produce mutations in the very structure and texture of lived experience and the social environment.

The postmodern turn in architecture involves a renunciation of modernist conceptions of stylistic purity, aesthetic elitism, rationalism, and universally based humanist and utopian political programs to beget a new humankind through architectural design. Against these principles, postmodernists like Robert Venturi, Philip Johnson, Christopher Jencks, Kenneth

Frampton, and Michael Graves renounced the abstract and ahistorical formalism of the International Style, embraced an eclectic mixture of historical styles, pursued an approach that respects both popular and professional tastes, and abandoned the utopian aspirations of modernism in favor of more "modest" goals. Let us, accordingly, engage the modern turn in architecture, examine the postmodern critique, and then explore the forms and theories of postmodern architecture. We shall argue that the turn from modern to postmodern architecture involves a transition from the regime of monopoly capital to a more aestheticized and transnational form of postmodern capital.

The Trajectories of Modern Architecture

> In lieu of cathedrals the machine for living in.
> —OSKAR SCHLEMMER

> The bourgeoisie . . . creates a world after its own image.
> —KARL MARX AND FRIEDRICH ENGELS

It is often argued that whereas the lines between the modern and the postmodern are hard to draw in literature, they are particularly clear in architecture. But in fact, the easy dichotomization between modern and postmodern architecture is arrived at only by equating modern architecture with the glass boxes of the International Style, which is just one version of architectural modernism, albeit one that dominated from the 1920s through the 1950s. It is typically ignored, however, that there were a profusion of modern styles, that there was considerable conflict among them, and that many architects of the International Style themselves developed different styles, some similar to the postmodern forms commonly opposed to it.

The architectural styles of the modern era include Renaissance (15th century), mannerism (16th), baroque (17th), rococo and neo-classicism (18th), expressionism (19th and 20th), art nouveau (19th and 20th), American industrial style, as in the skyscrapers of Louis Sullivan, and the organic regionalism of Frank Lloyd Wright. Architectural "modernism," however, is said in many standard postmodern accounts to begin with the genesis of the International Style, which appeared in the 1920s, was systematized by the early 1930s, and became dominant throughout the world by the 1950s.[8] Our argument in this chapter, however, is that the "modern architecture" constructed by the postmodern polemic is a reductive construct that collapses a great variety of modernist styles into a unitary category of the International Style, thus obscuring important differences. The postmodern turn in architecture is often celebrated and legitimated by a spurious conception of modern architecture that covers over its complexity, diversity, and richness, identifying it *tout court* with the high modernism of the International Style. We will argue for an architecture that draws upon both modern and postmodern styles to develop a mode that serves human needs and that produces a more livable and sustainable environment.

The International Style is largely the product of the Bauhaus School founded in Germany in 1919 when Walter Gropius was appointed director of two schools of Arts and Crafts in Saxe-Weimar which he amalgamated, changing the school's name to "Bauhaus," that is, "house of building." The school attracted top teachers and students, but its progressive ideals and plans scandalized the conservative citizens of Weimar, and so the school moved in 1925 to Dessau, Germany, where Gropius designed the edifice that became a prototype for the new style and developed the philosophy that would dominate architecture for the next several decades (Benevolo, 1977: 414ff.).

The International Style came to the United States when Gropius, Mies van der Rohe, and others emmigrated in the 1930s in order to escape from fascism. They were immediately embraced in the United States as architectural icons. Gropius was appointed head of the Harvard department of architecture in 1937 and Mies was designated director of architecture at the Illinois Institute of Technology in 1938, designing nearly all of the buildings on the new campus. Their students and followers became active in the major architectural firms in the United States, and their style of highly rationalized and functionalist buildings became the norm throughout the world.

The International Style is equated with "architectural modernism" because, like other modernist movements in the arts, it sought to make a clean sweep of the past, to be modern, to use new styles, materials, and technologies, and to advance new ideas. In the words of Gropius: "A breach has been made with the past, which allows us to envisage a new aspect of architecture corresponding to the technical civilization of the age in which we live; the morphology of dead styles has been destroyed; and we are returning to honesty of thought and feeling" (1965: 19). The reference to "honesty" is a critical attack on art nouveau, the prevailing aesthetic style that assaulted 19th-century eclecticism, neoclassicism above all, seeking a new "honesty" through imitation of natural forms such as trees and clouds, which it claimed to be the most valid sources of inspiration. Against art nouveau, the International Style asserted a superior "honesty" in the imitation of the geometric forms of the modern machine age and sought to replace nature with man-made environments, or at least to integrate nature into a massive new technoscape of dazzling proportions.

Although it presented itself as a break from the past, modernist architecture (hereafter, simply "modern architecture") was dependent upon the 19th-century technological innovations whose iron, steel, glass, and reinforced concrete allowed for innovations in building technique. Modern architecture, moreover, was heavily influenced by the values of dynamism and progress that dominated the 19th century, such as were celebrated in the Crystal Palace science and technology exhibit that opened in London in 1851. Attacking the philosophy of art for art's sake, modern architecture also embraced the utopian and humanist values of the Enlightenment by intending their architecture to contribute to the rationalization of the environment and the liberation of human beings from tradition. The Enlightenment emphasis on universal values was enthusiastically adopted by Bauhaus architects in their quest to construct a global, uniform architectural language that expressed rational values appropriate for the new "universal man."

The French–Swiss architect Le Corbusier also expressed the utopian ambitions of modern architecture in his cry of "architecture or revolution," believing that architecture—using new materials, new structural methods, and universal principles of design that were also sensitive to local conditions—could solve the major problems of contemporary urban life, such as housing, traffic, and the organization of cities, in ways that were both efficient and beautiful. Le Corbusier and the new breed of engineer–architects believed that architecture had to be reconceived to realize the new technological possibilities and to respond to the problems of the time.

Twentieth-century aesthetic movements also influenced modern architecture. Much of the architectural style of Bauhaus, as well as the work of Le Corbusier, is influenced by cubism, which featured abstract, geometric shapes. Like futurism, which concocted powerful images of the new machine age in excited motion, cubism expressed the dynamic aspects of the modern age through a multiperspectival method of representing various dimensions of objects simultaneously. Early in his career, Le Corbusier produced numerous paintings in the cubist style of Picasso and Georges Braque. He applauded the breakthrough of such painters in creating new forms that he believed had something to teach architecture: "Today painting has outsped the other arts. It is the first to have become attuned with our [industrial] epoch" (quoted in Blake, 1996: 23). Indeed, the cubist influence is vividly evident in the various homes that Le Corbusier designed in the 1920s and is consummately realized in his Villa Savoye (in Poissy, France, 1931), a rectangular boxlike house raised off the ground by concrete stilts, and his Weekend House (1935), comprised of diverse geometrical forms that Le Corbusier described as a form of painting in space.

In both his paintings and architecture, Le Corbusier followed the lead of cubism in the spatial arrangement of abstract geometric forms such as cylinders, cubes, and cones that signified nothing but their own formal qualities. Yet Le Corbusier felt that cubism had not completed the formal revolution it began. As Peter Blake put it: "Cubism had cleared the air by removing the most distracting elements of realism from painting, but it had . . . degenerated into a sort of playful, decorative movement" (1996: 28). In response, Le Corbusier and fellow painter Amédée Ozenfant produced a manifesto calling for a return to the geometric foundations of cubism and created a movement within the cubist tradition that they baptized "purism."

In October 1920, Le Corbusier and others published *L'Esprit nouveau*, a magazine dedicated to organizing all the arts around "*l'esthétique de la vie moderne*." In 1923, Le Corbusier—hitherto known by his birth name of Charles-Edouard Jeanneret—decided to focus on the architectural realization of his vision and resurrected himself as Le Corbusier, "the architect," and proceeded to revolutionize the discipline. Both Le Corbusier and the Bauhaus were also influenced by the avant-garde movement, which wanted to use art to transform everyday life, to devise an entire new modern world. Le Corbusier designed his first plan for an ideal city, *une ville contemporaine*, in 1922, and in 1925 he presented a plan to remake Paris, the *Plan Voisin*. He produced a housing complex at Pessac in 1925, and some decades later he designed a cityscape in Chandigarh, India (1952–1956).

The avant-garde aesthetic influences overlap with a crucial philosophical influence on Bauhaus and the International Style, namely, the utilitarian and functionalist ethos of capitalism and the modern machine age that it absorbed. Capitalist values of standardization and mechanization were embraced as liberating, with the result that architectural production became a factory assembly-line process: "We are approaching a state of technical proficiency when it will become possible to rationalize buildings and mass-produce them in factories by resolving their structure into a number of component parts" (Gropius, 1965: 39). Gropius calls for the transformation of architecture, once a trade subject to the seasons, the site, and "the arbitrary reproduction of historical styles," into "an organized industry" that prefabricates identical materials and standardized parts. This "will have the same sort of coordinating and sobering effect on the aspect of our towns as uniformity of type in modern attire has on social life . . . every house and block of flats will bear the unmistakable impress of our age" (40).

Such mass manufacture, Gropius insists, will allow sufficient range for architectural variety and free expression. Gropius rigorously trained a legion of students in the fundamentals of modern technique and design, encouraging them to take their place within the machine age, but the homogeneity of the International Style suggested that the pupils of Bauhaus did little but imitate their masters. Indeed, industrializing processes were so pervasive, so profound, that modern artists could not possibly ignore them. Every architect in particular felt compelled to respond to the machine age; this influenced how they built, what they built, and the philosophies informing their visions of architecture. They were intoxicated with the possibilities of the new technologies and in awe of the problems besetting the design of new urban environments.

Despite these technological, philosophical, and aesthetic continuities with the past, modernist architects generally sought to break from the history of architectural form, specifically the ornamental style that prevailed in Europe. They perceived such forms as the Gothic to be aesthetically unpleasing and bound up with authoritarian domination and social hierarchy, and they saw art nouveau as a slavish imitation of nature in contrast to the new industrial designs. For the Bauhaus theorists, massive churches, imposing government buildings, and public monuments projected power and authority, constructing testimonials of might for reactionary forces. Gothic cathedrals in this vision expressed the power and majesty of the church, and not just its aspiration for the divine. These insights are supplemented by Foucault (1965, 1977) who shows how mental institutions, hospitals, schools, and prisons helped construct the social space of modernity, a categorizing, separating, and incarcerating space that confines individuals who refuse to conform to existing norms and practices. This process produced a normalization of subjects through socially constructed definitions of "normal" and the "abnormal," in which individuals are classified and situated according to how they fit into the modern order. On this perspective, the construction of carceral and disciplinary space is a key aspect of the development and trajectories of modern societies.

For modernist architects, the reconstruction of space and the construction of a new type of architecture thus constituted an important part of a revolution against the past. For these modernists, architecture should free itself not only from tradition but also from the natural environment, in order to create its own utopian worlds of glass, steel, and concrete. Rather than integrating architecture with nature, a principle Frank Lloyd Wright adopted from art nouveau, many modernists argued that architecture should stand in bold contrast to the natural world. Modern architects sought a new, austere style that abandoned symbolism, ornamentation, and decoration in favor of unity, simplicity, and purity of form, paring down line and space to their bare essentials. In the words of Mies, "Less is more," and Adolph Loos even denounced decoration as "a crime."

Whereas historicist values and a plurality of historical styles and unique architectural personalities, such as Antonio Gaudi, flourished before the Bauhaus and would be recaptured afterwards in the movements leading toward postmodernism, the architects of the International Style intended to reduce this eclecticism and plurality to a single new style that they sought to make dominant throughout the world. The prototype for the new International Style was Gropius's design for the Bauhaus compound built in Dessau, Germany, in 1925–1926. The site is made up of a series of rectangular buildings joined together at perpendicular angles. In typical International Style, the design is uniform, geometrically precise, devoid of ornamental detail, and completely standardized, resembling a warehouse or prison yard. This minimalist style was duplicated, with equally uninspiring results, by Mies in his buildings at the Illinois Institute of Technology. In the building boom after World War II, the International Style high-rise skyscrapers came to dominate the urban environment throughout the world (although they are now being pushed out by new postmodern buildings). Buildings such as the tall glass and steel towers of Lake Point Tower, Lake Shore Apartments, the Federal Center, and the IBM Building in Chicago, and the Seagram Building in New York, all designed by Mies, exhibited the triumph of the International Style.

These phallic prisons, embodiments of Max Weber's iron cage, represent the ultramodernist philosophies that informed them. For the postmodernists, highmodernist architecture strives for purity of form and function, generally ignoring principles of communication (semiotics) and beauty. Aesthetic "beauty" is an obsolete principle that is jettisoned by many modern architects in favor of sheer functionality, whereby, in the words of Louis Sullivan, "form follows function." Beauty is superfluous when the emphasis is on utilitarian values, replicating the same bottom-line approach of capitalist society. Architectural modernists fetishized functional values and the technological culture that produced automobiles, highways, and factories. Except for futurism, no other modern art movement has been so conditioned by technological ideology and the modern mechanist worldview as International Style architecture.

But the standard postmodern narrative occludes the variety and diversity of architectural modernism. Certainly, not all architecture from the late 1920s to the 1950s had the look of the International Style. Not all modern architects were strict functionalists, and many sought aesthetic inspiration in machines and abstract shapes. The most striking exceptions to the International Style were the inventive modernist constructions of Le Corbusier and the poetic works of Frank Lloyd Wright, most notably the magnificent houses he designed throughout the United States, such as the Kaufmann House, Falling Water (Bear Run, Pennsylvania, 1936–1937), or Taliesin West (Phoenix, Arizona, 1938), each of which revel in the play of diverse forms and levels coherently linked. Wright, in fact, detested the International Style, Gropius's work in particular, for he witnessed how it came to dominate architecture in the United States and elsewhere, pushing out a diversity of more interesting styles, such as his own Japanese-inspired approach.

Other high modernist architects, most notably Le Corbusier, rejected the straitjacket of the glass box to produce highly interesting and innovative buildings. Le Corbusier's Church of Ronchamp (near Belfort, France, 1950–1953), his monumental series of governmental buildings at Chandigarh, India (1952–1956), and his Exhibition Hall (Zurich, 1967), show bold departures from the International Style. Le Corbusier's famous statement that the house is a "*machine à habiter,*" that is, "a machine for living in" is thus misleading and masks his fundamental concern with beauty and the transcendental qualities of abstract forms. "I compose with light," Le Corbusier said, as he meticulously broke up solid planes, on both exterior and interior surfaces, so as to absorb and reflect light and create mood. Le Corbusier speaks of his architecture as "poetry and lyricism brought forth by technics," thereby abolishing any rigid opposition between art and technology. He insisted that architecture "goes beyond utilitarian needs" such that the architect's passion can invent "drama out of stone" (quoted in Blake, 1996: 31)—as is vividly clear in his best works. Moreover, Le Corbusier explored tensions, ambiguities, and paradoxes that subvert oppositions such as those between machine and biology, between mathematical measure and lyricism, between engineering and aesthetics. He willfully combined mechanistic and biological metaphors; thus, his "*machine à habiter*," whether a house or a city, was composed of various "organs," and so he transcended merely mechanistic or formalist–functionalist conceptions.

Yet with the appointments of Gropius and Mies to major architectural chairs, the International Style literally became institutionalized and dominated the global scene, entrenching itself in Europe and the United States as a rigid orthodoxy that few architects dared to challenge if they wanted to work. Most, however, such as the Yale architects, produced endless variations of a highly limited form, seemingly unable to comprehend the possibility of color, locality, or irregular forms. An important incentive to remain within the orthodoxy was that International Style buildings, or so it was thought, were the most cost effective to produce. Those architects who did venture outside of the ideological limits and stylistic parameters of the glass box were condemned, ridiculed, and ostracized, and their designs were often prevented from being realized.

In fact, the modernist emphasis on change, innovation, and artistic autonomy was betrayed with the rigidification and repetition of the International Style. But modern architecture was not just formalist; it had a political content and philosophical vision that sought the rational transformation of human beings in conjunction with egalitarian values. The humanist and utopian values of modern architecture were most evident in its attempt to build a suitable form of mass housing for workers, the middle class, and low-income families. Yet these values were grossly belied when the housing projects became prisons befouled by drugs, crime, and graffiti. Apparently, the architects of these compounds presumed that the units were best designed as barren, small, and cramped. Many featured ceilings not to exceed eight feet in height, thin walls devoid of molding, casing, or baseboards, and narrow hallways. In contrast to other contemporary designs organized around patterned variety, such as those by Bruno Taut and Ernst May in Germany, Bauhaus-influenced designs in the United States were homogenized and oppressive.

Indeed, in Europe many more successful examples of workers' housing and planned urban environments were carried out by socialist governments. Ironically, Bauhaus architecture was originally conceived of as a socialist architecture, designed to meet the needs of working people. But the International Style became largely a corporate architecture, providing monuments for capitalism and housing units appropriate for its underclass. The utopian and egalitarian aspirations of earlier modern architecture were thus subverted in the capitalist appropriation of modern architecture to aggrandize capitalist corporations and to produce public housing designed more for social control and ghettoization than for creating new forms of democratic life.

Publisher Joseph Pulitzer erected the first major skyscraper in 1890 with his World Building in Lower Manhattan, which arose as the largest building in the world. It was soon surpassed by the Metropolitan Life Tower of 1909 (700 feet), the Woolworth Building of 1913 (792 feet), the Chrysler Building of 1930 (1,046 feet), and in the early 1930s the Empire State Building (1,250 feet) (Paul Goldberger, *The New York Times*, August 4, 1996: H30). Thus, whereas urban skyscrapers like the Empire State Building and the Chrysler Building were appropriate monuments to an earlier era of competitive capitalism, the glass and steel monoliths of the International Style are fitting emblems of a later stage of monopoly and state capitalism. The earlier urban skyscrapers designed by Louis Sullivan in Chicago and the first great high-rise buildings in New York expressed the architectural visions of their creators, with each having a distinctive style and look. Moreover, the spectacle of the earlier skyscrapers, with each striving to surpass previous heights, represented the ethos of a highly competitive and individualist capitalism. The skyscrapers paid homage to the entrepreneurs who dominated the early stage of market capitalism, generating the great economic fortunes and empires. Temples of the heroic stage of capitalism, icons of the market that made possible the amassing of great fortunes, these impressive and individualized edifices produced hieroglyphics of a competitive capitalism in which "Manhattan's great buildings were always happy enough to affront each other in a competitive verticality, the result of which is an architectural panorama in the image of the capitalist system: a pyramidal jungle, all the buildings attacking each other" (Baudrillard, 1983a: 135).[9]

High modernism in architecture fit perfectly with corporate capitalism and provided a useful ideology for its legitimation. The demand to restructure the environment, to destroy all obstacles in the path of modernization, was a perfect ideology for a relentless capitalist development. Small wonder that the movers and shakers of corporate capitalism were taken with the International Style. In a way, the avant-garde project of destroying tradition unwittingly abetted the agenda of a modernizing capital which itself was negating tradition and obliterating the past and thrived on the creation of perpetually new products and needs. The modernist project of abstraction in some ways furthered the capitalist project of abstracting from concrete human needs, tradition, and culture to promote its obliteration of traditional values and personality and creation of new subjectivities in the transition to consumer capitalism.

Moreover, as Tafuri argues (1976), the formalist project of the avant-garde and the International Style was parallel to Nietzsche's and Weber's demystification of the world, producing an ideology that could attack any obstacle in the way of capitalist expansion. Avant-garde formalism, which carried out an eradication of substance, individuality, and subjectivity, was parallel to the capitalist project of reducing the world to the pure stuff of domination (i.e., the worker as pure automaton, the citizen as object of manipulation). The avant-garde thus covertly aided the process of capitalist domination and massification in the production of a new regime of mass production, consumption, and culture no matter how much it attacked bourgeois culture and society. Indeed, the later postmodern turn to individual feeling, to aestheticization, to pleasure and indulgence, and difference and fragmentation, advances the contemporary capitalist agenda of generating a more aestheticized and eroticized world that will promote more individualized consumption, more segmented markets, with new choices, pleasures, and services, thus, once again, serving the agenda of capital that requires a new ideology convincingly served up by postmodernism.

The more restrained functionalism of the International Style, by contrast, is the proper representation of a state monopoly capitalism in which individuals submit to corporate control and their uniformity and homogeneity corresponds to the staid, ascetic, conformist, and conservative world of corporate capitalism that was dominant in the 1950s, with its organization men and women, its mass consumption, and its mass culture.[10] The International Style was thus appropriate to a homogenizing regime of capital that wanted to produce mass desires, tastes, and behavior. But the glass and steel high-rises of corporate capital can also be seen as monuments to their global power, with the same corporate style appearing everywhere, signifying the triumph of the giant corporations and their ability to remake the world in their own image.

The spectacle of high modernity was thus the U.S. corporation, the demiurge and progenitor of a new stage of global capitalism. The high-rise monoliths of the International Style were the temples of the megacorporation, the embodiment of the corporate vision, of the triumph and hegemony of global monopoly capital. Its imperatives meshed with those of modern architecture and its project to beget a new world that is clean, functional, efficient, and universal. But the modern city combined efficient and well-organized centers with regions of disorder, ugliness, violence, and chaos. The ideological abandonment of the modern project thus represents realist insight into the failures of the modernist–capitalist ambitions for a well-organized and functional corporate world and signals the transformation from modern to postmodern architecture, from the hegemony of the International Style to the eclecticism and pluralism of postmodern populism.

The shift from modern to postmodern architecture is thus not merely a mutation from one architectural style to another; it is, rather, a sign of a shift to a new regime of capital, a new social order. The fashionable postmodern architecture meets the needs of a transnational and global capital that valorizes difference, multiplicity, eclecticism, populism, and intensified consumerism. Thus, postmodern architecture, shopping malls, and spectacle became the promoters and palaces of a new stage of technocapitalism, the latest stage of capital, celebrating the postmodern image and consumer culture. Perhaps the emergent cyberspace of the Internet will be a new privileged domain of the infotainment society on the horizon. Following this logic, the spectacles of postmodern architecture are more appropriate to the contemporary forms of a highly aestheticized global consumer capitalism, as we shall see in the next sections.

New Directions

> The machine can no more adequately symbolize our culture than can a Greek temple or a Renaissance palace. . . . To persist in the religious cult of the machine, at this late day and date, is to betray an inability to interpret the challenges and dangers of our age.
>
> —Lewis Mumford

> I like elements which are hybrid rather than "pure," compromising rather than "clean." I prefer "both–and" to "either–or."
>
> —Robert Venturi

The monotonous cityscape of the International Style was indifferent to the needs of people, and though the initial modern housing projects sought to be responsive to these needs, they in fact were often oppressive. Thus, a growing chorus of dissent began to attack the International Style, and new architectural directions have been explored from the 1950s to the present. Charles Jencks, somewhat facetiously, claimed that modern architecture died in St. Louis, Missouri, on July 15, 1972, at exactly 3:32 P.M. (1977: 9)—for this is the place and time that the infamous Pruitt-Igoe housing project, a symbol of misguided modern visions, was dynamited, after millions of dollars had been wasted in futile efforts to renovate what the inhabitants vandalized and vilified. This failure, Jencks declares, revealed the bankruptcy of the utopian designs of modern architecture and the need to develop new postmodern concepts.

Critiques of architectural functionalism had already begun to surface in the 1950s. In a 1955 annual of the Yale Department of Architecture, Philip Johnson, a devotee of Mies and codesigner of the Seagram Building, stated: "Merely that a building works is not sufficient. You expect that it works. . . . Structure is a very dangerous thing to cling to. You can be led to believe that clear structure clearly expressed will end up being architecture by itself. You say I don't have to design anymore" (quoted in Klotz, 1988: 41). But during the 1950s and 1960s, several significant polemics against the International Style appeared, which paved the way for the postmodern turn in architecture in the writings of Lewis Mumford, Jane Jacobs's *The Death and Life of Great American Cities* (1961), and Robert Venturi's *Complexity and Contradiction in Architecture* (1966).

By the early 1950s, Lewis Mumford had become one of the leading social and architectural critics in the United States. Growing up in Manhattan, Mumford had extensive experience with both the stimulating and stifling aspects of modern urban life, and his work represents a growing dissatisfaction of many people with machine values and emphases on growth and centralization. In essays such as "The Case Against 'Modern Architecture,'" Mumford assailed the mechanistic paradigm that informed the work of Le Corbusier, Mies, Gropius, and others. He condemned their work as "the apotheosis of the compulsive, bureaucratic spirit" (in Miller, 1986: 78), as monuments to the modern fetish of size and growth. He saw that modern architecture had largely subordinated human beings to the circulation of commodities and traffic, ignoring their complex social, cultural, and psychological needs. He proclaimed that the vital function of both architecture and urban design was to realize these needs, to promote human interaction, stimulate creativity, and advance freedom and spontaneity. Decrying the reduction of the human to the mechanical, Mumford called for a new architecture that united art and technics, that promoted interaction among human beings, and that integrated culture and nature in an organic whole. He advocated a new architectural style that combined modern emphases on function with premodern emphases on decoration, symbolism, and expressive functions. In his emphasis on redesigning cities on a human scale organized around regional environments, he recaptured premodern emphases on decentralization and anticipated postmodern themes of a "critical regionalism" (Frampton, 1983) sensitive to locality, history, and tradition.

Jane Jacobs attacked modern urban planning and architecture that did not take into account the needs of the people and their actual communities. She argued that in the sprawl and disorder of the modern city, citizens had created a form of spontaneous order through communal interaction. Affirming the vibrancy and diversity of the urban neighborhood, Jacobs celebrated existing communities as sources of variety, liveliness, and cultural diversity that were being destroyed in favor of modernist visions of uniform order, structure, and homogeneity. By contrast, Jacobs valorized the intense variety and vitality of the city over what she saw as the sterile order projected by advocates of modern architecture. She believed that the self-organization of people in neighborhoods generated a form of ordered complexity that preserved life and community, writing: "There is a quality even meaner than outright ugliness or disorder, and this meaner quality is the dishonest mask of pretended order, achieved by ignoring or suppressing the real order that is struggling to exist and to be served" (1961: xx). Thus, she anticipated the postmodern celebration of diversity, otherness, and heterogeneity that would later be articulated in both postmodern theory and architecture.[12]

Mumford acknowledged the value of Jacobs's work and applauded her emphasis on the importance of family, neighborhood, and social diversity, but he argued that she had a limited conception of the modern city and was too uncritical of the worst features of urban life. Mumford claimed that Jacobs's vision of the ideal city life was merely that of a crime-free neighborhood, a goal attained through multiplying social and economic activity and people with watchful eyes. Mumford believed, however, that Jacobs's "cure" may be worse than the disease, leading her to affirm the values of the "megalopolis." With Jacobs upholding safety and activity above all other goals, Mumford argued that she failed to criticize urban crowding, overpopulation, excessive noise, the frantic pace of life, giganitsm in design, and excessive pollution. "Her simple formula [for eliminating crime] does not suggest that her eyes have ever been hurt by ugliness, sordor, confusion, or her ears offended by the roar of trucks smashing through a once quiet residential neighborhood, or her nose assaulted by the chronic odors of ill-ventilated, unsunned housing at the slum standards of congestion that alone meet her standards for residential density" (in Miller, 1986: 194).

Mumford agreed with Jacobs's affirmation of complexity and diversity, but thought she confused these with overcrowding. Overall, Jacobs ignored "the increasing pathology of the whole mode of life in the great metropolis, a pathology that is directly proportionate to its overgrowth, its purposeless materialism, its congestion, and its insensate disorder—the very conditions she upholds as marks of urban vitality" (in Miller, 1986: 192). Indeed, it could be argued that Jacobs was perceptive in her affirmative analyses of small, intimate urban areas like Greenwich Village, but that she lacked an adequate critical vision that sees the need for a radical reconstruction of urban life in its totality. Her work suffers from a contradiction between her love of the intimacy of neighborhood life and her embrace of the general structure of the megalopolis.

It is, however, Robert Venturi who is most centrally associated with the postmodern turn in architecture. Venturi (1966) established a series of principles in opposition to modernism, such as complexity and hybridity versus modernism's simplicity and purity. The very title of *Complexity and Contradiction in Architecture*, signifies opposition to Bauhaus minimalism, but the book was particularly influential because he coded his heterodox departures in language that paid homage to the masters, in the form of a "gentle" manifesto. Venturi also championed "the difficult unity of inclusion" over modernism's "easy unity of exclusion" (1966: 16). Thus, whereas high modernism systematically excluded ornamentation, decoration, and historical allusion, Venturi called for the inclusion of these elements, promoting a pluralist and eclectic aesthetic of inclusion that would embrace different styles, codings, and decorative elements banished by high modernism.

But it was his book *Learning from Las Vegas* (1972, coauthored with Denise Scott Brown and Steven Izenour), which signaled an explicit postmodern turn in architecture. Venturi and his colleagues aggressively celebrated the most distinctive forms of American commercial architecture, ranging from the Las Vegas Strip to billboards, neon, A&P parking lots, and suburbs. They distinguished between the "well-decorated shed," their metaphor for buildings suitably decorated and rich in signification, and the "duck," a metaphor for a building that merely signified itself, such as a doughnut shop in the shape of a doughnut or a boot shop in the shape of a boot, and they saw both as appropriate modes of architectural signification. Thus, in effect, Venturi and his colleagues provided an affirmative legitimation of dominant forms of U.S. architecture, which they defended as "all right" (i.e., "billboards are all right," "Main Street is almost all right," and ducks, decorated sheds, and other kinds of commercial U.S. architecture are all right too; the only thing that is not all right seems to be high modernism, the International Style).

Venturi and his associates saw architecture as a mode of communication and called for a rich architecture of allusion, comment, and ornamentation. Following this model, postmodern architecture returned to tradition, mixing traditional models, embellishment, and design with modern forms. For Klotz (1988), the primary characteristic of postmodern architecture is that meaning becomes as important as function, that buildings should "fiction" as well as function. Jencks (1977) also contributed to an understanding of how semiotic principles work in architecture and showed that postmodern architecture is part of the same linguistic turn that has influenced philosophy and social theory. His linguistic focus was influenced by the evolution in the 1960s and 1970s of structuralism and poststructuralism, each of which is informed by semiotic principles. But while academics and theoretical architects were debating problems with the International Style and searching for new vocabularies, many architects by the 1950s had already begun creating new styles that marked a clear departure from Bauhaus principles. As the International Style was enjoying a resurgence (under the command of large architectural firms such as Skidmore, Owings, and Merrill), dissident architects such as Eero Saarinen, Philip Johnson, Marcel Breuer, Louis Kahn, Venturi, and numerous others were returning to past styles for influence, recapturing prior emphases on ornament, and creating fascinating new shapes that exploded the limitations of the glass box.

That something new was afoot was clearly evident, for example, in Saarinen's airport terminals, beginning with the TWA Terminal at John F. Kennedy Airport (New York, 1956–1962). Saarinen created a curved structure that expressed the dynamism of air travel, with reinforced concrete that resembles a bird in flight. His Dulles International Airport Terminal (Chantilly, Virginia, 1962–1964) is a monumental curved roof, accented by a futuristic air tower composed of a series of rings anchored on a concrete base and topped by a round ball. Breuer's St. John's Abbey and University (Collegeville, Minnesota, 1961) features a huge, oblong auditorium with a honeycomb façade that is entered through diagonally set steps above which towers a concrete slab set into four arched supports that holds a belfry and a cross in a cubist-like depiction of an animal. Jorn Utzon's famous Opera House in Sydney (1956–1973) is an expressionistic assemblage of shell vaults set at odd angles.

By the time that John Portman's hotels had spread from Chicago to Los Angeles to Atlanta, beginning in the early 1970s, the postmodern style was well underway. Even when postmodern architects did build a modernist skyscraper, such as Kevin Roche's and John Dinkeloo's U.N. Plaza Hotel (New York, 1979), the phallic monotony was relieved with irregular shapes. That the new postmodern aesthetic was not welcomed by modernists who cut their teeth on the glass box, just as the modernism of Le Corbusier and others was often met with violent opposition and criticism, is evident by the reception of Michael Graves's Portland building (Portland, Oregon, 1982), which in the design competition was rejected as "a dog of a building," a "turkey," a "juke-box," and an "oversized, beribboned Christmas package" that should be set up in Las Vegas rather than Portland (quoted in Jencks, 1977: 7). Yet a look at any city's skyline today, such as Austin, Chicago, or Tokyo, shows an odd mixture of the new postmodern, neon-lit irregular shapes and the glass boxes of the International Style (both frequently surrounded by the ancient railroad tracks and run-down warehouse districts of 19th-century industrialism). Our contemporary cities, indeed, are aleatory, postmodern pastiches of different historical styles.

Postmodern Historicism, Eclecticism, and Pluralism

> I try to pick up what I like throughout history.
>
> —Philip Johnson

Postmodern architecture was a specific reaction to the aesthetic and philosophical assumptions of the International Style rather than to the wealth of architectural forms from the 15th century on or even to the various styles present in the 20th century—ranging from expressionism to the work of Frank Lloyd Wright—that were not associated with the Bauhaus School. In effect, postmodernism seeks a return to the historicism and eclecticism that prevailed in modern architecture before the stranglehold of the International Style. In direct opposition to the Bauhaus break with history to construct a fundamentally new aesthetic, postmodern architecture returns to the past to combine and play with different styles. Whereas modern architecture rejects history, postmodern architecture mines it for its rich symbolic, allegorical, and stylistic sources. The modern itself had become but a piece in the postmodern pastiche, as is evident in Philip Johnson's AT&T building (New York, 1980–1983), which rounds off a rectangular skyscraper with a Chippendale design.

The eclecticism of postmodern architecture was greatly enabled by the development of communications technologies that facilitated the rapid dissemination of ideas across the globe, allowing postmodern concepts and designs to quickly spread to places like India and Japan. Moreover, computers made possible mass production of "a variety of styles and almost personalized products" (Jencks, 1977: 5). Indeed, one could argue that modern and postmodern styles in architecture are each linked to specific technologies, one pertaining to industrial culture governed by the machine and the other to an emergent postindustrial culture governed by electronic mass media and the computer. They also correspond to different stages of urban culture, with high modernist architecture being appropriate to the well-planned, functional, homogenized world of state and monopoly capitalism while the more diverse, aestheticized, and multicultural "soft cities" of the contemporary era are appropriate to the universe of a global technocapitalism.[13]

The postmodern architectural spectacle is thus that of a global, postindustrial culture of consumerism, media culture, and the aestheticization of everyday life. The implosion of art into advertising, packaging, and design is preeminently visible in the highly aestheticized postmodern buildings that incorporate often gaudy and excessive historical allusions and neon ornamentation and coloring. Postmodern architecture thus breaks with the quest for purity and reintroduces symbolism, metaphor, color, and past historical styles. A new emphasis arises on decoration and the scenographic dimensions of architecture as opposed to abstract, compositional concerns, as tactile and surface properties are highlighted over abstract structural relations. Mies's claim that "less is more" is met with Venturi's rebuttal that "less is a bore."

The eclecticism of postmodern architecture entails a break with modern emphases on continuity of style in favor of willful discontinuity. Postmodern architecture juxtaposes different styles in discontinuous, often jarring ways. The clash between architectural styles in many postmodern buildings helps to foreground another key concern of postmodern style, the ambiguity of space. Postmodern buildings typically have unclear spatial boundaries, avoiding clear endings, beginnings, and climaxes, thus embracing principles of complexity and contradiction rather than of unity, harmony, and purity of form. This can produce the effect of a postmodern hyperspace that Jameson (1991: 44ff.) finds vertiginous, as well as the sense of clutter and disorder that some critics find characteristic of postmodern architecture.

Where modern architecture typically signifies little beyond its own abstraction and functionality, postmodern architecture seeks rich meanings and allusions; it is a consciously signifying form. As Jencks (1977) notes, modern architecture tried to suppress meaning in favor of an asignifying, formal purism, only to produce inadvertent, often embarrassing semiotics that reek of technocratic values. Jencks and other postmodernists condemn the bad communication of modern architecture. He sees the Chicago Civic Center, for example, a bland skyscraper distinguished only by a rusting Picasso statue in the outer courtyard, as an "inarticulate building" whose look fails to convey the important political functions that take place inside. Jencks deplores the fact that the buildings on Mies's Illinois Institute of Technology campus suffer from the same uniformity of style that makes the chapel and library indistinguishable from the boiler house. Similarly, Jencks condemns Gordon Bunshaft's Hirshhorn Museum (Washington, D.C., 1974), a pillbox-like structure that connotes "Public stay out!" and his Old Age Home (Amsterdam, 1975), which symbolizes white crosses on top of black coffins.

Such clumsy semiotics are sign crimes for the postmodernists. They much prefer buildings like Le Corbusier's Ronchamp Chapel, which is rich in metaphor and illusion, suggesting at once a floating duck, hands in prayer, a steamship, or a hat. For Jencks, the postmodern architect employs a double coding that seeks to overcome the opposition between elitism and populism. Double coding speaks both to the public and to "a concerned minority, usually architects" (Jencks, 1977: 6). Double-coded semiotics carries a different meaning for the architect and for the public, each signifier intended to be rich in allusions. For Jencks and others, postmodern architecture can appeal to the specialized interests of connoisseurs while breaking with elitism and appealing to a public who wants

> beauty, a traditional ambience, and a particular way of life. . . . Modern architecture suffered from elitism. Postmodernism is trying to get over the elitism not by dropping it, but rather by extending [through double coding] the language of architecture in many different ways—into the vernacular, towards tradition and the commercial slang of the street. (8)

Hence, the modern primacy of function gives way to the postmodern focus on semiotics, style, and communication. The emphasis on semiotics is so important for Jencks that the term "postmodern architecture" applies only to those buildings whose designers "are aware of architecture as a language" (1977: 6). For Klotz, "the very fact that we speak again of the meanings of architecture is the most decisive change in the architectural debate since 1945 . . . the new trends in architecture are predominantly marked by attempts to draw attention to other contents besides the functional qualities of a building" (1988: 3). Klotz refers to this postmodern liberation from the pure forms of modernism as the "fictionalization of architecture," and this endeavor to generate multivalent meanings has direct parallels with poststructuralist emphases on signification, ambiguity, open-ended meaning, dissemination, and so on.[14]

The concern for plural meanings also illustrates the postmodern attack on universals. Postmodern architecture repudiates the homogenization of the International Style and its contempt for place and favors the differences of regional styles and local cultures and traditions.

> Modern architecture has thus been judged by its natural product: the modern city, the suburbs without quality, the urban landscape devoid of collective values that has become an asphalt jungle and a dormitory; the loss of local character, of the connection with place; the terrible homologation that has made the outskirts of the cities of the whole world similar to one another, and whose inhabitants have a hard time recognizing an identity of their own. (Portoghesi, 1982: 7)

Such a concern for the local is evident in Frampton's manifesto (1983), "Toward a Critical Regionalism" (1983), as well as in Venturi et al.'s *Learning from Las Vegas* (1972). As in other postmodern fields, pluralism in architecture entails a relativism of values such that "no code is inherently better than any other" (Jencks, 1977: 87–88). Unlike modern architecture, which sought to embody fixed principles and a universal style, but similar to poststructuralism, postmodern architecture "acknowledges the all-important contingent nature of meanings (92), rooted in particular tastes and conventions. Often, postmodern architecture celebrates schlock, simulations, and "ersatz towns" (94). This is evident in Charles Moore's Piazza d'Italia, a public square in New Orleans (1976–1979), now decayed, that simulates an Italian piazza.[15] Postmodern architecture also implodes differences between public and private, or inner and outer, as in the Austin, Texas, Omni Hotel, which features rows of rooms that open upon the lobby and combines hotel rooms with offices, and nature with culture, with plants and trees decorated with lights. A similar implosion is found on the campus of the University of California at Santa Cruz, where dorm windows open to public walkways, exemplifying Baudrillard's (1983c) "ecstasy of communication" and "obscenity," where everything is transparent to those inside and outside. Alberta University, in Edmonton, Canada, the city with the world's largest mall, features a shopping mall in the center of the university campus, which combines restaurants and commercial venues with student housing, public spaces for study and conversations, and classrooms, erasing distinctions between learning, living, eating, and shopping.

The implosive function of postmodern architecture stands in bold contrast to the differentiating emphases of modern architects. Le Corbusier's monastery of La Tourette, for example, was divided into separate areas for study and prayer, spaces of solitude and collective life, and his urban designs divided metropolitan areas into different quarters for work, leisure, and government, as well as distinct zones for automobile traffic and pedestrian circulation. On Le Corbusier's vision, the city should be divided according to the "seven V's" (*voies*, "ways") composed of a regional road, a major urban highway, traffic circulating around residential apartment complexes, a shopping street, two streets providing access to individual dwellings, and one for pedestrian travel to schools, clubs, sport stadiums, and other places.

Postmodern architecture not only displays an awareness of the principles of implosion, simulation, and hyperreality as discussed by Baudrillard, it enacts them through ironic historical referencing to past styles. In its use of quotation, decoration, irony, and humor, postmodern architecture attacks the serious, rationalist, and utopian character of modern architecture. Postmodernists espouse a more "modest" approach to architecture and see modern ambitions as pompous and dangerous. In place of modern political and philosophical concerns, most revel in the innocent play of form. In place of modernist beliefs in "progressivist technology" and social regeneration through architecture, postmodernists accommodate themselves to popular culture and tastes. The "end of grand narratives" that Lyotard defines as central to the postmodern condition resonates in architecture as well as theory. In reaction to this loss of confidence in rationalism, innovation, progress, and monumentalism, postmodern architects abandon grand themes and heroic monuments and instead draw their inspiration from popular culture and common tastes, an attitude most blatantly on display in Venturi et al.'s *Learning from Las Vegas*.

The postmodern turn also opens up perspectives on the proliferating spaces of simulation, cyberspace, and other new terrains for architecture. Indeed, architecture is now rapidly imploding into simulation, deploying computer models of houses so that customers can enter virtual domains, change rooms and decor from simulation models, and envisage their homes in virtual space. Moreover, a new architecture of cyberspace is emerging that is producing an innovative spatial and cultural domain, an original configuration of space that is also a new mode of living, as more and more people spend increasing amounts of time in virtual communities and cyberspace. New virtual reality devices are on the horizon which will present yet more novel spaces to structure, organize, and live in, so that the construction of space will take on more and more postmodern, dematerialized forms as we head toward the new millenium.

As in the other aesthetic fields, we find that in architecture one can draw a distinction between ludic and oppositional postmodernism. Both groups celebrate difference and plurality, but to different ends: one to change art, the other to change life. We see Venturi as an example of a ludic postmodernist who indulges in stylistic play and irony for its own sake while holding uncritical attitudes toward consumer and media culture. On the other end, we see Kenneth Frampton (1983) as an example of a critical or engaged postmodernist in his advocacy of a regional architecture that provides a sense of place, interacts dialectically with nature, facilitates public interaction, resists the homogenizing forces of the megalopolis, engages more senses than just the visual, and ultimately seeks to advance a critical perception of reality. Building on these positions, in the next section we suggest ways that space, urban design, and architecture can be reconceived and reconstructed to create more livable and democratic public sites where individuals can create participatory communities and transform their societies.

A Requiem for Postmodern Architecture

> Democracy needs something basically better than a box.
> —Frank Lloyd Wright

During the 1970s and into the 1980s, postmodernism was the dominant form of architecture. But by the 1990s, for many critics, it had become as stale, boring, and pretentious as modern architecture, and just as irrelevant to human needs (see Blake, 1993). Just as the modernist steel and glass buildings became tedious clichés, so too did various forms of postmodern architecture become repetitive and trite. Hence, while the postmodern turn continues to be a strong force in theory and society, it has become somewhat passé in architecture, having been eclipsed in some quarters by "deconstructionism."

"Deconstructivist architecture" was formally defined and discussed in 1988, with the publication of Philip Johnson's and Mark Wigley's Museum of Modern Art catalogue *Deconstructivist Architecture* and a special issue of *Architecture Design Profile*. It is an application to architecture of poststructuralist theoretical principles, specifically as developed by Jacques Derrida, both to criticize architectural design and to concoct literal embodiments of Derrida's concepts. The result is asymmetrical, paradoxical forms of space that play binary oppositions off one another and construct and subvert spatial relations of hierarchy. While leading practitioners like Peter Eisenman uphold deconstructivist architecture as something entirely new, Michael Benedikt (1991a) claims that deconstructive attitudes and practices were already used in modern architecture and design pedagogy. What is new is the importation of a theoretical framework to discuss these principles in a sophisticated way and to exaggerate them in asymmetrical designs.

Benedikt argues that there is an intimate relation between Derrida's thought and architecture, since Derrida's work turns on the use of numerous architectural metaphors such as grooves, inside and outside, margin, center, and boundary. But Benedikt also displays an impatience with the highly theoretical, abstract, and semiotic nature of both postmodern and deconstructivist architectures, which abound in "the amoral delight of irony, pseudohistory, allusion, pyrotechnic self-reference, and fabulism" (1987: ix). Seeking a return to an "architecture of reality," Benedikt espouses a "High Realism" that conveys a strong sense of reality and allows one to have, to the extent possible, a "direct aesthetic experience" of an objective world that exists before linguistic framing and the contrived meanings of postmodern architecture and the media world (66ff.).[16]

Frank Gehry, conversely, combines dadaist techniques of using found materials with a form of deconstructionist architecture. His own house (1978) tends to look, in Ross King's description, "as if bombs had hit them" with "broken walls, bits pulled off and strewn around, crumpled bits often in cheap industrial materials, beams and columns seemingly in the wrong places" (1996: 165). Other examples of deconstructive architecture play with the pieces of traditional architecture but arrange them in eccentric ways, as in Eisenman's Wexner Center for the Visual Arts at Ohio State University, which links existing buildings to ones yet to be created, excavating the foundations of an armory once upon the site and with Mercator grids providing the scaffolding to a galleried street. Eisenman describes the structure as a palimpsest, "a place to write, erase, and rewrite," in other words, a deconstructive text à la Derrida (quoted in King, 1996: 164–165).

Although often semiotically sophisticated, much postmodern and deconstructive architecture is ludic in fatuous and offensive ways, replacing the iron cage with a neon cage, just as oppressive and indifferent to human needs as the worst modern architecture. In response to the asignifying, formalist nature of the International Style, postmodern architecture strives for "complexity," but often this is pursued for its own sake, yielding a meaningless chaos of forms, as most evident in Jean Renaudie's Jeanne Hatchette Center (Ivry, France, 1969–1972), where, in Klotz's vivid description, "a riot of prickly triangles and spiky trapezoids is unleashed for the mere purpose of dramatizing the sensationalist claims of originality" (1988: 34). Similarly, in reaction to the overly zealous attempts of certain modern architects to embody rational transformation through universal geometric forms, postmodernists spurned the idea that architecture should have social relevance. Against efforts to advance substantive social and political ideologies, many postmodern architects limit themselves to strictly stylistic and aesthetic concerns.

And yet architecture can only indulge in whimsy or pastiche in a decorative sense, for it must conform to the basic rules of design, function, and form lest it collapse. That is to say, it has a utilitarian dimension that the other arts lack. Moreover, it is a uniquely public art, and if the strict concern for style for its own sake is divorced from any substantive concerns with urban design and politics, the postmodern architect marginalizes his or her role and capitulates to larger political and economic interests. As Diane Ghirardo puts it:

> For nearly every project, the architect arrives last on the scene. Contemporary practice shrinks the role of the architect from that of an active agent in the construction of community and its structures to that of an exterior designer or an interiors specialist. Leasing agents, developers, commercial loan officers, planning and zoning commissions make the important decisions, leaving for the marginated architect the trivial task of selecting finishes and glosses inside and out. The utopian aspirations, social and political commitments, philosophical rigor and lofty self-confidence of the Modern Movement recede even further into the distance. Stylistic Post Modernism's exclusive preoccupation with style can be seen as a pathetic acceptance of the trivialization of the profession. (1984–1985: 190)

Hence, in architecture as in other fields, some forms of postmodernism entail the evacuation of normative, political, and utopian concerns. Under the hegemonic power of capital, one has to question the validity of the alleged "populism" of much postmodern architecture. If the architect is marginalized, how much more so must be the ordinary citizen, who Jencks claims plays a major role in design through competitions but whose contribution really seems limited, at best, to decoding "populist" signifieds that for most are probably unnoticed in the haste of everyday life.[17] Rather than being condemned to marginalization and irrelevance, while uncritically celebrating fashion, design, style, and consumerism, the architect should engage substantive social and political issues and design buildings that meet the needs of people. In a world of homelessness, joblessness, and hunger, the manifestos for today should be those of a pragmatics of human need concerned with producing decent housing and urban environments for all. But though postmodern architecture is clearly an advance over the International Style in its semiotic and stylistic richness, it is often just as indifferent to real human needs, providing more glitter and glitz than genuine enhancement of human life.

We should not move, therefore, from failed modern visions of new cities and housing complexes to a postmodern cynicism that dismisses utopian values as it fetishizes signification. Beginning in the mid-19th century, architects felt compelled to respond to the complex problems of urban housing, seeking to create affordable living units where human beings lived in harmony. Le Corbusier, for example, in 1922 developed plans for a *ville contemporaine* that would house around three million inhabitants, and throughout his life he proposed plans to reconstruct cities so as to improve the lives of the ordinary citizen. Acutely aware of the realities of the modern age, he rejected earlier visions of small, decentralized colonies and town villages. If "garden cities" are what the architect should build, Le Corbusier argued, these must be constructed vertically to maximize an efficient use of space in overcrowded cities. Hence, he designed large apartment blocks on the outskirts of Marseilles, Nantes–Rezé, Berlin, and elsewhere as models of modern urban life. His *Unité d'habitation* in Marseilles featured the innovative form of stacked duplex apartments with open terraces. In these complexes, Le Corbusier struggled to balance the need for individual privacy with the need for human community by integrating individual living units into a larger complex that provided services and places to shop.

Although his apartment complexes are far more subtle and sophisticated than the oppressive compounds built in the United States during the 1960s, Le Corbusier is often condemned as the inspiration for later architectural debacles, and his buildings and city designs are frequently criticized as too monotonous, sterile, linear, rigid, and overscaled. Lewis Mumford, for example, charged Le Corbusier with "Baroque insensitiveness to time, change, organic adaptation, functional fitness, ecological complexity" and cited his "sociological naiveté, his economic ignorance, and his political indifference" as major "deficiencies" that "reflected perfectly the financial, bureaucratic, and technological limitations of the present age" (1968: 114). With Mumford, Jane Jacobs argued that Le Corbusier's vision of urban life and architectural designs decisively influenced later models, proving to be "all but irresistible to planners, housers, designers, and to developers, lenders and mayors too" (1961: 23).

In many ways, Le Corbusier has been a scapegoat for the worst aspects of modern architecture, and critics often miss the nuances of his work, such as his attempts to account for regional conditions even when advancing a universal vision of design. Yet Le Corbusier clearly was too uncritical of modern values and was too optimistic about the role architecture could play in changing social life.[18] His visions anticipated what was to come, and the seeds sown in his designs and plans for the "radiant city" unfolded into massive empires of finance towering over congested, polluted, and dangerous streets, monumental realizations of his description of a city as "the grip of man on nature." His binary scheme of "architecture or revolution" evinces a naïveté regarding the colonizing power of capitalism, a failure to grasp how capital forms and deforms architecture, thwarts all rational visions, and produces environments suitable only for the maximization of profit while choking efforts to realize human needs.[19] As one critic puts it:

> Viewed in retrospect, Le Corbusier's visionary schemes seem to epitomize that optimistic peak of the modern movement marked by confidence in the splendors of the dawning machine age. As the honeymoon of man and machine persisted, it remained possible to derive a romantic excitement from automobiles, airplanes, and tall buildings. It was possible for architects to imagine that redesigning the city would accord with progress, and

> the new architectural imagery was accepted by many as a symbol of both technical and social advancement. (Evenson, 1987: 245–246)

Consequently, we are calling for an architecture between the modern and the postmodern, rejecting the excesses of both, building on strengths and overcoming weaknesses. Modern architecture is admirable in its utopian impulses, its emphasis on social relevance, and its drive to build a better world. Some of its best work by Wright, Le Corbusier, Gropius, and Mies is still extremely impressive, but it congealed into a deadening orthodoxy, in the service of corporate capital, that produced boring buildings and unlivable cities. This orthodoxy violated the very spirit of modernism, which extolled creativity, innovation, and a constant drive to "make it new." The emphasis on innovation in postmodern architecture, its historicism, eclecticism, and return to ornamentation, surely represents a positive step forward in comparison to the orthodoxy of the International Style. But postmodern architecture fell prey to ludic excess, to banal eclecticism and bad taste, and itself became repetitive, predictable, and boring, leading some to call for a revival of modernism.[20]

In our current postmodern climate, where the metanarratives of modernity are in shreds and the consequences of large-scale, uniform, and centralized planning are all too evident, the optimism of modern architecture is no longer credible. Postmodernists are rightly skeptical of facile visions of freedom, abstract universals, totalizing schemes of industrial rationalization, authoritarian modes of planning, and the notion of an artistic elite as the guardians of society, but their skepticism often unfolds as a cynicism about human amelioration in any form, retreating into a staid aestheticism that is no less elitist than modernism. Against the trivial pursuits of postmodern architecture, we claim that modern architecture is *an unfinished project*: We still need adequate public housing; democratic urban planning; innovative models for bringing together large numbers of people (where this is necessary) in homes, apartments, and neighborhoods that are clean and safe, as well as for decentralizing cities into a more human scale; and a harmonious integration of urban and natural environments. It was the historical task of modern architecture to make the first attempts at satisfying these needs, however problematically, but postmodern architecture is on the whole a sterile detour from the urgent problems of housing and urban planning. Of course, the real project is not architectural design but *social design*, the reconstruction of society from the standpoint of ecology and participatory democracy; the task is not "architecture or revolution" but architecture *and* revolution.

In particular, a new architecture could combine modern emphases on function and social relevance with the postmodern focus on style and meaning. At its best, not only is architecture functional and useful for living, working, playing, exercising, and studying—operating well on a technical and design level—but it can also inspire us, comfort us, and make us feel relaxed and at home. Buildings are definitely works of art in their own right, as well as environments for living. These principles did not go unnoticed by the Situationists, who advanced an important critique of the alienating architecture in modern homes and the urban environment. No environment, they argued, was phenomenologically neutral; rather, it affected the individual in distinct ways. The study of the lived impact of space and architecture on the emotions and behavior of individuals Situationists termed "psychogeography." "We are bored in the city . . . we really have to strive to discover mysteries" (Ivan Chtcheglov, quoted in Knabb, 1981: 2). The alienation of the modern condition, they argued, was as evident on our faces as in our buildings. The Bauhaus movement sought to provide the "education" artists needed to assume their place within the machine age; the Situationists sought to destroy the mechanization of life to liberate creativity.

In contrast both to modern functionalism and the towers of postmodern babble being erected, Situationists called for a poetic architecture that helped awaken imagination and desire. Through architecture, they argued, liberatory experiments with human nature and possibilities begin: "Architecture is the simplest means of *articulating* time and space, of modulating reality, of *engendering* dreams. It is a matter not only of a plastic modulation and articulation expressing an ephemeral beauty, but of a modulation producing influences in accordance with the eternal spectrum of human desires and the progress in realizing them" (Chtcheglov, quoted in Knabb, 1981: 2). The new architecture and space they envisage is to have an evocative power that, like a surrealist painting, can recreate and reawaken the fantastic, mysterious, and ambiguous aspects of existence. When the paintings of Giorgio de Chirico become architecture rather than just art, "there will be rooms more conducive to dreams than any drug" (3). To effect such changes, architecture will become modifiable, easy to change and manipulate; designs that incorporate rotating walls, for example, or that allow one to open the ceiling to gaze at the stars, will facilitate an active interaction with one's environment. The domestic space will stimulate human interaction, producing "houses where one cannot but love" (2).

The streets of the city, redesigned to facilitate the *dérive*, will be divided into "distinct psychic atmospheres" that stimulate various desires and moods. In general, the Situationists see architecture as the most important art form since it takes us far beyond the limitations of plastic art as a visual spectacle to the constructed situation where art is lived as a new form of space. "That which changes our way of seeing the streets is more important than what changes our way of seeing painting" (Chtcheglov, quoted in Knabb, 1981: 25). Their hope to revolutionize life through art makes them one of the last significant forms of an avant-garde movement. Thus, Chtcheglov suggested that "*the hacienda must be built*" (quoted in Knabb, 1981: 1), that new houses, cities, and sites for living must be constructed that will produce a new mode of everyday life.

At its best, modern architecture often achieved the poetic environments celebrated by the Situationists. The emotional effects of Le Corbusier's architecture, for example, are dramatically evident at La Tourette, where sunlight pours through large round circles onto a raised altar, at the church of Ronchamp, whose sloping walls are punctuated by irregularly shaped windows cut in stone, or in the magnificent lobby of the Assembly Hall at Chandigarh, where huge pillars capped with cone-shaped tops support the tall ceiling like mushrooms sprouting from the marbled floor. Le Corbusier was very concerned with both the texture of his materials and, in the case of Ronchamp and the Phillips Pavilion, with the acoustics of his rooms, whereby he attempted to devise a total audiovisual experience, or a "landscape acoustics." Through such considerations, Le Corbusier anticipated the multisensory architecture important to the Situationists and Kenneth Frampton.

Reconstructing social space requires insight into the social interests that control architecture, design, and urban planning. A more democratic politics of space would allow citizen input into the design of their communities and an appreciation for the specific contributions to the construction of space by women, youth, and people of different ethnic and subcultural groups. As Jane Jacobs pointed out, women have traditionally made important contributions to the construction of domestic space, community, and neighborliness, but their creativity has not been fully esteemed. The same could be said of youth subcultural groups, ethnic subcultures, and gay and lesbian cultures, which have often constructed imaginative and aesthetically pleasing habitats for members of their group, which provide a sense of comfort and belonging not found in the impersonal or oppressive public spaces of the contemporary city.

Dolores Hayden (1984), for instance, criticizes the ways that male and commercial culture deface the city (quasi-pornographic billboards, advertising, displays of macho violence, etc.) and discusses how the Los Angeles Woman's Building, an arts workshop, gets women involved in the construction of public spaces and public art. Jane Jacobs, Hayden, and others also endeavor to promote the appreciation of women's contributions to constructing community and to produce spaces and institutions, like daycare centers, rape crisis shelters, and community health centers, that meet women's needs. In King's view, a feminine spatial design will be very different from the male: "It will break things down, make them accessible; remove the steps and the podia; break the facades with flowers and scented fruits; reduce the scale; reinstate the tactile, the sounds of water and birds, the places of children's play (Nietzsche's 'play that calls new worlds to life'), the impermanent and the appropriable" (1996: 236–237). Mobilizing women to create community art and performance, to participate in the construction of alternative cultures, draws on women's abilities to achieve habitability and the even more extraordinary quality of the variety and freshness that flows from subtle change, adaptability, textures, and new sites and surfaces.

A postmodern philosophy of space would accordingly valorize the construction of domestic space and public space that would include not only new buildings and structures but new textures, sights, sounds, smells, and aesthetics, reinstating the tactile, the aural, the olfactory, and the auditory, thus affirming all of the senses as key constituents of the environment. This would involve an aestheticizing of everyday life and a reconstruction of the look, feel, and experience of social space with new buildings, public spaces, nature, and forms of art appropriate to specific local regions and sites. The reconstruction of social space would involve the reintegration of nature and the social, the resurrection of the senses, and new spaces to fulfill and cultivate the many-sided needs and potentials of the human being. The postmodern aestheticizing of the environment would thus realize the earlier avant-garde aspirations for the merger of art and life and would bring architecture, sculpture, paintings, and the other forms of visual culture into a closer relationship in a reconstruction of culture and society.

THE POSTMODERN TURN IN PAINTING

> The question of imitation, the gestural look of Abstract Expressionism, and of the words that had been hurled as insults for as long as we could remember—illusionistic, theatrical, decorative, literary—were resurrected, as art became once again ornamental or moral, grandiose or miniaturized. . . . It was defying all the proscriptions of modernist purity.
>
> —KIM LEVIN

As with architecture, the postmodern turn in painting similarly endorsed connecting visual culture with everyday life and renounced the trends toward abstraction in high modernism, returning to representation and to popular and historical images and references. Although the postmodern turn in culture was first advanced in discourse and practice in architecture and literature, a distinctive anticipation of postmodernism in the arts is found in the works of Robert Rauschenberg, Jasper Johns, and the New York avant-garde of the 1950s, and in our narrative, postmodernism in the arts appears vividly in the works of Andy Warhol and in pop art. By the 1980s, the postmodern turn toward historicism, eclecticism, populism, and ludic play with tradition—evident in architecture—constitutes the dominant trend of the visual arts, reversing the modernist emphasis on formalism, aesthetic purity, innovation, and negation of tradition.

The Modernist Moment in Painting

> A picture—before being a warhorse, a nude woman, or some sort of anecdote—is essentially a surface covered with colours arranged in a certain order.
>
> —MAURICE DENIS

> Picasso, Braque, Mondrian, Miro, Kandinksy, Brancusi, even Klee, Matisse and Cezanne derive their chief inspiration from the medium they work in. The excitement of their art seems to lie most of all in its pure preoccupation with the invention and arrangement of spaces, surfaces, shapes, colors, etc., to the exclusion of whatever is not necessarily implicated in these factors.
>
> —CLEMENT GREENBERG

The modernist insurrection in the visual arts erupted in 19th-century Europe as a rebellion against traditional academic and classical forms. Painters like Monet and Manet in France broke with the conventions of French academic painting, interjecting new subject matter and forms into their work. The modernist artist, like Gauguin and van Gogh, was often an outsider, seeking novel experiences, visions, and modes of painting in isolation from cliques and official art institutions. The result was a series of modernist aesthetic revolts in which painters created their own unique styles and forms and innovative works and artistic breakthroughs.

European modernists in the 19th and 20th centuries also formed avant-garde movements, with each new group claiming to be the vanguard of art, the most advanced art of its day (hence the military metaphor of the "avant-garde"). In the field of painting, modernist groups promoting French impressionism, German expressionism, Italian futurism, Russian constructivism, and cubism crossed national borders and spawned new movements, like dadaism, surrealism, and vorticism, that called for revolutions in art and life (Burger, 1984). "Modern art" therefore became both a movement and a slogan to attack traditionalism and dominant academic forms while advocating the creation of new ones.

The histories of painting and visual culture most in favor during the modernist era tended to celebrate modernism as a successful revolution against traditional art and to view the history of art from the perspectives of high modernism, especially abstract expressionism, which was seen as the contemporary realization of the modernist program.[21] One of the insights of postmodern theory is that histories are constructs written from the perspectives of specific standpoints, with all their defining biases, assumptions, and prejudices. In the following sections, we will provide a rereading of the history of modern visual art from the perspectives of the postmodern turn, which produces a different artistic pantheon and appraises the major modernist movements and painters quite differently from the way modernist perspectives do.

Histories of modern painting focus on France, seeing Manet, the impressionists, Cézanne, cubism, and others as the heroes of modernism. In these histories, Paris is considered to have been the world capital of painting during the first decades of the 20th century, and indeed most of the great modernist painters moved and worked there, including Picasso, Braque, Matisse, Gris, Leger, and countless others. French modernism, especially cubism, sought a world of painting autonomous from everyday reality and moved relentlessly toward abstraction, striving to establish a new world of art as a realm unto itself. Breaking with the illusionism of Renaissance perspective and classical efforts to create three-dimensional surfaces, cubism reduced paintings to abstract lines and planes, rejecting the traditional concept of depth and seeking to flatten the surface of art to a depthless physical field. The cubists, following the pathway of Cézanne, also abstracted objects into "cubes" or geometric shapes, while painters like Picasso or Braque emphasized relationships among these forms, producing new surfaces that accented the interrelatedness of objects and provided a multiperspectival view. Such paintings existed in their own space, creating their own world, a field that abstracted from the denseness and heaviness of the social world in order to construct a purified realm of forms, images, and abstract representations of the real. Of course, cubist paintings also tended to be iconic signs of the everyday world; thus, their representations of guitars or apples, for example, alluded to real ones, standing as signs of everyday objects.

Cubist paintings also included actual elements of everyday life such as newspapers or labels of wine bottles in their paintings, although the formalist readings, which dominated during the succeeding decades of high modernism, stressed that the cubist formal experiments and collages were carried out in order to provide purer, more geometrical forms, shapes, and lines and thus to advance the modernist agenda of purifying art into its essential constituents. For the histories of art that dominated during the following decades of high modernism, formalist painters like Cézanne, those in the cubist and constructivist movements, and other stylistic innovators were seen as key figures in 20th-century art history. Yet every era must reconstruct its own history, and from the perspective of the postmodern turn in art, the collage elements in cubism, the bringing of everyday objects into painting, was an important anticipation of the postmodern efforts to collapse art into life, to subvert its autonomy, to create new aesthetic forms, strategies, and effects.

In fact, the cubist works of Juan Gris anticipate the postmodern turn more than do the abstract earlier "analytical cubist" works of Braque and Picasso. Gris's cubist paintings of objects were more decorative, colorful, and stylized than his predecessors as were the works of German expressionists and other painters influenced by cubism.[22] From the perspective of the postmodern turn, it was thus the innovations in the decorative, the development of new collage forms, the aestheticizing of objects of everyday life, and the ludic play with appearances of Gris, the fauvists, and Matisse that anticipated the postmodern turn that began in the 1950s.

Moreover, the dada movement and iconoclasts like Marcel Duchamp take on new importance as anticipations of the postmodern turn. The dadaists' mocking assaults on the institution of art, on the religion of art and the artist, and on belief in the autonomy of art, combined with their attempts to redefine art and to greatly expand its field, prefigured later postmodern projects. These motifs were taken up by Duchamp, whose appropriations of classical images and objects of everyday life would impress later artists involved in the postmodern turn. Yet cubism plays a very important role as well in our story of 20th-century art. Not only are its stylistic innovations important for later formalist movements like abstract expressionism, but its use of collage, its spurning of signification, and its blending of art and the objects of everyday life anticipate the postmodern turn in art, which we will map in this chapter.

Abstract Expressionism: The Final Ecstasy of High Modernism

> We had a lot to say and we did pick up the cudgel of abstract art, which was left behind by the Europeans. They dropped it, and we picked it up and carried it on.
>
> —PHILIP PAVIA

> Hell, it's not just about painting!
> —ABSTRACT EXPRESSIONIST CREDO

The drive toward abstraction and nonrepresentational art was significantly advanced by U.S. painters of the New York scene of the 1940s and 1950s, who produced what in retrospect can be seen as one of the last flowerings of high modernism in art.[23] Previously, the United States was something of a backwater in international art, with U.S. painters achieving few breakthroughs and little international renown. The abstract expressionists, however, took up the challenge of European modernism, to continually revolutionize art and to produce new aesthetic forms and styles. In the 1940s and 1950s, they created the first internationally significant U.S. art and catapulted New York to the center of the art world—where pop art and postmodernism would follow as internationally important art forms and movements.

New York was the preferred site for European émigrés who were fleeing from German fascism and World War II. Such luminaries as André Breton and Duchamp from France, the German émigrés Hans Hofmann and Josef Albers, Willem de Kooning from Holland, and others brought the impulses of European modernist art to the United States. The synergy of the European-born painters and the U.S.-bred talent created a dynamic cultural matrix that generated a vast wealth of world-class art. The cosmopolitanism of the New York scene anticipated the coming global culture, though the commitment to high cultural purity and the aversion to mass culture sharply distinguished the ethos and work of high modernism from the coming postmodernism.

The New York-based abstract expressionists were determined to create a new modern art that embodied their own experience, vision, and aesthetic form. In Emile de Antonio's documentary film *Painters Painting* (1973),[24] Barnett Newman tells how painting was dead for him and his colleagues in the 1940s and how they sought to produce a new modernist art. In the somber experience of the Great Depression of the 1930s and the carnage of World War II, Newman and his colleagues could not think of producing a beautiful art. Newman himself did not want to create a world of pure forms either, in the idealized style of cubism or the perfect geometric forms of Mondrian. Thus, the painter had to start from scratch, Newman insisted, which for him meant abstracting from the world completely, totally negating both the everyday world of objects and an ideal world of purified natural forms, in order to explore a new subject matter of color, shapes, lines, and forms.

The move within abstract expressionism toward a more abstract and nonrepresentational art constituted something of a political rebellion and rejection of the existing society and world. The "ugliness" (as it initially appeared) of a work by Pollock, Kline, de Kooning, or Newman is registered from this perspective as a renunciation of the existing world, as a refusal to beautify or idealize it, while attempting to produce an alternative vision and new forms of art. The abstract expressionist painters thus opposed the official world of art and the society it beautified, even if they did not make explicitly political statements themselves in their work.

Jackson Pollock (1912–1956) is frequently presented as the initiator of the dramatic breakthrough toward abstraction in post-World War II American art.[25] Pollock emphasized the expressive aspects of his painting, that he was rendering into concrete form the turmoil of his inner life, expressing his feelings and subjective vision. Pollock's "action painting" helped launch a movement that came to be called "abstract expressionism." It was more abstract and nonrepresentational than any previous form of American art, but it was also expressive and highly subjective in the tradition of German expressionism.[26] As with the German expressionists, the abstract expressionists were angry artists who refused the existing bourgeois world and who created alternative worlds and visions with their distorted images, violent colors, and sometimes bizarre shapes and forms. But Pollock and the abstract expressionists went much further toward a nonrepresentational art than did their German predecessors.

The group of painters who produced new aesthetic breakthroughs by creating a variety of forms of nonrepresentational art came to be known as the "New York School," or the "American School." These artists formed various groups and coteries, deeply influenced each other, and undertook a combined research program into the possibilities of abstract painting and the project of creating art free from traditional forms, conventions, and style. In a sense, the abstract expressionists embodied perfectly the modernist ethos and ideology. They rebelled against aesthetic and societal conventions, created highly innovative works, and took art deadly serious, dedicating their lives to the creation of monumental artifacts. They thus shared the modernist religion of art, the concept of the artist as genius and hero, and undertook the quest to develop lasting and monumental works that had societal and spiritual importance and universal meaning.

A 1939 essay, "Modern Art and Kitsch," by art critic Clement Greenberg enunciated the modernist principles that legitimated the artistic revolution being carried out by the abstract expressionists (collected in Greenberg, 1961). His essay is as important for modernist aesthetics as Jameson's 1984 essay "Postmodernism: The Cultural Logic of Late Capitalism" is for articulating the postmodern turn. Greenberg envisaged art as a purely autonomous activity that had its own norms, techniques, and goals. His aesthetic was strictly formalist, with the goal of art being to delineate a pure world of form, autonomous from extraneous considerations. "Avoid content like the plague," Greenberg prescribed, and in turn he championed the contemporary modes of painting that he saw as embodying his formalist aesthetic, which was congruent with the modernist project in architecture represented by the International Style.

A coterie of abstract expressionist painters carried out formal innovations in the arts, advancing the program of modernism sketched by Greenberg; they included the German émigré Josef Albers, who experimented with form and the perception of color, and the Dutch émigré Willem de Kooning, who combined stunning color, expressive abstraction, and impressive innovation. Franz Kline explored abstract form and empty surface space, Helen Frankenthaler developed color-field painting with her pigment-dyed canvases reworked with abstract designs, and Robert Motherwell experimented with scale, massing, design, and gesture. Barnett Newman created nonrepresentational canvases that were divided by vertical stripes into vibrant fields of color, while Mark Rothko produced abstract and enigmatic paintings that exuded spirituality. Color-field painters like Kenneth Noland, Jules Olitski, and Larry Poons painted abstract fields of color. Frank Stella in turn constructed works exhibiting a flat, nonrepresentational surface without subject matter, content, or theme.

Pollock, Newman, Gottlieb, Motherwell, and other abstract expressionists exhibited a heroic impulse to create paintings larger than those accommodated in the usual home or gallery space. The violence, aggressivity, daring, innovativeness, monumentality, risk, and extreme individuality in abstract expressionism is typically American and created the first American art movement that generated worldwide acclaim. In a sense, abstract expressionism was the art form of the hegemony of U.S. global capitalism, expressing the energies, drive, and mastery of of U.S. global power. U.S. capital generated the wealth to purchase and inflate the prices of the masters of abstract expressionism and to exhibit their monumentally large art in large corporate spaces. Thus, although abstract expressionism in ideology was anti-corporate and anti-mass culture and had its spiritual pretensions, it fit all too well into the triumph of U.S. capitalism in the global market.

Between Modernism and Postmodernism

> As the paintings changed, the printed material became as much of a subject as the paint (I began using newsprint in my work) causing changes of focus. . . . There is no poor subject (Any incentive to paint is as good as any other).
>
> —Robert Rauschenberg

Abstract expressionism manifested the modernist urge for the monumental, grandiose, and highly original work that would project the exalted vision of the modern artist–hero. High modernism in painting had reached its apogee by the mid-1950s. A new generation of abstract expressionists tended merely to imitate the previous masters. Some New York painters—most prominently Robert Rauschenberg and Jasper Johns—began moving beyond abstract expressionism and toward incorporating objects of everyday life into painting. Influenced by composer John Cage and painter Marcel Duchamp, both of whom wished to redefine art to present the material of everyday life as art objects, Rauschenberg combined the formal experiments of abstract expressionism with the development of a new type of collage art.

Making art out of his bed, newspapers, advertisements, found objects, and other debris of the consumer and media society, Rauschenberg overcame the separation between high art and lowly everyday life that was one of the key features of high modernism. He utilized the collage form to expand the field of painting to include both the abstract images and rebellious use of color and form monumentalized by the earlier generation of abstract painters and quotations of classical paintings and ordinary objects and images, all thrown together into his canvases in a seemingly random order and often exhibiting an exuberant dadaist spirit of humor and rebellion. His jokes and dadaist irony, such as erasing a drawing of de Kooning and then exhibiting it, broke with the seriousness of high modernism, and his use of quotation and pastiche rejected the modernist tendency toward innovation, originality, and purity of image, mixing a wide range of materials and styles in his heterogeneous works.

For instance, one of Rauschenberg's collages, *Bed*, was made from his bedspread, sheets, and pillow, and his *Monogram* featured a stuffed goat, which he had bought at a secondhand shop, standing in a tire and serving as the center of a collage of everyday objects. *Factum I* and *Factum II* juxtaposed seemingly identical collages to undermine the notion of authenticity and to call attention to the constructed nature of the artifact. His collages were thus significantly different from the more formalist cubist collages, with his scraps of newspapers, found objects, and images of traditional art (dis)assembled into postmodern work that scattered objects and images in a random and seemingly disorganized and unruly fashion. Whereas modernist collages organized materials into an aesthetic unity and harmony, Rauschenberg's and later postmodern work would present collage material in its very fragmentation without attempting to make a coherent statement or to organize, unify, and produce an aesthetic effect of harmony.

Jasper Johns also collapsed distinctions between high art and mass culture in his paintings of targets and U.S. flags and in his sculptures made of beer cans and coffee cans. In addition, Johns made a central feature of his painting the sort of flat, one-dimensional, nonexpressive image that would be, in Jameson's (1991) view, characteristic of postmodern culture as a whole. Deeply concerned with the formal problems of painting, Johns abstracted from the concreteness of flags to create a depthless, formal image. Representations of the flag or targets were thus chosen for their formal qualities rather than for their symbolic associations, thus flattening the image and striving to drain it of meaning and symbolism. The goal was not to paint his impression of flags or to faithfully mirror and reproduce the flag in ultrarealist fashion. Rather, Johns wished to create the flag as a pure art object, as a self-enclosed form, as a pure iconic image.

Obviously, this painterly and formalist goal puts Johns in tune with the motifs of modernist abstract art, but he also developed postmodern themes. Johns often noted the "detachment" and "distance" from his subject matter in his painting of targets, numbers, and other objects from everyday life. In a 1989 documentary, he describes his efforts to "feel removed" from the objects of his work and to be "detached and not involved."[27] Johns's self-image instantiates the "absence of affect" that Jameson would eventually identify with the postmodern sensibility. Thus, where the modernist Pollock attempted to project his feelings into his work and to express himself with exuberance, Johns coolly chooses to display the objects themselves, abstracted into aesthetic form.

Reacting against purism and formalism in painting, Rauschenberg utilized recognizable figures of consumer and media culture in collage paintings, and Johns made bronze sculptures of everyday objects like beer cans. This move constituted a rebellion against the tendencies of high modernism to abstract itself altogether from mass culture and its commercial flora and fauna. Rauschenberg's work with EAT (Experiments with Art and Technology) also imploded distinctions between art and technology (see Tomkins, 1975). Working frequently with other avant-garde performers in theater, happenings, and concerts in the 1960s, Rauschenberg constantly erased boundaries between the arts. Here he was influenced by his friend John Cage, and, indeed, Rauschenberg and Johns worked with Cage, Merce Cunningham, and others in the New York avant-garde throughout the 1960s in performances that would later be called "postmodern" with their erasure of traditional aesthetic boundaries and subversion of traditional codes, combining dance, music, theater, and spectacle.

Thus, whereas modernists sought to separate the various arts and to pursue each one's own autonomous logic, postmodernists created multimedia works and mixed together artistic genres. The 1960s New York avant-garde wanted to bring the audience into the artworks themselves, stressing how audiences created meanings. Rauschenberg believed that viewer perception of his white paintings, with the shadows and modalities of light projected on them, helped create their meaning and experience, making the audience participants (Tomkins, 1965: 203). In fact, he, Cage, and others anticipated the turn to the audience as the creator of meaning that has been developed by cultural studies (see the discussion in Kellner, 1995a). John Cage has said on the album cover to *Variations IV*:

> Most people mistakenly think that when they hear a piece of music, that they're not doing anything but that something is being done to them. Now this is not true, and we must arrange our music, we must arrange our Art, we must arrange everything, I believe, so that people realize that they themselves are doing it, and not that something is being done to them.

On the same album cover, Joseph Byrd cites aesthetician Morse Peckham's position in *Man's Rage for Chaos* that art constitutes "any perceptual field which an individual uses as an occasion for performing the role of art perceiver." In other words, it is the perceiver, the art audience, who constitutes the object as "art," not any apparatus of critics, any coterie of artists, or any consensus of the art establishment. This position was also taken by Duchamp, who, according to Calvin Tomkins's summary of a 1957 seminar in Houston, Texas,

> proposed a somewhat surprising definition of the spectator's role in that mysterious process known as the creative act. The artist, Duchamp said, is a "mediumistic being" who does not really know what he is doing or why he is doing it. It is the spectator who, through a kind of "inner osmosis," deciphers and interprets the work's inner qualifications, relates them to the external world, and thus completes the creative cycle. The spectator's contribution is consequently equal in importance to the artist's, and per-

haps in the long run even greater, for, as Duchamp remarked in another context, "it is posterity that makes the masterpiece." (Tomkins, 1965: 9)

The "happenings" staged by the New York avant-garde literally involved the participation of audiences in multimedia performances, thus undoing the separation both of the audience from the work and of the various media, like theater or film, from each other. This form of artistic rebellion and redefinition of art appealed to Rauschenberg, who enthusiastically carried out moves toward a postmodern turn in painting and participated in various multimedia creations. Rauschenberg pastiched his own previous paintings, redoing his notorious white paintings of the early 1950s in 1968. Other postmodern painters would also aggressively quote previous art forms; for example, Larry Rivers cited Dutch masters and cigars; Alfred Leslie replicated Italian masters and reproduced Thomas Cole's Hudson River valley paintings; Tano Festa pastiched in a photomat series Michelangelo's painting of Adam in the Sistine Chapel; and there were numerous other appropriations of classical paintings in postmodern art.

These artists who began the postmodern turn were carrying forth some of the program of dada and were in turn labeled "neo-dada" during the 1950s and early 1960s.[28] In retrospect, dada can be seen as a virus that entered the body of modernism and affected many later modern artists, causing it to mutate into postmodernism. With the exception of some abstract expressionists who maintained a religion of art, the spirit of dada entered into contemporary painting, creating a critical distance from tradition and an iconoclastic attitude toward bourgeois culture. Via the mediation of Duchamp and several important dada and surrealism exhibits in the 1950s (see Craft, 1996), the dada virus was especially active in the New York intellectual scene, affecting Rauschenberg, Johns, Cage, Warhol, and others. Thus, the spirit of dada is constitutive for the postmodern appropriation, hyperirony, and play with tradition, although much ludic postmodernism renounces the earlier dada emphasis on art as a vehicle of social change and the avant-gardist pretensions of many dada artists.

It was thus Duchamp's more ironic and apolitical version of dada that most influenced the later artists who would begin the postmodern turn away from high modernism in the arts. Duchamp himself began coming to New York during World War I, frequently visiting the city, including a long interlude during World War II. He had successful museum and gallery shows and influenced many young painters associated with the postmodern turn (see Tomkins, 1965). His readymades redefined the art object and broke down barriers between art and everyday life. His appropriations of previous aesthetic images, such as drawing a mustache on an image of the *Mona Lisa* and labeling it with a provocative title (*L.H.O.O.Q*, which pronounced phonetically in French becomes "Elle a chaud au cul," meaning "She has a hot ass").

These aesthetic games anticipated key postmodern strategies, as did his experiments with photography, film, and other media. In addition, Duchamp's adoption of a fictive name (Rrose Selavy), doubling his identity as an artist, prefigured the postmodern insight that identity is constructed, flexible, and multiple, pointing to the overlaps between postmodern theory and art and to the complex ways that they interact and mutually influence each other.

Within the field of painting, Duchamp's hard-edged abstractions anticipated the work of American abstractionists like Frank Stella and color-field painters such as Noland and Olitski, who moved toward the sort of flat, euphoric, nonsignifying images that would be associated with postmodernism. Duchamp's cover design of a 1936 issue of *Cahiers d'art* "superimposed red and blue hearts (*Coeurs Volants*) in which the juxtapositions of the two colors set up chromatic vibrations that created the illusion of depth—a discovery that preceded by thirty years the 'optical' art of Morris Louis, Kenneth Noland, and a number of others" (Tomkins, 1965: 59). Yet the endeavors of the later American high modernists to develop nonrepresentational and nonsignifying images were accompanied by moves toward the formal purity and autonomy of art celebrated by modernism.

The artists influenced by Duchamp whom we discussed in this section are between modernism and postmodernism, as they combine motifs from both. Although they brought the objects of everyday life into art and participated in the multimedia happenings and artistic experiments characteristic of the postmodern turn, they maintained important elements of modernism in their work. Rauschenberg combined the abstract motifs and painterly gestures of high modernism with the detritus of the consumer society in his works, and although Johns pictured targets, flags, and other everyday objects, he also had a rigorously formalist aspect to his painting and soon turned to more abstract and nonrepresentational work. Both Rauschenberg and Johns thus had a highly developed painterly dimension to their work that would distinguish them from many later postmodern artists who employed the technologies of serial reproduction, the techniques of hyperrealism, or the slick mechanical look of commercial art. Thus, we must look to the pop art of Andy Warhol and others for more consequent moves toward a postmodern turn in painting that decisively broke with abstract expressionist nonrepresentational art and that returned to the materials of everyday life as the substance and form of art, imploding art into objects of everyday life, commodities, and hyperreality.

Pop Art and the Postmodern Turn

> The canvas is an absolutely everyday object, on the same plane as this chair or this poster.
>
> —Andy Warhol

> The philosophy of representation—of the original, the first time, resemblance, imitation, faithfulness—is dissolving; and the arrow of the simulacrum . . . is headed in our direction.
>
> —Michel Foucault

In those artists labeled as representatives of pop art (Warhol, Roy Lichtenstein, Claes Oldenburg, Jim Dine, et al.), there was less rage, less alienation, and a more comfortable attitude toward U.S. society than in the previous generation of abstract expressionists. The public persona of Andy Warhol, who surfaced as the Prince of Pop, is especially revealing of how the concept of the artist had changed from the earlier days of abstract expressionism, which had continued the romantic mythology of the artist as heroic outsider at war with society. Emile de Antonio's filming of his longtime friend Warhol in *Painters Painting* caught him in characteristic poses in his downtown studio. Filmed in a mirror sitting beside his assistant Brigid Polk, Warhol declares that painting is over, claims that he hasn't done any painting in three years and that Polk has been doing his work, and is generally self-deprecating. What is significant about Warhol's self-presentation is the way that he deflates the usual aura of pretension that surrounds major artists. Indeed, Warhol constantly comments in the form of put-ons, highlighting his commercial aims and downplaying the aesthetic significance of his work, thus utilizing masks and simulating indifference to the traditional aesthetic goals and values that had been apotheosized in high modernism (see also the interviews with Warhol in the film *Andy Warhol Superstar*).

It is also very significant for delineating a postmodern turn in the arts that Warhol enthusiastically embraced commercial art and the artifacts of media culture in order to produce pop art. Whereas Rauschenberg and Johns did commercial art in the 1950s to support themselves, they quit when they achieved success in the New York art world. For them, it was either/or, and they choose high, or serious, art. Warhol, by contrast, continued doing commercial art even after he won approval as an officially sanctioned artist and arguably collapsed the distinction, turning commercial art into museum art and highlighting the commodity status of art itself—indeed presenting art as the supercommodity.

Whereas Jasper Johns painted representations of the real objects of everyday life, turning them into art objects with an intensely painterly dimension, Warhol tended to work with images of images, drawing on photography, advertisements, and various reproductions of "real" objects. Warhol's pop art thus produced simulacra of images of commercial objects, as well as replicas of representations of stars like Marilyn Monroe or Elvis Presley, political figures like Mao Tse-tung, or newspaper photos of such things as an electric chair or traffic accident. Warhol's pop art was thus an art of simulacra, second-order images, representations of representations, and was in this sense more abstract than his predecessors, even though his objects looked more "real." But Warhol's "real" was a hyperreal, a realer-than-real, producing purified images from preexisting representations whose ultimate "origins" in the social world may be difficult to detect.[29]

Warhol's use of silk screens, which literally reproduced a series of images from photography or other sources, proliferated his art of simulacra. Furthermore, his attempt to produce totally bland, nonsignifying art that renounced deeper meanings beyond the images themselves and his presenting himself as an empty one-dimensional character ("I want to be a machine") anticipates postmodernism, which renounces the real in favor of depthless collages of images and words. Warhol thus erased the very concepts of the artist, artwork, and creativity, leveling the significance of the art object to a mere sign among other signs, undermining a depth hermeneutic whereby one sought meaning behind, within, and beyond the artwork. For Warhol, the creator was just a technician without anything to say, and his style was simply that of the existing image world.

Although there remains a residue of tongue-in-cheek dadaist irony and aesthetic subversion in Warhol, it is without the political bite of earlier dada and is more affirmative of its objects than destructive. On the whole, pop art not only depicted the objects of the consumer and media society as the subject matter of art but utilized commercial methods and techniques to produce art. Whereas Johns and Rauschenberg continued to make painterly pictures, with signs and traces of their work as artists, Warhol and other pop artists tried to be as impersonal and objective as possible, seeking to erase all personal and stylistic elements from their work. They thus enacted the theme of the disappearance of the subject and "end of the author" central to some versions of postmodern theory (e.g., Barthes, Baudrillard, Foucault, and Derrida). Moreover, the pop artists returned to the representation of images and objects of everyday life and eschewed the nonrepresentational elements still evident in the work of Johns and Rauschenberg.

Thus, the pop artists repudiated major tenets of modernism, and certainly Andy Warhol represents a totally different figure of the artist from the modernist concept, played to the hilt by the abstract expressionists, of the artist as hero and cultural warrior. Instead, for Warhol and pop art, the artist is merely a chronicler of the images of the day, a participant in the media and consumer society, a producer of images within the society of promotion. The successful artist, on Warhol's example, is the artist who most effectively circulates his or her work, and Warhol once said that "publicity" is the greatest art of the 20th century. Indeed, Warhol himself was a master of publicity and became a superstar celebrity as well as an internationally renowned artist.

The pop artist also renounces the spiritual and universalist concerns of high modernism, found at the core of abstract expressionism, and abandons transcendence for total immersion in immanence, in the existing society. The artist is thus no longer an outsider, no longer the representative of alienation and nonconformity, no longer the cultural warrior against the crass bourgeois society. Instead, for Warhol and pop, the artist is merely a player in the game of contemporary commerce, publicity, and image construction, a member of the hip, chic, and—if successful—rich.

If abstract expressionism represents the triumph of heroic individualism, pop art reveals the triumph of the spectacle of the consumer society, its colonization of every aspect of life, including art. Pop puts on display the commodification of art, the reduction of art to commodity, but also art as a commodity spectacle of the affluent consumer and media society. Pop art exposes the role of image in reproducing capitalist culture and the reduction of art, culture, style, and identity to image, to a hyperreal simulacrum, to an infinite Möbius-like play of signifiers without a real or stable referent. The interminable precession of images in pop art highlights the spectacles of the media and the commercial culture of glitz, where everything is for sale. Its meticulous reproductions dramatize the spectacle of new technologies that can produce exact replicas and reproductions of the "real," moving art into the sphere of the hyperreal.

Warhol's forte was that he understood that the new media culture was a culture of images and their technological reproduction and dissemination, and he produced and reproduced images of the newly dominant media culture in his work. He thoroughly deconstructs the modernist notions of authorship, creativity, originality, authenticity, and auratic art, creating a new art form based on serial reproduction. His sculptures of Brillo boxes, pictures of Campbell's Soup cans, and silk screens of celebrities such as Jackie Kennedy and Marilyn Monroe themselves became defining images of media culture. Such work is emblematic of the loss of aura, the erasure of uniqueness and authenticity, in the postmodern image culture and the ways that images are commodified, mass-produced, and circulated in the infotainment society.

Warhol's work therefore is of essential diagnostic value in dissecting the nature and values of contemporary culture. His work puts on display the icons, celebrities, and forms of a culture that was increasingly becoming a media culture, based on mass-produced images, commercialism, and spectacle. Warhol saw that images permeated and constituted every realm of society, from economics to politics to culture and everyday life. He faithfully reproduced the images of advertising, politics, consumer culture, and the icons of everyday life *as images*, as simulacra of commodity and media culture. Warhol thus perceived the move toward a culture and politics of the image in U.S. life, in which all of social life is filtered through the media, politics becomes a battle of images, and identity is mediated through image and look (Kellner, 1990, 1995a). Warhol also saw that the media culture was becoming a culture of celebrity and that anyone could become a celebrity—at least for 15 minutes. And finally, Warhol grasped that art itself was a question of images, of producing pleasing and resonant images that were attractive and appealing. Warhol's own work—especially his painting—did indeed produce compelling images that resonated with cultural experience.

Crucially, Warhol showed that art is a commodity and a commercial business, that the commodity is the fetish of late-capitalist society, that commodity fetishism is its organizing principle, and that art has itself become the fetish of fetishes and a thoroughly commercial activity in a fashion- and publicity-driven business culture. Yet although Warhol could have provided a demystification of art and the banality of commercial culture with these insights, instead he remystified it, providing an aesthetic aura to the commodity, sanctifying art as a commercial activity, affirming and celebrating its commodity status, and thus ultimately advancing the values of capitalism. The same is true of his insights into media culture and hyperreality: Warhol could have put on display the mechanisms of celebrity fetishism, demystified the culture of the image, and shown the hyperreal to be a degraded and illusory form of the real. Instead, he chose to revel in the world of celebrity, image, and promotion, assiduously circulating his own image as pop artist and himself becoming a major celebrity and figure of the pop scene in the process. Warhol's work thus inhabits a space between a possible demystification of the commodity and media world of contemporary capitalism and its remystification in aestheticized images of the commodity and the hyperreal. With Venturi and others, Warhol belongs to the camp of conservative, ludic postmodernists.

Warhol's work could be compared with that of Duane Hanson, whose sculptures of shoppers in a market mock and attack commodity fetishism, or with Oldenburg's works, which by exaggerating the fetishistic aspects of commodities, media images, and art create a critical distance and tension lacking in the work of the Prince of Pop, who did not distance himself from commodity culture but, rather, immersed himself in it. Indeed, Warhol thrived on the commodity fetish and made his fortune within it, thoroughly integrating his art into the marketplace and commodity aesthetic. Much postmodern painting replicates this ludic play with existing capitalist and media culture, although, as we shall see in the next section, there are also striking examples of a postmodernism of resistance in the visual arts. Thus, as we shall see in the next section, postmodern art is a contested terrain between opposing tendencies and forces.

THE VICISSITUDES OF POSTMODERN ART

> Generally speaking, the play on quotations is boring for me. The infinite nesting of box within box, the play of second and third degree quotes, I think that is a pathological form of the end of art, a sentimental form.
>
> —Jean Baudrillard

> In the illusory babels of language, an artist might advance specifically to get lost, and to intoxicate himself in dizzying syntaxes, seeking odd intersections of meaning, strange corridors of history, unexpected echoes, unknown humors, or voids of knowledge . . . but this quest is risky, full of bottomless fictions and endless architectures and counter-architectures . . . at the end, if there is an end, are perhaps only meaningless reverbations.
>
> —ROBERT SMITHSON

Warhol himself became an image, a trademark, a celebrity, an icon of the pop culture that he chronicled, tracked, and immersed himself in. As Venturi et al. (1972) remarked, pop art had a tremendous impact on what we can now see as a postmodern turn in architecture, and it influenced fashion, design, advertising, TV, film, and other forms of media culture as well. And, significantly, in the spirit of Warhol, the art world itself fell under the sway of its publicity machines, which generated excitement over the emergent trends of the 1960s, when new movements attracted attention one after another—op art, conceptual art, minimalism, photorealism, earth art, and the like. Indeed, after the moment of pop in the early 1960s, aesthetic pluralism seemed to be the dominant tendency in U.S. art, with an explosion of forms, reappropriation of past images and styles, frenetic revivalism, and hype that would later be identified with the postmodern turn.

The art world thus reflected the new world of commodity spectacle and media culture dominated by publicity, specialized markets, and intense commercialization. The art object became part of a circuit of global capital and media culture in which fashion and consumer trends rapidly circulated from country to country. Obviously, the financial wealth of an expanding U.S. geopolitical global empire made possible both the financing and inflation of prices of the U.S. art of the period, as well as its promotion (see Guilbaut, 1983). For a brief moment, the United States ruled the world of art, producing the most significant artists and aesthetic breakthroughs, from abstract expressionism to pop art and the proliferating movements of the 1960s. We now live in a more decentered world, with no artistic core, in which a plurality of styles uneasily coexist, and there is no aesthetic consensus as to what is quality or advanced art, as there was in the 1960s when New York reigned as the art capital of the world. Instead, the art world, like postmodern culture in general, is ruled by fragmentation and the hype of media and consumer culture, with its cacophony of competing trends, works, and artists.

In the 1960s, there was a rapid proliferation of the last modernist movements, with all logical possibilities seemingly exhausted. With minimalism, conceptualism, and earth art, anything could be an artwork, or at least part of one, thus bringing modernism to an end with a whimper and not a bang. Postmodernism in turn absorbed the characteristics and strategies of all modernist movements, which it either revived in a neo or retro mode or combined, undoing boundaries between genres and styles, and between art and life. The postmodern moment in the visual arts had thus arrived.

By the 1970s, the label "postmodern" was frequently applied to new trends and forms in the visual arts. In popular journalism and art journals, voices were beginning to suggest that modernism in the arts was finished and that a new postmodern art had arrived, although no major manifestos *for* postmodern visual art and against modernism appeared comparable to Jencks's or Venturi's interventions in architecture, or John Barth's essays legitimating a postmodern turn in literature.[30] Brian O'Doherty (1971: 19) noted that the term "post-modernism came into common usage in the late sixties" but indicated that "though it is our diagnosis for what surrounds us, one never hears it defined." Moreover, he presented it rather negatively as an "angry dumbness . . . fundamentally an anger at being forced to complete its own obsolescence." Articulating what would later become a standard conception of the term in the arts, Kim Levin pointed to a postmodern turn in articles written in the 1970s and 1980s, and collected in a 1988 volume *Beyond Modernism*. She pointed out that terms that were insults in the modernist vocabulary, such as "illusionistic, theatrical, decorative, literary," were resurrected as positive terms in new postmodern art that renounced modernist purity and reveled in ornamentation, quotation, and resurrections of tradition (1988: 3).

A variety of radical critics associated with the journal *October* and other art journals developed a wealth of accounts of a postmodernism of resistance in the late 1970s and the 1980s. Douglas Crimp (1980) valorized a more political and oppositional form of postmodern art that renounced the aesthetic purity legislated by high modernism and generated works that crossed disciplinary boundaries and produced mixed-media forms. In a similar vein, Rosalind Krauss (1983) presented the postmodern turn as a positive renunciation of the fetish of originality and an undoing of boundaries that expanded the aesthetic field. Craig Owens (1983) claimed that postmodern appropriation, hybridization, eclecticism, and other strategies described an "allegorical impulse" that defined the postmodern turn and that Hal Foster characterized as a series of "recodings" that could be given critical and oppositional inflections (Foster, 1983, 1985).[31]

Every account of postmodern art constructs new genealogies of contemporary art, with different histories, evaluations and canons, and emphases and judgments. In our view, the postmodern turn in art mobilized motifs from cubism, dada, and the works of Duchamp into a new (anti)aesthetic that rejected key tenets of modernism in order to create new types and styles of art for the contemporary era. In our rereading of the history of contemporary art from the perspective of the postmodern turn, we see postmodern art as radicalizing the anti-modernist tradition of Duchamp, whose readymades subverted the canons of modernist style, and as extending the principles of dada and cubist collage to go beyond modernism. Postmodern visual art ranges from the transitional collage and representations of objects of everyday life of Rauschenberg and Johns, to the pop art of Lichtenstein and Warhol, which replicates the icons of consumer society, to neo-geo and simulation art, which seeks to capture the new forms and experiences of the computer society, to the works of Jenny Holzer, Hans Haacke, Barbara Kruger, Laurie Anderson, and David Wojnarowicz, which use the strategies of postmodern art to advance critiques of the media and consumer society.

As do postmodern architecture and literature, postmodern visual art opposes the key trends and tenets of modernism and embodies a new aesthetic against much of what modernism stood for. The postmodern pastiche and quotation in painting, for instance, was not only a joke or a commentary on the circulation and proliferation of images in a media society; it also signaled a return to tradition, the vernacular, and reference, which were spurned in the modernist aesthetics of originality, innovation, purity, and formalism. Its return to past forms and styles consciously enacted a form of historicism that renounced modernist notions of progression in art and its cult of originality. Deployment of pastiche and quotation repudiated the notion of the author as original creator and the notion of the authenticity and auratic uniqueness of the art object.[32]

Furthermore, postmodern forms like fluxus, happenings, environmental and earth art, and various modes of interactive art and performance art undid the boundaries between art and audience, between art object and world, thus merging art with its environment and everyday life and contesting the conservative function of the museum. Happenings and plays by the Living Theater and Performance Group implicated the audience in participation in dramatic spectacle; the earth art of Michael Heizer, Walter De Maria, and Robert Smithson involved the spectator in the art environment and collapsed boundaries between art and nature; Christo's projects blended art, commerce, and politics, bringing "art" to a diversity of sites, urban and rural; and sculpture merged into buildings, landscapes, and urban environments as never before, becoming part of an expanded aesthetic field that undid the boundaries and aesthetic of previously classical and modernist art. Rosalind Krauss claimed that such postmodern sculpture extended and reconstituted the field, undoing the modernist isolation of sculpture from environment and architecture (in Foster, 1983), thus in effect integrating art with everyday life.

Joseph Beuys, for example, referred to his performance pieces as "social sculpture" because he intended his art as a political force to promote thought and social change, in effect helping to create and shape a new society and humanity, thus imploding art into education and politics. The implosion that we have seen is central to postmodern theory is thus also a key to depicting postmodern culture as a whole.[33] Implosion in the arts involves an undoing of boundaries among the arts, as well as, in some cases, the merger between text and audience and between art and life. The postmodern turn thus sub-

Your
w o r d
(verbs)
the
side
of my
(noun)

verts classical conceptions of art as independent from mass culture and its everyday world. Rejecting notions of the autonomy of art, it questions the museum and other spaces in which art is publicly represented, thus attacking the institution of art and its field of exhibition, publicity, and distribution (see Crimp, 1983)—though few postmodern artists have been able to resist exhibiting in museum shows when offered under favorable circumstances.

Within the field of painting and the visual arts, the term "postmodern" is generally applied to artifacts that plunder the entire history of art for styles that range from neo-realism to neo-expressionism. Much postmodernist painting consists of playful pastiche and repetition, quotation, and allusion to painting of the past. Indeed, in the 1970s and 1980s there was a frenzy of revivalism, with Chia and Cucchi rekindling the forms of neo-primitivism, and a variety of American and European artists generated a neo-expressionism, while Julian Schnabel and Francesco Clemente appropriated a variety of neo-romantic forms and Jeff Koons engaged in reappropriations of baroque and rococo styles. Even more recent trends such as pop art and op art were revived, with Philip Taaffe, for instance, redoing Bridget Riley's 1964 op art work *Crest* as *Brest* in 1985. Thus, specific iconography and images of previous painters were appropriated, reproduced, and quoted à la Rauschenberg.

Against complex modernist meaning machines, ludic postmodernism plays with cultural forms or deploys nonsignifying surface images, randomly juxtaposed to resist interpretation, preferring, à la Warhol, an aesthetic of the surface rather than depth, form rather than meaning, and randomness and discontinuities over well-wrought aesthetic artifacts and unities. The title of the film *Stop Making Sense* by David Byrne and the Talking Heads enunciates the tendency of this form of postmodern art, perhaps first articulated by Susan Sontag (1967), to seek pleasure in aesthetic forms and surfaces and to eschew systems of meaning, polysemic and multilayered complex artifacts that demand depth hermeneutics, and works that intend to make personal or political statements.

Many artists labeled "postmodern" play with historical and traditional images and forms, in accord with the descriptions of postmodern theory that evoke a depthless, one-dimensional, and nonsignifying image culture as characteristic of the present moment (Jameson, 1984, 1991). Painters like David Salle utilize flat, vivid surfaces and broken images to present a schizophrenic fragmentation of images with no clear narrative or structural organization. Frequently appropriating images from other painters, Salle provides cold depictions of fragments of contemporary life, disconnected and without context. His *Sextet in Dogtown*, for instance, juxtaposes 10 images that do not connect in any obvious way, presenting fragments of a disconnected postmodern condition from the media, urban life, and the consumer society.

The fragmentation and disorder in much postmodern art of the 1980s can be read as exhibiting a loss of ability to contextualize, narrate, and provide order to experience (Taylor, 1985). The fragmented and inexplicable juxtapositions in the paintings of Salle, Robert Longo, Schnabel, and others has been interpreted as the dissolution of the creative and synthesizing artistic ego into a neurotic, narcissistic, and schizophrenic consciousness lost in private fantasies, disparate images, and desperate attempts to salvage beauty or transcendence from the fragments of an otherwise barren world, bereft of community, shared traditions, or aesthetic norms. The renunciation of meaning and interpretation privileges the retinal over the conceptual, the eye over the mind, thus reversing one of the major currents of modernism that had run from cubism through the work of Duchamp and abstract expressionism. Hyperrealism and the revivals of pop, op, and other recent trends can also be seen as an escape from self, interpretation, and meaning.

Other artists, like Jeff Halley, utilize the forms of computer art to depict the complexity of new communications environments and the new aesthetic of a simulated world, while Nam June Paik deploys video to highlight the new media image world and its fragmented flow and never-ending proliferation of images. Jeff Koons exhibits objects from everyday life and pays artisans and commercial firms to replicate kitsch ornaments, objects, and advertisements in neo-realist verisimilitude. Halley and others belong to the neo-geo revival of abstraction, specifically as influenced by Baudrillard's theory of simulation. An example of such "simulation art" is found in Allan McCollum's 1983 installation of hundreds of "generic paintings" featuring different sizes of black squares framed in white squares, which presents a commentary on how all works of art are ultimately reduced to the abstraction and homogeneity of exchange value in a commodity market.

Yet one can distinguish between a ludic and "conservative–pluralist" posture of "anything goes" in art, characterized by pastiche, quotation, and play, and a more "critical–oppositional" strain of postmodern art (Foster, 1983; Wallis, 1984; Conner, 1989). This more deconstructive and socially critical art retains the same openness to a wide range of form and media of its more ludic–conservative variant, but in addition it attempts to develop a new postmodern vision, as in the work of Erich Fischl, who represents contemporary forms of alienation, anguish, and worse (see Kroker and Cook, 1986), or Thomas Lawson, who in *Don't Hit Her Again* (1981) presents a shockingly large face of a battered small child as an emblem of child abuse.

Political and oppositional forms of postmodern art refuse the antihermeneutic approach of ludic postmodernism in favor of a densely signifying practice that endeavors to subvert dominant meanings. A variety of radical critics, such as Crimp, Krauss, Owens, Foster, and Hutcheon, valorize different forms of political postmodern art that offer critical visions of culture and society and that advance oppositional practice. Several of these critics identify one such postmodern practice, "postmodern photography," as a model of political postmodernism, a "political art of the first order" (Hutcheon, 1989: 13). They see postmodern photography as generating a critique of realism and mimesis, "showing photography to be always a *re*presentation, always-already-seen. Their images are purloined, confiscated, appropriated, *stolen*. In their work, the original cannot be located, is always deferred; even the self which might have generated an original is shown to be itself a copy" (Crimp, 1980: 98). All images—including those of photography, which seem to present an objective view of the world—are challenged as arbitrary, conventional, and coded interpretations and constructions rather than objective snapshots of reality.

As Owens (1983) shows, many women artists have made use of postmodern strategies to advance radical and feminist critiques and politics. Martha Rosler's *The Bowery in Two Inadequate Descriptive Systems* problematizes images and languages as means of depicting social reality and raises questions concerning how such things as the Bowery and drunkenness can be represented. Sherrie Levine's *Photograph after Edward Weston* crops a photo of Weston's son near the top of the penis, signaling how art idealizes the male body and that women can deploy artistic forms and strategies to attack patriarchy. Cindy Sherman's *Untitled Film Still* presents the artist in a pastiche of a Marilyn Monroe figure, alone in a dark night in a film noir setting, evoking the dangers to women in predatory urban environments—and from the distortions of stereotyped systems of representation. In turn, Barbara Kruger's montages of images and discourse problematize systems of representation, and her billboards and public art break down boundaries between art and environment, politics and aesthetics (images from these artists are reproduced in Foster, 1983: 68–76).

Similarly, Richard Prince deconstructs the images of advertising, and Hans Haacke exposes the lies and deceptions of corporate and political advertising and of the art establishment. As Debord emphasized two decades earlier, these postmodern artists are suggesting that contemporary life is thoroughly mediated by representations. The Situationists, in fact, anticipated this parodic "photo-piracy" by a couple of decades in their technique of *détournement* (see Chapter 3), and their critical image art was anticipated by dadaist montage artists such as John Heartfield, who made photomontages exposing the corruption of German capitalism and its complicity with Hitler and the Nazis (i.e., Heartfield's famous montage of the meaning of the Hitler salute shows Hitler in salute taking money bags from German capitalists).

Thus, postmodern image appropriation can be an ex-propriation, seizing already-existing cultural images to use against the grain to reveal the artificiality, derivativeness, and conventionality of dominant images, to criticize hegemonic forms of representation and social ideologies. It can advance a "politics of representation" that challenges dominant images, ideologies, and cultural stereotypes and suggests their role in constituting subjectivity. Following a modernist strategy advanced by the Russian formalists, the surrealists, Brecht, and others, oppositional postmodern image artists such as Barbara Kruger, Negativland, and the Tape Beatles seek to "denaturalize" the familiar image environment, to expose it as a contingent construction rather than an icon of the real, and to prompt critical awareness of the mechanisms that produce allegedly natural signs and images.[34] Subverting realist notions that representations are objective and transparent pictures of reality, oppositional postmodern artists intend to show how all meanings are socially and historically constructed.

At the same time, by blatantly stealing images, they are debunking the humanist ideology informing modernist notions of artistic genius, originality, and autonomous language, arguing that no language ever escapes from a historically constituted web of intertextuality. The main weapon of appropriation art is parody, which, as Hutcheon (1989) rightly argues, is not simply a detached irony or nostalgia for the past but, rather, a key strategy for contesting both realist epistemology and modernist autonomy. As postmodern, this art is much more ambivalent and contradictory than is modernist art and, like postmodern art in general, it simultaneously installs and subverts ideologies, being both critical and complicit, making us aware of the power of the image and the ways that representations constitute our subjectivity and modes of seeing the world. As suggested by the title of one of Victor Burgin's works (1986), postmodern photography is "between" various boundaries, operating on the fault lines between art, politics, and theory, undoing oppositions between text and image, high art and mass media, and artistic practice and theory. Indeed, as Burgin's work shows, the work of postmodern image artists is extremely sophisticated and theoretical, drawing on Marxism, feminism, semiotics, psychoanalysis, and deconstruction.

Women, gays and lesbians, and people of color, often excluded from official art worlds, also used new art forms like film and video as means of aesthetic subversion and cultural critique. Martha Rosler's political videos attack oppression and the complicity of intellectuals therein; Joan Braderman's videos take apart the forms and ways of reading women's magazines and tabloids; Yvonne Rainer's films deconstruct traditional forms of cinema and attempt to discover new forms and aesthetic strategies to advance feminist critique; Marlon Riggs's films and videos critically analyze images of blackness and gayness and depict the tensions between these conflicting aspects of his identity; Julia Dash uses film to interrogate black identity and history; and Laurie Anderson deploys performance art, film, and video to play with and comment on contemporary technological culture and society.

Interestingly, following the lead of Anderson and others who began using video and film, superstar postmodern artists such as Longo, Salle, and Schnabel all turned to film in the mid-1990s as vehicles of artistic expression, taking conventional genres of Hollywood film as a medium for aesthetic practice (e.g., Longo utilizing the SF–cyberpunk genre in *Johnny Mnemonic*,

Salle deploying the crime–gangster genre in *Search and Destroy*, and Schnabel producing an artist biopic in *Basquiat*). Film and video can capture aesthetic experience and through techniques of mechanical reproduction can distribute artworks through much broader channels than can traditional art, which requires presence and immediacy of response; and these technological mediations are increased even more by the possibility of Internet circulation of artistic materials.

Indeed, the visual arts are especially important in a postmodern media and consumer society because the image is the semiurgic force that generates thought, behavior, and the very ethos of everyday life. In a world saturated with images, the Situationists use image to fight image. Painting and visual art freezes representations into crystallized icons of experience and vision. In a highly speeded-up media and consumer society, the still and quiet images of painting or photography can serve revelatory functions, showing the role of image in the media and consumer society and producing counterimages that can generate critical insights. The Situationists and postmodern artists like Kruger and Holzer juxtapose image and text to promote critical reflection on the forms and messages of the media and consumer society.

But a postmodern culture involves sound as well as sight, discourse as well as images, music as well as visual art. Indeed, juxtapositions between word and image, sight and sound, are characteristic of postmodern culture, which brings spectacle into musical performance and sounds into many visual art installations, as well as making use of video and computer art. The spoken word—as well as graphics presenting concepts and texts—is very important in the works of Martha Rosler and Joan Braderman, who use the power of the word to demystify and deconstruct dominant images and to encourage thought and reflection. Music is central to many forms of postmodern avant-garde art ranging from the works of John Cage to Laurie Anderson. In *Home of the Brave* (1986) and succeeding performance art, Anderson mixes media, genres, and aesthetic forms to provide critical commentary on the media and consumer society. Reviving the art of storytelling, dormant in a media culture, Anderson tells stories about the postmodern technoculture, punctuated and illustrated by a cornucopia of sights, sounds, and spectacle.

With the postmodern turn in art, critics are now more inclined to look toward women's art, the art of people of color, non-Western art, and art from sources previously excluded from the established pantheons for new and exciting developments than they were during the first decades of the postwar period when there was still something of an aesthetic consensus and establishment, with its pantheon of largely white male artists. Now aesthetic values are up for grabs, and there are continual redefinitions of art and controversies concerning the most appropriate and advanced art of our time. There is, as of yet, no established history, genealogy, tradition, or canon of postmodern art. Rather, there continue to be intense controversies between defenders of the modern and the postmodern and an overwhelmingly diversity of new postmodern artifacts, with heated controversy over their significance and value.

The variety and diversity of postmodern forms of art is by now bewildering, and it is impossible to survey here the multiplicity of contemporary postmodern interventions in the arts. During the past decade, the postmodern turn has not produced many new art heroes, canons, or monuments. It seems that perhaps there is too much irony, renunciation of originality, and revival of the past to win acclaim for new works and artists. Yet by the mid-1990s, the postmodern turn in the arts seems to have facilitated more exhibits in galleries and museums by younger artists, and the contemporary scene appears to be ever more multicultural and international in flavor, tending toward postmodern eclecticism and pluralism (see the articles in *The New York Times*, May 17, 1996: B1, and May 26, 1996: section 2: 1). Many younger and established artists are turning to multimedia productions, and one imagines that the computer, the Internet, and new technologies will be crucial to new postmodern developments within art.

It has also becoming strikingly evident that the postmodern turn is helping to generate a new global culture, with postmodern forms in architecture, painting, media, computer, and consumer culture, and other spheres currently traversing the globe with incredible speed. New information and media technologies make possible instantaneous communication and circulation of images, ideas, and artifacts from one corner of the world to another. Thus, one sees similar architecture arising in Tokyo, London, Sydney, and Los Angeles. Japanese architects like Tange Kenzo, Isozaki Arata, and Ando Tadao are among the most advanced promoters of the postmodern style, and neighborhoods in Japan, such as the Aoyama and West Shinjuku districts of Tokyo, exhibit some of the most striking postmodern architecture.[35] Indeed, cities today are increasingly global and cosmopolitan (see Sassen, 1991), and both the International Style and postmodern architecture are visible everywhere.

Media culture too is increasingly global and a bearer of postmodern forms that seem to travel easily across borders. MTV is now international, with different versions on practically every continent of the world. Forms of popular music such as rap, heavy metal, and grunge have their analogues all over the planet: Japanese youth tan themselves, Westernize their eyes, and cultivate dreadlocks, affecting the mannerisms of African American rap singers, and imitations of rap, heavy metal, and grunge music are found everywhere. And now the Internet is instantaneously conveying global culture from one side of the earth to the other, making accessible the latest ideas, forms of culture, and modes of interaction. The global village is clearly the habitat for postmodern selves, and the ideas of the postmodern itself have gained currency and force from the new global culture and modes of communication.

Thus, there is no question of the significance or global nature of postmodern cultural forms, but it is not certain that the postmodern is yet *the* dominant as Jameson (1984, 1991) claims. Rather, much of the world lives in a nonsynchronous, overdetermined mixture of the traditional, modern, and postmodern, with these cultural forms often synthesized and overlapping. Indeed, we have been arguing that even in more advanced capitalist countries, we are currently between the modern and the postmodern, in an interregnum between paradigms, experiencing the breakdown of modern theory, culture, and society and the emergence of new postmodern forms. This ferment in the contemporary situation, this transition from the modern to the postmodern, is dramatically apparent in the field of science, in which new postmodern paradigms are assaulting previous modern concepts, leading to controversies within and about contemporary science. Indeed, as we show in the next chapter, the emergence of postmodern forms of science is contributing to the genesis of a new postmodern paradigm that we delineate in Chapter 6.

NOTES

1. There are, of course, a tremendous number of different constructions of modernism and of its origins, history, canons, and effects. We follow the broad historical vistas described by Marshall Berman (1982), who sees modernist art emerging out of modernity, appearing first in mid-19th-century Europe with Charles Baudelaire and his followers in poetry; Edward Manet, the impressionists, Paul Cézanne, and others in painting; Gustave Flaubert, Fyodor Dostoyevsky, and myriad literary figures; and with similar formal innovations in the other arts. For other constructions of modernism, see the collection edited by Bradbury and McFarlane (1976) and Berman (1994). In this chapter, we will critically analyze various constructions of the history of modernism, architecture, and painting and the visual arts in the transition from modernism to postmodernism. We will indicate the biases and assumptions of some dominant approaches that we will contest. We suggest that modernism looks quite different from the vantage point of the postmodern turn and that we accordingly need to reconceptualize the history and trajectory of modern art in the light of contemporary postmodern developments.

2. See Baudelaire's prose poem "One O'Clock in the Morning," which concludes with the poet returning to his cherished solitude after a maddening day in the city, praying that God will grant him a few good verses to prove that he is not inferior to those he despises (1970: 16). In another vignette in *Paris Spleen*, "Beat Up the Poor," Baudelaire describes his attempt to batter a beggar after being disgusted by moralizing literature. Such passages combine irony and an articulation of the elite attitudes that the modernist artist was superior to the urban rabble and that art was the only significant activity, a credo memorialized in the doctrine of *l'art pour l'art*, "art for art's sake."

3. On the distinction between modernism and the "historical avant-garde," see Burger (1984) and Calinescu (1987).

4. On the last, see Woolf (1929) for a manifesto on women and modernism; Rosemont (1997) on women and surrealism; and Benstock (1986) for women in the Paris literary scene. On women in painting, see the studies in Broude and Garrard (1982), and on women and modernism in general see Jardine (1985) and Martin (1991).

5. The term "avant-garde" is, of course, a military metaphor that implies that artists are the "front line" in bringing change to an oppressive and decadent bourgeois society. As Calinescu (1987: 101ff.) notes, Saint-Simon first used the term in reference to artists in 1825, in his vision of a new wave of leaders led by industrialists and engineers who would bring progressive change to society. Lenin appropriated the metaphor to designate Marxist intellectuals as the front guard and relegated artists to the rear guard, privileging politics over art. Since then, art has been subsumed to one political system or another, while avant-garde groups and artists attempt to be in the cultural forefront of an assault on existing culture and society in the project of creating new forms of art and life—a project abandoned by most postmodernists.

6. For some useful surveys of postmodern art, see Tomkins (1975); Foster (1983); Trachtenberg (1985); Connor (1989); Kearney (1989); Levin (1988); Perloff (1989); and Hutcheon (1989).

7. This distinction between oppositional and apolitical postmodernism seems to be a replay in a different register of the distinction made by Peter Burger (1984) between an apolitical high modernism and a political "historical avant-garde" that renounces the aloof formalism of high modernism and stands in vigorous opposition to existing culture and society. The difference, however, is that high modernism, however abstract, was deadly serious in its ambition to create novel and monumental forms of culture, while postmodernism is more ironic and playful. Moreover, even an oppositional postmodernism gives up some of the grandiose dreams of social revolution that animated the historical avant-garde and is content with more modest opposition (a difference that could, of course, be appraised either negatively or positively depending on whether one subscribes to modern or postmodern politics). Andreas Huyssen, by contrast, finds grandiose and megalomaniacal trends in the postmodern (1986: 179ff.), whereas we are arguing that the postmodern sensibility is overall more ironic and playful than the modernist sensibility and renounces its drive for monumentality and originality, along with its avant-garde pretensions. The remnents of grandiosity and pretension that Huyssen finds in the allegedly postmodern Kassel Documenta Seven art exhibit, the work of Anselm Kiefer, or films like Werner Herzog's *Fitzcarraldo* are, we argue, aftermaths of the modern, a hypermodernism in its moment of decline, and are not, properly speaking, the products of a full-blown postmodernism that revels in irony, quotation, and play.

8. The 1932 Museum of Modern Art exhibit on the International Style and the oft-printed and updated book, originally publised in 1932, by Henry Russell Hitchcock and Philip Johnson (1966) presented the general principles of the style and canonized it as *the* modern form of architecture.

9. Baudrillard claims that the two towers of the World Trade Center represent an image of the stabilized bipolar world of corporate capitalism and state communism, signifying the replication of the two systems in "a vertigo of duplication" and the stabilization of a bipolar global system (1983a: 135ff.). The collapse of communism deflated this fantasy, and we suggest that contemporary architecture mirrors the

transition from a system of competitive capitalism reflected in the early individualized high-rises, to a stage of state and monopoly capitalism articulated in the International Style, to a new global stage of postmodern capitalism reflected in the eclectic and historicized postmodern architecture.

10. The periodization that we are suggesting corresponds to a distinction made by the Frankfurt School between competitive, market capitalism and the stage of state and monopoly capitalism that appeared in the 1930s; see the texts in Bronner and Kellner (1989) and the discussion in Kellner (1989a). We are also suggesting correlations between the International Style and state monopoly capitalism from the 1930s into the 1960s and between a postmodern architecture and the new stage of more individualized and aestheticized consumption in the era of transnational consumer capitalism, with its "soft cities," information and entertainment society, and postmodern culture.

11. Bertens notes that British architectural historial Nikolaus Pevsner used the term "post-modern" in a discussion of the return of historicism in 1961 and in subsequent articles in which he criticized forms of architectural revivalism that he saw as "a sign of weakness" (1995: 79–80). Thus, although the term was in circulation in the 1960s, it only became a positive banner for the transformation of architecture in the 1970s.

12. Jacobs also articulated the importance of women's urban experiences and the ways that urban environments nourished women who in turn helped produce urban community—a point that we shall return to later in this chapter. It is perhaps no accident that the profession of architecture is so dominated by men in a world in which money, technology, and power determine what will or will not get built. Buildings stand as secular and phallic monuments to male business culture and connote authority and social hierarchy. It is thus not surprising that the dominant class, gender, and race controls the field, though postmodern pluralization might create openings in this sphere for individuals excluded from public power in the previous organization of society and culture.

13. David Harvey describes how Jonathan Raban's *Soft City* (1974) presents a postmodern view of the contemporary city "as an 'encyclopedia' or 'emporium of styles' in which all sense of hierarchy or even homogeneity of values was in the course of dissolution: (Harvey, 1989: 3–4). This view saw the city as a "labyrinth" of diverse networks of social interaction, images and spectacles, and pleasures that allowed "soft" multiple identities and interactions. Rejecting the rationalist and massified vision of the corporate city, Raban affirmed the contemporary city as a site of postmodern diversity and heterogeneity that is fun oriented, variegated, multicultural, and eclectic—a clearly middle-class view that overlooks tendencies toward urban disintegration.

14. Jencks's double coding has become something of a contemporary popular wisdom: for example, the *Harvard Architectural Review* writes in an editorial: "Postmodernism recognizes that the abstract play of masses in light may not be enough to retain the involvement of the observer, that private dialects directed at a professional audience may not alone satisfy the requirements of a civic art" (Spring 1980: 6). See also King (1996: 140).

15. It is ironic that this obviously failed example of urban architecture has been enshrined as exemplary of postmodern architecture in glossy pictures in Jencks

(1977) and Portoghesi (1982) and in positive citations in Hutcheon (1989: 12) and Bertens (1995: 65). Perhaps its neglect and decay can then be taken as symptomatic of the limitations and failure of the postmodern turn in architecture.

16. More recently, Benedikt (1991b) has been exploring the architecture of cyberspace, which operates in the virtual realm of computer networks. Seeking to visualize the nonphysical and to design electronic edifices, cybertecture (i.e., cyberspace architecture) represents "the coming dematerialization of architecture" (Benedikt, 1991b: back cover), and it is perhaps here where postmodern architecture will come into its own.

17. Jencks claims that "the public becomes more engaged in architectural issues over which it has increasing choice: matters of taste, matters of influencing an architect through competitions" (1977: 6). But in fact dominant social and economic forces decide what buildings are to be made, in what style, and for what purpose. See, for example, the studies in Dutton and Mann (1996).

18. Consider, for example, Le Corbusier's celebration of Henry Ford's automobile factory in Detroit where "everyone works to one end, all are in agreement, all have the same objective, and their thoughts and actions flow along the same channel" (quoted in Blake, 1996: 96), or his remark in 1935 that "New York's skyscrapers are too small" (92).

19. Many critics see Oscar Niemeyer's design of Brasília, the new capital of Brazil, as a monument of alienation that is true to Le Corbusier's principles (see Evenson, 1987), while others condemn his own governmental complex at Chandigarh as a debacle wholly unsuited to the area (see Correa, 1987).

20. Peter Blake (1993), for instance, who earlier had criticized the failures of modern architecture (1974), now calls for a revival of the modernist spirit of social relevance, of the desire to create a better world, though in an era ruled by capital he is skeptical of whether this can happen. We would argue that the creative spirit of modern architecture is an important resource, but it needs to overcome the constricting dogmatism of the International Style to continue to be a positive force in the present.

21. Most (modernist-inspired) histories see modernism as the triumph over sterile academicism and classicism. This is especially true of the histories and theories of art that see abstract expressionism as the highest achievement of contemporary American and even global art, including Greenberg (1961); Rose (1967); Geldzahler (1969); Sandler (1970); and Ashton (1973). Most of these histories privilege Cézanne, cubism, and other abstract and formalist precursors of the program of abstract expressionism as the key aesthetic currents of the century and see modernism as the triumph and model of authentic art, thus registering the hegemony of modernism within the academy and official art establishments. We will suggest in this chapter that the history of 20th-century art will look different when written from the perspective of the postmodern turn, in which different precursors, schools, and aspects of important movements like cubism, dada, and Duchamp will appear salient to postmodernist developments in painting.

22. See the illustrations in Cooper (1970); there are also many excellent reproductions of Gris's work in the "WebMuseum" (http://www.oir.ucf.edu/wm/). Such sites are a postmodern realization of Malraux's concept of a "museum without walls."

23. Curiously, Butler (1980) sees abstract expressionism as part of the postmod-

ern turn in art, whereas we see it as the final ecstasy of high modernism, interpreting Rauschenberg, Johns, Cage, and those who produced conceptual art as significant transitional figures between modernism and postmodernism. Thus, whereas we agree that there is a transitional period and figures between high modernism and the postmodern turn, we narrativize the transition differently from the way Butler and previous critics do.

24. In the following discussion of American painting, we draw upon Douglas Kellner's research for work on a CD-ROM, released by Voyager in 1996, reproducing Emile de Antonio's 1973 film on American art from 1945–1970, *Painters Painting*. This project involved over 700 pages of interviews with major U.S. painters, critics, and dealers and collectors that de Antonio carried out for his film. For other perspectives on the art of the period, see the sources in Note 21.

25. Pollock's centrality to the explosion of high modernism is stressed in the texts cited in Note 21 and is highlighted in de Antonio's *Painters Painting*. Guilbaut (1983) indicates how Pollock's work solved the essential contradictions of the time (i.e., that art was to be revolutionary, politically progressive, abstract, and formally innovative at the same time) and was promoted by the media and the art establishment as the finally achieved breakthrough in U.S. painting.

26. On German expressionism, see Bronner and Kellner (1989). Willem de Kooning also exemplified the dual aspects of expressivism and abstraction found in Pollock, embodying as well the ethos of the modernist artist working to create new forms and styles.

27. See the 1989 documentary *Jasper Johns* made by RM Productions.

28. See the study by Craft (1996) of the category of "neo-dada" and how it functioned in art criticism through 1965. We would suggest that neo-dadaism became a key element in the postmodern turn in the visual arts.

29. We therefore disagree with Irving Sandler (1988) and other critics who see pop art as a return to the object, as an art of the concrete, of the real. We are suggesting that Warhol's pop art is really concerned with images of images and is thus a form of what Baudrillard (1981, 1993, 1994) calls second-order simulacra, or the hyperreal. We also disagree with Kim Levin, who claims that "Pop Art was thoroughly within the Modernist tradition in its acceptance of technology, and its formal use of contemporary objects from everyday life as abstract signs was not essentially unlike that of the Cubists" (1988: 10). We see, however, Warhol's replication of commodities, obsessive reproduction of media images, use of silk screen and other multiple-reproductive techniques, hyperrealism, hyperirony, and self-deprecatory concept of the artist as quite antithetical to cubism and to modernism in general and thus as carrying out a postmodern turn in painting.

30. For studies of some of the key postmodern visual art forms in the 1970s and 1980s, see Tomkins (1976); Davis (1977); the studies collected in Foster (1983, 1985) and Wallis (1984); and Levin (1988). Tomkins subtitles his journalistic studies of Henry Geldzahler, Warhol, lithography, EAT, earth art, the new film culture, video art, and experimental theater *Reports on Post-Modern Art*, but he does not theorize the term; nonetheless, his work is valuable for its evocation of the variety and diversity of 1970s art, the moment when the postmodern turn arguably became dominant—or at least symptomatic of the most interesting new work. Davis titled his 1977 book *Art-*

culture but admits that it was his editor who foisted the subtitle *Essays on the Post-Modern* on him (1980: 12), and he does not engage the concept of the postmodern either.

31. The essays just cited of those and other critics who seek to develop a more oppositional postmodernism are collected in Wallis (1984); see also Foster (1983, 1985); Krauss (1985); and Owens (1992).

32. For examples of postmodern quoting of past paintings and engaging in an art of appropriation, see some of the studies collected in Foster (1983); Wallis (1984); and Levin (1988).

33. Otherwise disparate and sometimes conflicting accounts (e.g., Baudrillard, 1983a, 1983b, 1993; Jameson, 1984, 1991; Huyssen, 1986; Hassan, 1987; Hutcheon, 1988, 1989; Lash, 1990) agree that postmodern culture crosses the "great divide" between modernism and media culture and undoes a series of boundaries that separate high from low culture, leading its proponents to celebrate postmodern populism or an oppositional postmodernism, while its more traditionalist opponents attack the collapse of high culture or loss of aesthetic standards and values.

34. See the documentary *Sonic Outlaws* (1995), which depicts a variety of contemporary oppositional appropriations of media images and is itself an example of a critical postmodernism.

35. On postmodernism and Japan, see the articles collected in Miyoshi and Harootunian (1989), and on postmodern architecture in Japan see Bognar (1995). Indeed, as Darrell Hamamoto has suggested to us in correspondence: " 'International Style' has an authoritarian ring to it. 'International' is in fact synonymous with Euroamerica." By extension, then, postmodern style is more eclectic and pluralistic.

CHAPTER FIVE

Entropy, Chaos, and Organism in Postmodern Science

> The carapace of science that has more or less imprisoned our century . . . is showing signs of cracking.
>
> —HUSTON SMITH

> *Bona fide* sciences . . . have broken sharply with the ideals and assumptions long identified with modern science. These postmodern sciences are the really exciting developments in the evolution of science, culture, and value today.
>
> —FREDERICK FERRÉ

In the 1970s and 1980s, after a long and complex movement through various aesthetic and theoretical discourses and practices, the postmodern turn emerged on the scene of science and social theory. In the realm of theory, since the 19th century, various critiques assaulted Enlightenment rationality, scientific mechanism and reductionism, and a progressivist view of history, leading to the postmodern attack on humanism, grand narratives, and the search for universal or transcendental values and new emphases on difference, language, history, and contingency (see Best and Kellner, 1991, and Chapters 1 and 2). In the arts, postmodernists broke with the modernist drive toward monumentality, originality, and aesthetic purity and progress, generating a wealth of new aesthetic practices that recycled traditional forms and broke down boundaries between the various arts, as well as between art and life (see Chapter 4).

In science, the postmodern turn emerged as a break from the mechanistic, reductionist, naive realist, and determinist worldview of Newtonian physics. Advocates of postmodern science claim that the modern scientific paradigm is giving way in the 20th century to a new mode of scientific thinking based on concepts such as entropy, evolution, organism, indeterminacy, probability, relativity, complementarity, interpretation, chaos, complexity, and self-organization.[1] In significant ways, this new mode of thought is congruent with changes that have occurred in social theory, and it also overlaps with recent shifts in the arts, suggesting that the postmodern turn is not merely a sign game, struggle for cultural capital, or frivolous fad but, rather, concerns the construction of a new transdisciplinary paradigm, one that we continue to map and assess in this and the succeeding chapter.

As with conceptual breaks in other fields, the changes leading to the construction of a "postmodern science" have a long history and numerous points of origin. Postmodern science has at least five major sources of influence: thermodynamics, which emerged in the 19th century; evolutionary biology and ecology, which have been developing throughout the 19th and 20th centuries; quantum mechanics and relativity theory, which appeared at the beginning of the 20th century; cybernetics and information theory, first cultivated during the 1940s; and chaos and complexity theory, which surfaced in the 1970s and 1980s. Together, these discourses have generated a ferment of new thinking, producing a variety of critiques of modern science and proposals for a new postmodern science.

In this and the next chapter, we argue that recent scientific developments are part of a larger postmodern paradigm shift and assess the contributions and limitations of these new discourses. As we shall see, arguments for a postmodern turn in science draw on the most advanced scientific and philosophical theories, which attempt to generate new scientific paradigms and concepts, leading to radically different understandings of the natural and social worlds. We discuss central differences between modern and postmodern epistemologies and worldviews and examine the social, political, ecological, and ethical implications of postmodern science. We show that postmodern science is not a simple monolithic assault on modern scientific norms but constitutes a plurality of positions, the most salient of which are positive and reconstructive. We also discuss the recent eruption of "science wars" between advocates and enemies of a social constructionist view that sees science as a sociologically shaped body of thought rather than as a pristine, objective representation of the world. With positivistic defenders of science attacking postmodern discourse as a new form of irrationalism seeking to destroy critical reason and its achievements, the science wars demonstrate the highly contested nature of the postmodern turn.

THE MODERN SCIENTIFIC WORLDVIEW

> Let the human race recover that right over Nature which belongs to it by divine bequest.
>
> —FRANCIS BACON

> Give me extension and motion, and I will construct the universe.
>
> —RENÉ DESCARTES

> I esteem the universe all the more since I have known that it is like a watch. It is surprising that nature, admirable as it is, is based on such simple things.
>
> —BERNARD LE BOVIER DE FONTENELLE

> The universe is . . . one, infinite, immobile. . . . It does not move itself locally. . . . It does not generate itself. . . . It is not corruptible. . . . It is not alterable.
>
> —GIORDANO BRUNO

In the transition to modernity, reason awakens to its potential power and embarks on the task of the theoretical and practical mastery of the world. This became possible only with the dethronement of God as the locus of knowledge and value and the construction of a new epistemology in which mathematics and the experimental method of science are the keys to unlock the mysteries of the universe. The new science was validated by the impressive achievements of celestial mechanics that began with Nicolaus Copernicus's overturning a geocentric universe for a heliocentric one and culminated with Newton's discoveries of the laws of gravity and with tremendous advances in technological knowledge. For the major architects of the modern worldview—Galileo, Bacon, Descartes, and Newton—the cosmos is a vast machine governed by universal and invariable laws that function in a stable and orderly way that can be comprehended and controlled by the rational mind. Beginning in the 16th century, scientific explanations of the world replace theological explanations; knowledge is used no longer to serve God and shore up faith but, rather, to serve the needs of human beings and to expand their power over nature.

For modern science to develop, it had to disenchant the world and eradicate from it all influences that saw nature to be infused with living or spiritual forces. This required a frontal attack on the notion that the mind participates in the world and on all animistic and religious ideologies—from the preSocratics to Renaissance alchemists—that believed nature to be magical, divine, or suffused with spirit and intelligence (Berman, 1981). Through advancing strictly mathematical and physical explanations of the universe, modern science presided over the "death of nature" (Merchant, 1980) and transformed a living natural world into a dead machine. This perspective greatly facilitates the project of dominating nature, for it assumes that lifeless matter neither is violated nor resists manipulation.

For modern thinkers like Galileo, who rejected Aristotle as a master of the sciences and committed himself to a direct study of nature apart from the dogmas of scholasticism and scripture, the investigation shifted from a *qualitative* study of the essence of things, their "whatness," their inner tendencies and purposes, to a *quantitative* analysis of their weight, solidity, and size and thus to the external, physical characteristics of matter that can be measured in fixed mathematical laws by a detached observer. Aristotle's fourfold notion of causation, modeled on living organisms and including the goals and purposes of animate beings as they unfold in their developmental impulses, was rejected for a mechanistic explanation of cause and effect. Hence, "teleology" became the bête noire of modern science, repudiated as a metaphysical relic from the past, and the universe was emptied not only of purposes and directional tendencies but also of value, meaning, and change. With the distinction between primary qualities (shape, size, weight) and secondary qualities (taste, texture, smell, and so on), the natural world was reduced to little but "a dull affair, soundless, senseless, colorless; merely the hurrying of material, endlessly, meaninglessly" (Whitehead, 1967: 54).

Modern science, therefore, allowed human beings to gain more certainty in their knowledge of the world, but at the cost of feeling comfortable or at home within it. The estrangement of human beings from nature is well represented in Bacon's *Novum Organum*, published in 1620. Here we find the preeminent expression of the new confidence in reason whose task was to master and control a natural world defined in opposition to human life. *Novum Organum* is a veritable manifesto for the Western anthropocentric outlook, which holds that human beings stand at the apex of creation and that the earth and its sundry life forms have value only insofar as they serve human needs. As Horkheimer and Adorno (1972) argued, the domination of nature included domination of human beings in a subject–object dialectic in which men were positioned as the Subjects, the Lords and Sovereigns of Being, while nature, women, children, and slaves were posited as objects of domination and body was subordinated to mind. Whereas philosophers in the premodern world believed that the purpose of knowledge was to contemplate eternal truths, Bacon and the modern world exalt applied knowledge and hold that the function of knowledge is to gain control over nature. For Bacon, praised by his contemporaries as "the secretary of nature," the purpose of knowledge is "to extend more widely the limits of power and greatness of man, [to command natural forces for] the relief of man's estate." Centuries before Foucault, Bacon stated that "knowledge is power" which allows human beings to control natural and social processes.

While Descartes approached the study of the world from a rationalist perspective grounded in the clarification and systematization of *a priori* ideas, rather than the data of empirical observation, like Bacon he understood nature in mechanistic terms and sought to use science to extend the domain of human power. On Descartes's dualistic logic, so decisive for later Western development, the mind was rigidly divorced from the body, and human beings (essentially soul, consciousness, or "thinking substance") were ontologically estranged from nature and physical bodies ("extended substance"), in order to be upheld as the "lords and possessors of nature." Descartes dreamed of "The Project of a Universal Science Destined to Raise Our Nature to Its Highest Degree of Perfection," and in his *Discourse on*

Method he sketched out a conception in which space or extension became the defining feature of nature and bodies in motion became the substance of the new science, with mathematical physics as the one method that would provide certain knowledge of matter (1960 [1637]).

Pursuing the logical implications of Descartes's mechanistic vision of the universe, Hobbes, Spinoza, and other thinkers of the 17th century rejected his dualistic ontology to create a more consistent mechanistic monism. Whereas Descartes exempted human consciousness from mechanistic laws, Hobbes, in a direct critical reply, claimed that the human mind itself was a machine: "If this be so, reasoning will depend on names, names on the imagination, and imagination . . . on the motion of the corporeal organs. Thus mind will be nothing but the motions in certain parts of an organic body" (1947: 5). Similarly, Spinoza boasted, "I shall consider human actions and desires in exactly the same manner as though I were concerned with lines, planes, and solids" (quoted in Randall, 1976: 247). Hence, both denied that the *res cogitans* was exempted from mechanistic explanation, and both sought to exorcise the "ghost in the machine" for a more rigorous and consistent materialism and deterministic logic.

This reductionist program informed the work of John Locke and the British empiricists and spread throughout Europe via the Enlightenment champions of Newton and mechanistic materialism, attracting figures in France such as d'Holbach, d'Alembert, and, most notoriously, La Mettrie, who stated in 1748: "Let us conclude boldly then, that man is a machine and that there is only one substance, differently modified, in the whole world" (quoted in Matson, 1966: 13). Subsuming human beings to the natural world had the advantage of subverting Cartesian dualism, which posited an ontological gulf between human beings and nature, but only to obliterate the differences between animate and inanimate life, mind and matter, humanity and the natural world. Hence, the mechanistic paradigm made human beings themselves nothing but matter in motion, pawns of natural forces, and denied them freedom and spontaneity. "If man believes himself free," said d'Holbach, "he is merely exhibiting a dangerous delusion and an intellectual weakness. It is the structure of the atoms that forms him, and their motions propel him forward; conditions not dependent on him determine his nature and direct his fate" (quoted in Cassirer, 1951: 65).

Although the scientific revolution brought tremendous intellectual, technological, and medical gains, feeding into the Enlightenment notion of progress, these were developed through an antagonistic relation to the natural world. As many have argued (Horkheimer and Adorno, 1972 [1947]; Marcuse, 1964; Leiss, 1974; Merchant, 1980; Bordo, 1987; Harding, 1986, 1991), the Baconian–Cartesian mind-set is one of conquest and control. The violence implicit in the modern attitude toward nature is overtly evident in Bacon's constant use of sexual metaphors and rape images. Nature is portrayed as a female to be captured by the male mind, tortured through mechanical inventions, "bound into service," put into "constraint," and made a "slave" to rational knowledge, which will "penetrate" its hidden secrets (Bacon, 1960). For feminist critics of modern science, the modern scientific worldview is thus highly anthropocentric and patriarchal, with the "human" project of the domination of nature really the extension of patriarchal attitudes.

To the modern mind, nature is conceived of as something that can be understood, mastered, and put to human service. With the publication of *Mathematical Principles of Natural Philosophy* in 1687, Newton vindicated Bacon and Descartes by discovering the universal laws of gravity and motion and the mathematical methods to describe them. After Newton, what the premodern world took to be ineffable mysteries of nature were revealed to be the orderly and precise mechanisms of a well-regulated clock or machine. The universe was thought to be law abiding, orderly, universal, and fully predictable. All events—past, present, and future—were determined by the same laws; chance and indeterminacy played no role in the smoothly running gears of nature. From Descartes to Newton and Hobbes to La Mettrie, the machine metaphor dominated the early modern mind, such that not only the physical universe but also society, animals, and even human beings were seen as different kinds of machines, devoid of any life impulse.

The modern world thus became the first civilization to be organized around mechanistic science and quantitative reasoning. The abstraction process generated by science, in which the natural world was emptied of meaning and reduced to quantitative value, is paralleled by the abstraction process created by capitalism, in which all objects, including labor itself, are subsumed to exchange value as mediated by money. In both cases, a reductionism takes over placing the entire world within the frame of technological manipulation for power and profit. The kind of knowledge employed in both cases—*Herrschaftswissen*, knowledge for the sake of domination, or what has been termed "instrumental knowledge"—is only one kind of knowledge, but according to the ideology of "scientism" or "positivism" it is the only or supreme mode of knowledge in the modern world.[2] Instrumental knowledge is based on prediction and control, and it attains this goal by linking science to technology, by employing sophisticated mathematical methods of measurement, and by abstracting itself from all other concerns, often disparaged as "nonscientific," "subjective," or "inefficient." Modern science, in its classical self-conception, sharply separates "fact" from "value," thereby pursuing a "value-free" study of natural systems apart from ethics and metaphysics, just as capitalism bifurcates the public and private sectors, disburdening private enterprise of any public or moral obligations.

Carrying through the Newtonian program in the late 18th century, Laplace (1951 [1814]) hypothesized that a sufficiently intelligent being, or "demon," could grasp any past or future event from an adequate comprehension of the present. Laplace drew out the implicit syllogism informing the Newtonian paradigm: The present state of events can be grasped with full precision; all future events are identical to present events; therefore all future events can be exactly predicted. As a kind of machine or automaton, reality is basically static and tautological. Time is "reversible," such that one could go forward or backward at any point and the same unchanging laws would be in operation.

Yet Laplace was describing a highly idealized universe, and he and other followers of Newton understated or ignored problems in Newton's allegedly universal theory, such as his inability to adequately explain the movements of the moon in its complex gravitational relationship with the sun and the earth. On this very problem, Peterson argues, determinism began to founder, for it was "in the moon's astonishingly complicated movements that astronomers first glimpsed dynamical chaos and truly began to learn the limits of mathematical prediction" (1993: 119). Newton himself had observed irregularities in the movements of the planets that he suspected might disrupt the harmony of the solar system, but he argued that God would intervene periodically to adjust his great clock, thereby maintaining harmony and stability. Such glimpses of irregularity that defy deterministic models of predictability would later fuel chaos theory and provide a key transition to a postmodern scientific mind.

Yet Newtonian natural science proved itself in practice, with its theories producing impressive results, and its success had a dramatic impact on the new social sciences emerging in the 18th and 19th centuries. Social scientists and historians, awaiting the arrival of their own Newton, began to search for laws that controlled the social and psychological world in the same way that laws controlled the natural world. Before Comte's definitive vision in the 19th century, the idea of the social sciences emerged with 18th century thinkers, like Hume and Montesquieu, who set out to find the laws of society and human nature, that is, to construct a "social mathematics" or "social physics" based on alleged laws of behavior and human nature. The repressive implications of this mechanistic vision were made explicit in the ideas of Saint-Simon and of his student Comte, who envisaged a society controlled by scientific elites that manipulated human affairs to conform to the natural order of things. In Comte's frightening words: "True liberty is nothing else than a rational submission to the preponderance of the laws of nature, in release from all arbitrary personal dictation" (1850: 432).

Consequently, in the 19th and 20th centuries, positivist concepts and methods dominated study not only of the inanimate world of physical objects but also of the animate worlds of biology, economics, society, history, philosophy, and psychology. Various fields of human behavior were marked out, discursively defined in terms of laws, regularities, and norms, and social scientists embarked on the kind of normalizing and disciplinary programs described by Foucault. Even the great thinkers in opposition to the main dynamics of modern society had assimilated core aspects of the mechanistic paradigm. According to Darwin's theory of natural selection, organisms passively respond to their environment, not shaping it in turn or having any self-determining powers of their own (see below); Marx often couched his theories in mechanistic terms, appealing to the "iron laws" of capitalism or to the "laws of motion of modern society"; and Freud practiced psychoanalysis as a scientific discipline designed to master the inner world and extensively employed mechanistic metaphors relating to libidinal energy systems and hydraulic operations.

Thus, from the 17th to the twentieth centuries, a new *modern paradigm* emerged, organized around the logic of determinism and rooted in the objectifying, mechanistic, abstract, and atemporal mode of thought stemming from the natural sciences. The reduction by behaviorism of purposeful consciousness to observable signs and by logical positivism of ethics, art, and metaphysics to meaningless ("unverifiable") discourse demonstrate the hegemony of scientific materialism and determinism over every area of human thought. Scientism became a modern faith, promoting the belief that the scientific method alone provided the royal road to truth, that there was one legitimate logic and one reliable methodology, and that eventually all sciences and fields of intellectual endeavor could be unified within the same nomological, reductionist, and materialist framework.

Though science became the new God and science-driven technological change in turn advanced the Gospel of Progress, a handful of modern thinkers—ranging from romantics like Blake and hermeneuticists like Dilthey, to existentialists like Kierkegaard and Nietzsche and philosophers such as Bergson and Whitehead—did step forth to challenge the methods and goals of reductionism and mechanism. By the early 20th century, the Newtonian dream of exact measurement, precise predictability, absolute certainty, and total knowledge was challenged by an increasing number of thinkers in philosophy and in science and mathematics themselves. These heretics chipped away at the epistemological cornerstone of the modern worldview, undermining both the "quest for certainty" and the representational or "spectator theory" of knowledge (see Dewey, 1979). Modern scientific epistemology involves a rigid subject–object bifurcation whereby the unbiased scientific mind confronts the cosmos in a detached mode of observation, passively receiving sense impressions rather than actively shaping the understanding of the world, and translucently bringing this knowledge to consciousness through an exact process of linguistic designation.

In the disenchanted world of modern science, the perceiving subject is a neutral observer and the object a pure datum of perception, each separated from the other by a chasm of nonparticipation. Defined as a "mirror of nature," the mind was thought capable of representing the world through objective knowledge that was stable, certain, and accurate (see the critique in Rorty, 1979). As described in Foucault's (1973) archaeology of the classical theory of representation, language, once an inextricable part of the divine web of things, came to be seen in the modern era as distinct from the world of objects and thus able to represent them in objective terms. But once both the perceiving subject and the object were understood in terms of time, history, change, and becoming—a development that Foucault rightly observes occurred within the modern era, specifically, the 19th century—the movement to the postmodern was underway.

THERMODYNAMICS, ENTROPY, AND EVOLUTION: THE EMERGENCE OF POSTMODERN SCIENCE

The world-order is based on a lie.
—Franz Kafka

Ignis mutat res [Fire Transforms All Things].
—Ancient Saying

"Our vision of nature is undergoing a radical change toward the multiple, the temporal, and the complex." With these words Nobel Prize-winning physicist Ilya Prigogine and chemist and historian of science Isabelle Stengers begin their influential book *Order Out of Chaos* (1984), a landmark of postmodern science.[3] This work charts a key transition from mechanical dynamics to thermodynamics, from a static and deterministic view of life to a new theory of "dissipative structures" based on principles of complexity, self-organization, and order emerging from the "chaos" of nonequilibrium conditions. The "new science," which Prigogine and Stengers distinguish from "modern science," overturns the static deterministic view to reinterpret the universe as being constituted by forces of diversity, evolution, and instability, and by a complex dialectic of order and disorder. Change and time introduce instability and disorder into the world, but these in turn create new and more complex forms of order.

As described by Prigogine and Stengers, the first challenge to the Newtonian paradigm was the introduction of temporality into the understanding of natural processes. Time and becoming, expunged from the old picture, play fundamental roles in the new theories of science. Prigogine and Stengers date the birth of a "new type of science," the "science of complexity" based on dynamic processes, to 1811, when Jean-Baptiste-Joseph Fourier won the

French Academy of Sciences prize for his rigorous mathematical description of the propagation of heat in solids. The first form of the new science was thermodynamics, the dynamics of heat properties. The rapid spread of steam-engine technology in Britain generated a new interest in the mechanical effects of heat, and thermodynamics emerged out of a concern with its possibilities of producing mechanical energy. By describing the process whereby heat propagation gradually leads to a homogenization and equilibrium of thermal properties, which involved qualitative changes in the nature of heat and the states of energy, Fourier dealt the first major blow to the Laplacian school dominating European science.

Whereas classical dynamics discussed system modification in static terms of energy conservation, thermodynamics emphasized the active properties of natural systems that involve energy degradation and increased entropy. There were both crucial continuities and discontinuities involved in this change of emphasis: "A physical theory had been created that was every bit as mathematically rigorous as the mechanical laws of motion but that remained completely alien to the Newtonian world. From this time on, mathematics, physics, and Newtonian science ceased to be synonymous" (Prigogine and Stengers, 1984: 104). The difference lies in the nature of the heat engine, as opposed to the mechanical engine, as well as in the discontinuity between life processes and the operations of machines:

> A *mechanical engine* gives back in the form of work the potential energy it has received from the outside world. Both cause and effect are of the same nature and, at least ideally, equivalent. In contrast, the *heat engine* implies the material changes of states, including the transformation of the system's mechanical properties, dilatation, and expansion. The mechanical work produced must be seen as the result of a true process of transformation and not only as a [continuous and unaltered] transmission of movement. (106)

According to thermodynamics, heat can only do work when it descends from higher to lower temperatures. Not all heat, however, can be converted into work without energy loss in some systems, and all energy on earth ultimately degrades into heat. Energy conservation and equivalences, the emphases appropriate to the mechanistic paradigm, are only part of the story of nature; they must be supplemented by another aspect involving the boilers of steam engines, chemical transformations, and life-and-death processes, all of which are bound up with *irreversible changes*. This new theory that substituted irreversible for reversible time, such that one cannot go backward or forward in time without significant transformations in the world, escapes the gravitational influences of the Newtonian paradigm to establish a new outlook on reality: No event or natural state in the present moment is exactly similar to any other event or state in the past or future; change, becoming, and transformation are inherent aspects of life.

Fourier's findings were later consolidated by other physicists and were formulated as the second law of thermodynamics by William Thomson in 1852. Thomson was the first to extend the results of thermodynamics from the limited case of engine technology to a cosmic principle applying to the fundamental dynamics of the universe itself. If the world is indeed a machine, then it is one moving in the direction of increasing waste and disorder. In 1865, Rudolf Clausius characterized the second law of thermodynamics as the "law of entropy" (from the Greek word meaning "transformation") and, with Thomson, emphasized that the energy of the universe is increasingly passing from useful to useless states. The more transformations energy undergoes, the more "disordered" it becomes and the less available it is for further use. The same amount of energy will always exist (the first law of thermodynamics), but the degree of order and availability changes (the second law). While the quantity of energy never varies, the quality of energy, the amount available for useful work, changes irreversibly. If one burns a lump of coal, for example, it is dissipated in the form of heat, smoke, and ashes and thereby becomes unavailable for human use. Its "free" energy becomes degraded into "bound" energy that is disordered.

Any closed system, one that does not exchange heat or energy with its surroundings, tends toward equalization of temperature, pressure, and other physical characteristics. Entropy is a mathematical measure of the disorder and unavailability of energy, the result of irreversible, qualitative changes in the nature of energy. In his book *The Runaway Universe* (1978), Paul Davies drew the pessimistic implications of the cosmic application of the law of entropy: "The unpalatable truth appears to be that the inexorable disintegration of the universe as we know it seems assured, the organization which sustains all ordered activity, from men to galaxies, is slowly but inevitably running down, and may even be overtaken by gravitational collapse into oblivion" (1978: 197). According to this view, the universe is fated to perish through a massive loss of heat and subsequent freezing.

But though some theorists like Clausius and Davies accepted this pessimistic idea, others rejected it. As early as 1931, in the work of E. A. Milne, doubt was cast upon the validity of applying the concept of entropy to the universe as whole. Milne argued that there is no way to assess the change in entropy for the universe as a whole because, hypothetically, it is limitless.[4] In this vein, Toulmin (1982b) has argued that the cosmic interpretation of entropy is a scientific myth based on a disanalogy between heat engines and the universe as a whole. At most, Toulmin claims, entropy implies something about the universe, but no one has established that the universe is a thermally isolated system and hence that the "law" of entropy applies universally. The pessimistic interpretation of entropy is relevant only for closed systems; in open systems there is the possibility of continuous exchange of matter and energy with their surroundings, such that growth and evolution may occur, entropy does not have to accumulate, and systems can become "self-organizing."

With the introduction of evolution and temporality into the conceptual framework of science, the concept of matter itself was revised; matter is "no longer the passive substance described in the mechanistic world view but is associated with spontaneous activity," or self-organization (Prigogine and Stengers, 1984: 9). For Prigogine, whom Jantsch (1980: v) refers to as the "catalyst of the self-organization paradigm," open systems exist in conditions that are "far from equilibrium." Unlike stable systems with little fluctuation (such as are described by the general laws of physics), open systems are dynamic, spontaneous, changing, and evolving. Out of these fluctuating processes arise highly complex forms of order, which mutate into ever-new patterns and conditions. Accordingly, Prigogine and Stengers distinguish between reversible "equilibrium structures" and irreversible "nonequilibrium" or "dissipative structures," which, as in chemical processes, are self-organizing and have abrupt, unexpected, and simultaneous changes in behavior that lead to new forms of order.

The active concept of matter has spawned the notion of "self-organization" or "autopoiesis," an element of a new scientific paradigm—emerging in fields such as physics, chemistry, and evolutionary biology—that sharply distinguishes life processes from mechanistic operations. On the postmodern paradigm, life forms are viewed not as machines concerned with the uniform manufacture of a product but, rather, as dynamic processes that are self-organizing, as living systems that spontaneously and autonomously create their own conditions of self-renewal through evolving complexity (see Varela, 1979; Jantsch, 1980; and Kauffman, 1991, 1995). For Jantsch, the principles of self-organization, dissipative structures, nonequilibrium, and complex order through fluctuation provide "an emergent unifying paradigm" of life that seeks to grasp the interconnection of natural dynamics on all levels, from cosmic gases to human intelligence, without annulling their differences. With the development of this new paradigm, Jantsch argues, "science is about to recognize these principles as general laws of the dynamics of nature" (1980: 18) and supersede the mechanistic paradigm. As an example of this recognition British scientist James Lovelock formulated the "Gaia hypothesis" (1987) in the mid-1960s, which emphasizes that "Gaia," the Greek word for "earth," is a vast, living superorganism with complex self-regulating dynamics that adjust toward optimum conditions of self-reproduction.

Of course, Darwin's theory, with its emphasis on "descent by modification," played a major role in changing the static scientific conception of reality, overturning the essentialist view of species as "natural types" accepted both by religion (creationism) and science (natural theology). But theorists of the self-organization paradigm reject Darwinian theory for its mechanistic reduction of life to forces of blind struggle and chance and for its deterministic negation of the dynamic interplay between organism and environment, making living organisms nothing but the result of external, random influences. Lovelock (1987), for example, emphasizes that biospheric life forms actively contribute to the constitution of the atmosphere, soil, and other components of the environment. As Stuart Kauffman puts it:

> Darwin could not have suspected the existence of self-organization, a recently discovered, innate property of some complex systems. It is possible that biological order reflects in part a spontaneous order on which selection has acted. Selection has molded, but was not compelled to invent, the narrative coherence of ontogeny, or biological development. Indeed, the capacity to evolve and adapt may itself be an achievement of evolution. (1991: 78)

According to the new paradigm, natural selection remains an important explanatory principle for the emergence and development of life, but it must be *complemented* with the new perspective of active matter: "We may have begun to understand evolution as the marriage of selection and self-organization" (Kauffman, 1991: 78). Indeed, for Jantsch, "the ultimate principle of evolution does not seem to be adaptation, but transformation and the creative diversification of evolution" (1980: 144).

Entropy and Information Theory: The Dialectic of Noise and Meaning

> Two dangers threaten the world: order and disorder.
> —Paul Valéry

> The highest destiny of mathematics [is] the discovery of order among disorder.
> —Norbert Wiener

The concept of entropy is a key link between modern and postmodern science, and its saliency in both discourses suggests that we are in a borderland between the modern and the postmodern. The concept has seeped into the postmodern zeitgeist, linking thermodynamics and physics with the life sciences and ecology, with information theory and communication, and with postmodern literature and social theory, since entropy plays a key role in the works of Pynchon, Baudrillard, and others. Despite differences between the deployment of the concept in different fields and thinkers, it has been associated with postmodern notions of disorder and indeterminacy that undermine mechanistic and deterministic conceptions of modern science. This occurs in two ways: first, by introducing time and becoming into static and deterministic modern theories; and second, by challenging the modern dogma of progress and the reckless optimism that science and technology are the panaceas for all problems in a world to be bent to human will. Introducing the discordant notes of resource limits and energy degradation into the modern symphony of progress, the concept of entropy highlights the importance of indeterminacy and probability for postmodern science, has important implications for ecology, and is articulated with a general cultural mood of decline and crisis.

The anti-deterministic motifs of the entropy framework were further advanced by the new sciences of cybernetics, systems theory, and information theory, all of which developed the notion of entropy in a new communications paradigm. In this paradigm, it is argued that all complex systems can be seen as systems of information and communication governed by an interplay of order and disorder. Any system or relation whereby a message of some kind is sent from one person or place to another is seen to be an information system governed by universal laws. DNA, for example, is a coded system that communicates information required to create specific life forms. According to the new paradigm, nature is interpreted not only as matter and energy, as modern science believed, but also as information. The proper interpretation of nature, therefore, requires not only knowledge of physics, chemistry, and biology but also a theory of communication.[5]

Cybernetics, developed by MIT scientist Norbert Wiener and others, is part of the new information theory paradigm that belongs to the postindustrial, communications revolution just as entropy derives from the Industrial Revolution of the machine age. "It became clear to me almost at the very beginning," Wiener wrote in his memoirs, "that these new concepts of communication and control involved a new interpretation of man, of man's knowledge of the universe, and of society" (1966: 325). During World War II, Wiener and other scientists undertook an intense study of the problem of making radar screens as precise as possible in order to maximize the ability of pilots to predict the movement of enemy planes and shoot them down. Facing the task of separating orderly messages from the "noise" of electrical interference that corrupts the integrity of the message, Wiener saw that the equations describing entropy could be used to model information and communication systems just as well as they could describe engines and physical systems.

Wiener rejected formalistic theories that sought completeness and certainty within a deterministic logic to proclaim the ubiquity of random behavior, chance forces, and entropy, but he also sought ways of reducing entropy and increasing order. This was the task of cybernetics. The word "cybernetics" is derived from the Greek word meaning "steersman" and therefore connoting order, stability, and proper functioning; cybernetics is the science of maintaining order in any system, whether physical, chemical, electrical, economic, or informational, as a thermostat maintains temperature stability in a room in response to constant fluctuations. The human body, for example, is an ordered system that works according to the principle of homeostasis that strives to preserve health and order. Illness represents an entropic breakdown of physical order, and healing is a cybernetic response. Similarly, according to Gaia theory, the earth labors to ward off blows to its physical integrity to reestablish ecological balance.

With his seminal publications in 1948, Claude Shannon, researcher at Bell Telephone Laboratories, formalized information theory. According to Shannon, the laws of information hold good for any system in which a message is sent from one place or person to another. For such information systems theorists, increase of entropy represents a progressive loss of information, whether it be the weakening of radio pulses or verbal communication distortion. The higher the entropy of a system, therefore, the less information about it is possible. Or, the more entropy, the more random the messages and the greater the possibility of misunderstanding and distortion. For communication theorists influenced by Gregory Bateson, however, the model of information developed by Shannon, Weaver, Wiener, and other pioneers of information theory required supplementation with more sophisticated conceptions of organism, environment, and communicative interaction with the natural and social world (see the studies in Wilder and Weakland, 1981). Information theory often remained too immersed in the premises of scientific determinism and reductionism, leaving out the contextual pragmatics of communication including feedback and the ways that self-regulating systems—ranging from the individual to the environment—are involved in patterns of communicative action. Such an approach overcame Cartesian dualism and scientific determinism by placing the individual in an interactive relationship with an environment in which processes of information, meaning, and communication were seen to characterize both the natural world and the social world. This transdisciplinary perspective made possible a postmodern science that would help integrate the natural and social sciences.

Extending the metaphors of communication to human beings and nature not only makes possible new transdisciplinary perspectives, but also helps enable a new understanding of the universe in which patterns of order and structure are seen in apparent disorder, thus helping prepare the way for the chaos theory that we discuss in a following section. Such perspectives also pose a strong challenge to the entropic interpretation of the universe. As Jeremy Campbell states:

> Information theory shows that there are good reasons why the forces of antichance are as universal as the forces of chance, even though entropy has been presented as the overwhelmingly more powerful principle. The proper metaphor for the life process may not be a pair of rolling dice or a spinning roulette wheel, but the sentences of a language, conveying information that is partly predictable and partly unpredictable. (1982: 12)

Thus, whereas entropy theory in thermodynamics may in some cases be an overreaction to the modernist belief in a perfectly structured and harmonious and deterministic universe, information theory seeks a dialectical interpretation in a world believed to be governed by forces of chance and nonchance, order and entropy. According to Chomsky (1975), there are innate, universal principles of grammar within the brain that allow communicative competence, invariably structuring noise into meaning and thereby countering information entropy. Similarly, although evolutionists often emphasize that natural selection is nonteleological and random, advantageous traits are selected for species reproduction, giving life a specific structure, and natural selection therefore can be seen as an anti-entropic force. Moreover, as we have seen above, evolutionary biologists and theorists of the new science believe that life has innate tendencies to evolve in specific ways, which are not solely determined by the environmental influences of natural selection and which can be seen as constant and form giving. Prigogine argues for the existence of a force that pushes living organisms into states of ever-greater complexity, a principle "that is totally against the classical thermodynamic view that information must always degrade. It is, if you will, something profoundly optimistic" (quoted in Campbell, 1982: 101).

From the postmodern perspective, modern science enacts an illicit imperialism via the application of inorganic models to the entire universe. For postmodern science, the principles of life and evolution hold sway, replacing mechanism, with the result that the same methods, models, and concepts used to describe inorganic systems do not hold for organic systems. While inorganic systems are subject to entropy and ultimately demise, organic systems are open to countervailing and counterentropic forces. Life according to this view is self-perpetuating and self-organizing, seeking to expand, develop, and unfold, often in surprising and novel ways. Indeed, novelty and creativity are anomalies for the mechanistic worldview, whereas a more organic postmodern view sees emergent orders developing out of previous elements, positing a multidimensional universe with a multiplicity of self-organizing systems within it, including consciousness, the human immune system, nature, and the cosmos itself (Varela, 1979; Jantsch, 1980).

Certainly, there is a tension between the principle of evolution and entropy since the former emphasizes life as an expanding, orderly, and increasingly complex process and the latter posits energy systems as running down, becoming ever more disorderly and simplified, but this disparity is less a matter of contradiction than complementarity. As some theorists argue, both principles are operating simultaneously, and so we must think in terms of a both/and rather than an either/or logic. According to Harvard astronomer David Layzer (1975), there is not, in Sir Arthur Eddington's phrase, only one "arrow of time," the irreversible arrow of entropy by which energy is transformed into less orderly and available form as time passes; there are also arrows of cosmic expansion, evolution, and history, which involve movement toward greater order and complexity.

For Prigogine, Jantsch, and others, the tendency toward greater forms of order and complexity is a property of open rather than closed systems. Unlike closed systems, open systems exchange matter, energy, and information with their surroundings and hence are far from states of equilibrium that characterize closed systems. The principles of entropy and evolution, then, are relative to different kinds of systems, and the notion of complementarity allows postmodern science to deal with such tensions better than modern science with its rigid logic of noncontradiction.[6]

THE MULTIPLE WORLDS OF RELATIVITY THEORY AND QUANTUM MECHANICS

> It had not been possible to see what could be wrong with the fundamental concepts like matter, space, time and causality that had been so extremely successful in the history of science. Only experimental research itself, carried out with all the refined equipment that technical science could offer, and its mathematical interpretation, proved the basis for a critical analysis—or, one may say, enforced the critical analysis—of these concepts and finally resulted in the dissolution of the rigid frame.
>
> —WERNER HEISENBERG

> The fact that relativity and quantum [theory] together overturned the Newtonian physics shows the danger of complacency about our worldview. It shows that we constantly must look at our worldview as provisional, as exploratory, and to inquire. We must have a worldview, but we must not make it an absolute thing that leaves on room for inquiry and change. We must avoid dogmatism.
>
> —DAVID BOHM

In the Newtonian framework, space and time were seen as absolute and separate realities; correct measurements of each would never vary, neither affected the other, and both were independent and outside of human consciousness. The speed of light could be measured from a position at absolute rest, in the ether of the cosmos. Whereas thermodynamics disturbed this absolute order by introducing irreversible time, Einstein sent it reeling through his rejection of space and time as rigid or separate constructs.

Einstein's theories boldly depart from commonsense notions of reality to create a strange universe of curved space, black holes, simultaneity, and a fourth dimension of reality based on a space–time continuum. Above all, Einstein introduced relativity into the scientific framework and broke down the firm divide between subject and object. Because Einstein believed that there were constants in the universe, most notably the speed of light, the term "relativity theory" is misleading; a better term might be "perspectival theory" since he argues that observations of the physical world vary according to the speed at which an observer is traveling.

For Einstein, there is no inertial point in the universe that is absolute, constant, or privileged; rather, everything is moving relative to everything else. As formulated in his famous thought experiments involving an observer looking out the window of a moving train, the perception of time is relative to the speed of movement, such that the faster the observer travels relative to any clock external to the train, the slower the movement of time recorded by that clock for the observer. Past, present, and future and causal sequence, as well as the shape and color of objects, vary according to the speed of the observer. As one approaches the speed of light, space and time are not separate coordinates but, rather, become intertwined in a fourth dimension, a space–time continuum, where space contracts as time dilates and space dilates as time contracts.

Einstein can be seen as a transitional figure toward a new postmodern consciousness. He broke with the mechanistic conception that space is composed of separate particles to see it as a unified field of strong and weak regions, thereby anticipating the holistic logic of quantum theory. While not abandoning norms of objective measurement, he introduced elements of subjectivity and relativity into the scientific framework, such that the "objective world" (including the solid and invariant "primary qualities" described by modern science and philosophy) changed size, form, color, and sequentiality as an observer changed speed and direction relative to it. By introducing the notions of relativity and energy (seeing energy and matter as ultimately the same thing), Einstein shattered the unified stabilities of the Newtonian universe. As Matson puts it:

> Not "matter" but "energy" was now the basic dictum of science; no reliance could henceforth be placed upon actions-at-a-distance, nor upon mechanical conceptions of force or of quasi-solid ethers, nor upon the integrity and stability of Space and Time as familiarly conceived—nor, indeed, upon the bedrock axioms of Euclidean geometry. (1966: 120)

According to Bohm, however, Einstein did not make a complete break with atomism because he rejected the theory of nonlocal connection, which holds that things can be linked at any distance without any apparent connecting force: "Relativity theory and quantum physics agree in suggesting unbroken wholeness, although they disagree on everything else. That is, relativity requires strict continuity, strict determinism, and strict locality, while quantum mechanics requires just the opposite—discontinuity, indeterminism, and nonlocality" (1988: 65).

Einstein played a major role both in developing quantum theory and criticizing some versions of it. While (special) relativity theory broke with the classical theory of objectivity, Einstein shared the Newtonian ambition to produce a complete description of nature (the general theory of relativity) and rejected theories of indeterminacy, arguing against Niels Bohr that "God does not play dice with the universe." Einstein believed that there was an unknown property, a hidden variable, that, once discovered, would restore determinacy to physics. Moreover, Einstein was repelled by the implication of his theories, which pointed to a changing, expanding universe, and altered his equations to make them match his faith in a static cosmos. As Brian Swimme (1988) notes, when Edwin Hubble provided the empirical support for an expanding universe, Einstein repudiated his alteration of the field equations as the "biggest blunder of my life."

Despite the attempt of this scientific giant to plug the holes in the dam of determinism, the forces of indeterminacy and chaos were breaking through in unstoppable waves of change unleashed by quantum mechanics. Although quantum mechanics acknowledges the validity of Newtonian laws of motion and gravity in the macroscopic world of stars, planets, and motion visible to the eye and telescope, it explores a new subatomic world where indeterminacy prevails and the behavior of matter fails to conform to the familiar laws of the large-scale world. Quantum mechanics followed Einstein in producing counterintuitive and paradoxical scientific notions, but these were developed in an altogether different framework. The radicality of quantum mechanics leads Smith (1982) and Bohm (1988) to speak of it as a "postmodern physics" that is significantly different from the classical physics informed by a mechanistic logic. Whereas Einstein's vision of the world is between the modern and postmodern, Bohm and others cross over into the postmodern, carrying out a decisive break with mechanism, determinism, and absolutism and championing a postmodern science that is more organic, indeterministic, probabilistic, and multiperspectival than the previous modern paradigm.

On Matson's account, quantum physics is the culmination of a postmodern movement against mechanistic science, but he finds important changes earlier in the 19th century with the thermodynamic theory of probability and with James Clerk Maxwell's equations in electromagnetic theory. These formulas "clearly repudiated the entire foundation of Newtonian mechanics" (Matson, 1966: 115) by admitting uncertainty in our ability to know the true nature of electricity and by demonstrating the inadequacy of the Newtonian principle of action at a distance to explain interaction between fields of electromagnetic force. In what is probably the first reference to "postmodern science," Matson states: "The success of the electromagnetic theory was significant not only as the first breach in the great wall of mechanism, but also as a premonitory clue into what might be termed the 'postmodern' conception of the nature of scientific knowledge" (115). Insights into the limitations of Newtonian science were advanced further at the end of the 19th century by Ernst Mach, Heinrich Hertz, and Karl Pearson, and there was also a movement to reject the "ether theory" that, paradoxically, provided immaterial foundations for mechanism. Einstein's theory of relativity added even more cumulative weight to these changes. But despite the originality of these developments, Matson argues, all "may be fairly regarded as only a prologue to the real revolution in twentieth-century science: that of quantum physics" (122).

Established as a coherent theory by 1927, the quantum revolution began in 1900, the year Max Planck reluctantly presented his paper on energy radiation, whose implications went far beyond the Newtonian paradigm. Where Aristotelian and Newtonian theories had long held that matter moves in a smooth and continuous way, Planck discovered that atomic oscillators emit energy in abrupt, discontinuous bursts, involving packets of energy he called "quanta." This assertion of discontinuity at the fundamental level of reality, Matson argues, "rudely violated the cardinal faith of modern science in the uniformity and continuity in nature—specifically, in its belief that the evolution of every self-contained physical system is constituted by a continuous chain of causally related events" (1966: 122).

In the subatomic world, the most basic components of reality—which a standard interpretation of quantum mechanics understands in terms of relations rather than as independent entities—cannot be isolated, precisely identified or predicted, or grasped "as they really are." A fundamental insight of quantum mechanics is that in the process of perceiving and analyzing subatomic particles, the scientist unavoidably influences their behavior through the use of measuring instruments (e.g., gamma ray microscopes that alter the behavior of particles). As Niels Bohr emphasized, the scientist cannot identify both the position and velocity of a subatomic particle but must choose which aspect to measure in the most accurate manner. In a strict inverse relation, the more one focuses on one characteristic, the less one can discern about the other. Thus, precisely the conditions assumed by classical "physics" as prerequisites for exact prediction—for example, the stability of the objective world and the neutrality of the observing subject—are unattainable in microphysics.

Werner Heisenberg argued that there is a strong element of "uncertainty" involved in quantum physics that cannot be eliminated. Like Newtonian science, quantum mechanics attempts to predict the behavior of matter, but it understands that the element of indeterminacy in the subatomic world prevents exact understanding and that the predictions it makes involve only probabilities, statistical regularities, and not certainties. At best, scientists can predict only what an ensemble of phenomena will do and not the behavior of individual elements. Consequently, as a radically new way of modeling reality, the philosophical implications of quantum mechanics are profound, and all of them register in postmodern theory. Quantum mechanics challenges modern representational epistemology on all possible grounds by theorizing a realm of being in which the perceiving subject cannot adequately grasp the objects of perception. Against the certainties of the Newtonian world, quantum mechanics discovered that the most fundamental level of reality is unrepresentable in any exact manner. Indeed, since no one has ever actually seen an electron without the aid of measuring devices, quantum scientists admit that they are modeling reality rather than actually mirroring it and they understand our conception of the "objective world" to be a cognitive construction.

This leads physicists such as Heisenberg to argue that we should not even try to visualize what particles are like. We should "abandon all attempts to construct perceptual models of atomic processes" (Heisenberg, 1971: 76). Physicist P. W. Bridgman suggests that "the structure of nature may eventually be such that our processes of thought do not correspond to it sufficiently to permit us to think about it at all. . . . The world fades out and eludes us. . . . We have reached the limit of the vision of the great pioneers of science, the vision, namely, that we live in a sympathetic world in that it is comprehensible by our minds" (quoted in Smith, 1982: 8). Similarly, the "Copenhagen interpretation" of quantum mechanics advanced by Bohr operates not with a correspondence theory of truth that tries to match theory with reality but, rather, with a pragmatic theory of truth that seeks results in experimental situations. Bohr believed that indeterminacy was fundamental to nature, but he did not think it suggested any inadequacy in the nature of scientific knowledge. He argued that science should learn to accept indeterminacy without trying to explain it and should base its theories on what *can* be observed and measured experimentally.

Where one cannot precisely determine relations or predict future events, the mechanistic concept of causality is unhinged. As Bohr wrote, quantum mechanics entails "the necessity of a final renunciation of the classical ideal of causality and a radical revision of our attitude toward the problem of physical reality" (1958: 60). Fundamental to this theoretical revision is the rethinking of another major tenet of representational epistemology. Quantum mechanics problematizes the very distinction between subject and the object and undermines spectator epistemology by implicating the observer in the behavior of the observed. Where the subject is both observer and participant in the perceptual process, absolute scientific detachment becomes a chimera. As Heisenberg put it: "What we observe is not nature itself, but nature exposed to our method of questioning" (1958: 58). In quantum mechanics, both the subject and object are seen as active rather than passive and the human perception and understanding of the world "external" to it are inevitably mediated by assumptions, biases, technologies, and practices. This hermeneutic approach to observation would be developed elaborately by phenomenologists, Marxists, feminists, pragmatists, and philosophers of science, becoming a central aspect of postmodern science (see Toulmin, 1982a, 1982b) and a key point of contestation in the "science wars" that erupted in the mid-1990s.

Quantum mechanics has also challenged the Aristotelian logic of identity that informs all Western metaphysics by showing that light is both particle and wave, in that it manifests different characteristics according to the kind of experiments designed to analyze it.[7] Consequently, the "many worlds interpretation" of quantum mechanics holds that there are different aspects of reality that come into being due to the experimental devices one employs. Thus, quantum mechanics problematizes the modernist search for a single, universally valid language of reality. According to Bohr's theory of complementarity, reality is irreducibly plural and complex, and no single theoretical description can exhaust it. Rather, various languages and perspectives are needed—mutually exclusive if pursued monologically, complementary if taken together.

CHAOS THEORY: INDETERMINACY AND THE LIMITS OF PREDICTION

> Consequently: he who wants to have right and wrong,
> Order without disorder,
> Does not understand the principles
> Of heaven and earth.
> He does not know how
> Things hang together.
>
> —Chuang Tzu

> Where chaos begins, classical science stops.
>
> —James Gleick

After the development of quantum mechanics in the 1920s, the seeds of uncertainty continued to bear fruit in the 1930s and 1940s. In his undecidability theorem, first published in 1931, Kurt Gödel argued that no formal system of arithmetic can be complete, that no axiomization of arithmetic can entail all its truths, and that the consistency of a formal system cannot be demonstrated within the logic of the system. This result shattered the dream, first advanced by Frege in 1879 and continued in the 20th century by Whitehead, Russell, Hilbert, and others, that mathematics can be completely and consistently formalized in one system through axioms, deduction, and proofs.[8] If a proposition and its negation alternatively can be taken as axioms, however, there is no guarantee that a deduction will not lead to a contradiction and inconsistency. Because mathematics is the formal language of science and the standard of rational knowledge and certainty, Gödel's theorems had disturbing implications for the ideals of formalization and rigorous science.

These implications were embraced and developed by chaos theorists in the 1970s. Some writers see the most recent assault on modern scientific theory, chaos theory, as the third great scientific revolution of this century. As

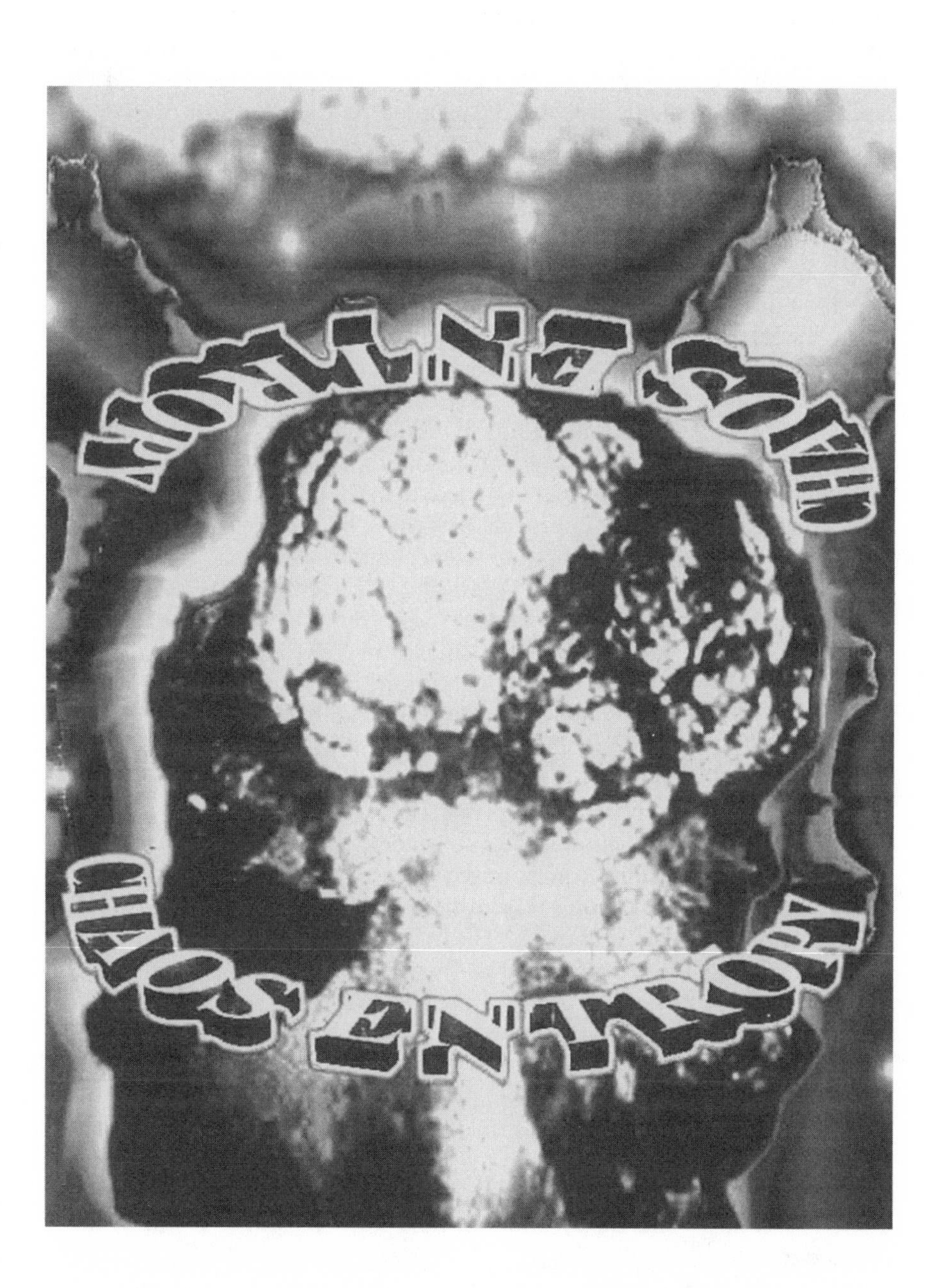
CHAOS ENTROPY
CHAOS ENTROPY

physicist Joseph Ford puts it: "Relativity eliminated the Newtonian illusion of absolute time and space; quantum theory eliminated the Newtonian dream of a controllable measurement process; and chaos eliminates the Laplacian fantasy of deterministic probability" (quoted in Gleick, 1987: 6). Though chaos theory is a rupture from the Newtonian paradigm, it is perhaps an exaggeration to speak of it as a third scientific revolution, given its continuities with quantum mechanics. Yet many of the tenets that quantum mechanics applies to the microscopic realm, chaos theory extends to the macroscopic realm and to physical processes in general.

Despite the conceptual revolution achieved by quantum mechanics, chaos theory had to fight its own battles for legitimacy. In *Chaos* (1987), James Gleick describes how this new view of reality, initially met with scorn and rejection, has become increasingly influential. Once pariahs, chaos theorists are now well ensconced within academic bureaucracies; once obscure and underground, chaos journals and conferences now proliferate. Chaos theory has challenged long-standing barriers between disciplines, thus creating a new interdisciplinary approach, breaching boundaries between physics and biology and between the natural sciences and life sciences.

According to Peterson's (1993) genealogy, the term "chaos" was first applied to mathematics in 1975 by James Yourke, but it has a much older pedigree. Interestingly, Peterson roots this lineage in Newtonian mathematics itself, which provided the deterministic foil to chaos theory. Newton's calculations, Peterson argues, employed "two-faced equations [that] have hidden within them not only the rare instances of order that Newton so brilliantly illuminated but also the darkness of ubiquitous chaos" (1993: 97). Newton selected areas where total precision was possible, and his equations were valid for these contexts, but he also generated tools to analyze irregularity and nonpredictability, which posed special problems. The suppressed chaos within Newton's calculus was brought out first in the late 19th and early 20th centuries by Henri Poincaré. His innovative work in differential equations led him to the conclusion that there was room for unpredictability within deterministic systems. After correcting an error in a prize-winning paper, he changed his conclusions about the stability of the solar system and argued in a revised paper that geometric surfaces reveal complexity, not stability. In his 1903 essay "Science and Method," he gave a classical definition of what has become known as chaos theory: "It may happen that small differences in the initial condition produce very great ones in the final phenomena. A small error in the former will produce an enormous error in the latter. Prediction becomes impossible, and we have the fortuitous phenomenon" (quoted in Peterson, 1993: 167).

Poincaré did not overthrow Newton's laws of motion, but he uncovered new subtleties and complexities, previously hidden from view, that were worked out over the next decades by a handful of mathematicians, astronomers, and scientists who continued his unorthodox approach. Because ignorance of a system at the level of initial conditions will make future predictions of its behavior impossible, Laplace's dream becoming nightmare, the fantasy of fully known present conditions as the basis for infallible retrospective and prospective knowledge founders on the reef of quantum indeterminacy at the starting point of reflection. The kind of "chaos" that results from sensitive dependence on initial conditions is easily demonstrated by hitting two balls at only slightly different angles on a bumper table and following their very different ricochet trajectories. Complex systems' sensitivity to initial conditions is vividly described in Edward Lorenz's concept of the "butterfly effect": The flapping of a butterfly's wings in Tokyo theoretically can alter weather patterns in New York, which helps to explain the impossibility of accurate long-range forecasting—as meteorologists often discover to their chagrin.

Like quantum mechanics, chaos theory is a dynamic view of reality, one that understands the behavior of matter to be often complex and unpredictable. Chaos theory applies to any system with unpredictability, whether it be the stock market, populations, chemical reaction rates, the movement of ants, brain waves, or cardiac rhythms. Using sophisticated equations worked out by computers, chaos theory attempts to construct mathematical models of these systems in order to provide insight into how they operate.[9] For chaos theorists, complex systems have an underlying order, and conversely, even the most simple deterministic systems can be highly complex and unpredictable in their operation.

Thus, through the impact of quantum mechanics and chaos theory, scientists are beginning to revise their vision of the world in terms of a heterogeneous system of forces that interact in complex, random, and irregular ways. Important aspects of postmodern science, therefore, are that it complexifies, diversifies, pluralizes, and temporalizes the old views and that it is considerably more open to change, becoming, indeterminacy, and randomness than science operating within the Newtonian paradigm. In the postmodern paradigm, however, the older views of reality are not necessarily demolished; rather, they are recontextualized in complementary relations with the new views. Prigogine, Kauffman, Jantsch, and others attempt to construct a synthetic theory where reversibility and irreversibility, chance and necessity, dynamics and thermodynamics, entropy and evolution, natural selection and self-organization, Newtonian and post-Newtonian outlooks are synthesized but in which the old paradigm applies only to certain closed subsystems of nature and in which chance, irregularities, and the dynamic movement of open systems are the rule rather than the exception.[10]

In this vein, it should be emphasized that the "chaos" scientists are finding everywhere today is not perceived as the logical antithesis of order, for that is the type of rigid, binaristic thinking that postmodern science attempts to displace. As Prigogine and Stengers hold, echoing the deconstructive sentiments of postmodern social theory: "The epoch of certainties and absolute oppositions is over" (1984: 299). Consequently, new forms of order emerge out of chaos, considerably more complex, intricate, and irregular in nature than the forms of order that previously were thought to exist. Thus, chaos theory spawned a related branch of study known as "complexity theory."[11] Benoit Mandelbrot discovered, for instance, that there was an unexpected degree of order within the most disorderly realms of data. Such "irregular regularities," as we might term them, are evident in the famous Lorenz attractor, whose swirling lines of aleatory patterns resemble the mysterious eyes of an owl. While the name "chaos" suggests wild and erratic behavior to the layperson, scientists understand chaos in terms of complex order and "limited predictability."

A distinction can be drawn, therefore, between sheer "noise" and chaos, with the latter yielding complex patterns. But as a kind of order emerges out of chaos, so chaos underlies order, discovered by chaos theorists in the simplest of physical systems. Such a scientific breakthrough is a matter of revisioning, a new way of seeing. Previously chaos and complexity were seen as negative, as limits to good science, as noise to filter out or to overcome. Modern science previously looked for simple and repeatable regularities, laws, and causal mechanisms. But comprehending complexity and chaos created not only a new optic, but new ways of conceptualizing and interpreting natural processes. Examining indeterminacies, seeming randomness, chance, and disorder reveals new forms of order, as well as how disorder and order could coexist. Thus, precisely the focus on chaos and complexity enabled scientists to see hitherto unknown patterns, new structures, new forms of order, as well as the ways that disorder contained implicit order. Accordingly, the turn toward chaos and complexity theory refocused scientists on order and disorder, structure and dissipation, regularities and nonregularities, patterns and chaos. Some saw deeper patterns and structures behind the disorder, while others saw the disorder, the chaos, producing new patterns of order: order out of chaos. Thus, though deterministic systems contain the seeds of chaos, uncertainty, and nonpredictability, that which is uncertain and nondeterministic does not necessarily stand outside of a field of quantification and predictability; it does, however, in many cases involve a logic of statistical probability rather than of absolute certainty.[12]

The concepts of chaos and entropy, moreover, should not be seen as antithetical; although entropy, the passing of a system from an ordered to a disordered state, suggests chaos, "chaos" in actuality can give rise to new and complex forms of order. If, as Hawking argues, the universe will continue to expand forever, then chaos—form-giving dynamic change—rather than entropy becomes the principle for the universe as a whole. Indeed, complexity theorists like Stuart Kauffman (1991, 1995) of the Sante Fe Institute see the "self-organization" of complex systems as a corrective to the principle of entropy, as an evolutionary force driving organisms toward increasingly complex states. In dynamic states of equilibrium, we find a constant breakdown of order and dissolution of structure *and* the emergence of order and complexity. Ultimately, entropy and chaos can be seen as discrete moments in a social and cosmic dialectic of disorder–order, understood differently according to one's perspective or contextual interpretation. The disordering of an ordered state could allow new forms of order to emerge in more differentiated ways. Furthermore, simplistic oppositions such as "Disorder is bad and order is good" should be eschewed, since disorder can bring greater complexity, creativity, and ultimately order itself, and not all forms of order are valuable or good. Although in open systems the cycle of disorder and order seems potentially endless, the principle of entropy suggests that in closed systems this dialectic disintegrates into nothingness or uselessness.

Thus, values that modern thought takes as opposites polarized in rigid binary distinctions, such as between fact and value, determinism and chaos, and subjectivity and objectivity, implode in postmodern thought, but rarely with the effect of nullifying the relative truth of modern theories and concepts or totally effacing the differences. In postmodern science, as in postmodern thought in general, these "oppositions" can be rearticulated in a looser way, preserving differences, or they can be occluded. The distinctions between fact and value, or between objectivity and subjectivity, for example, can be rethought in a way that reconstructs and salvages the concept of truth as a hermeneutic construct to be verified pragmatically within a scientific community, or it can be dissolved in the relativism of Feyerabend or Baudrillard. Hawking, to give another example of a reconstructive postmodern approach, tries to maintain strong concepts of truth, objectivity, and prediction *within* the context of uncertainty. The task of science, as he redefines it, is "the discovery of laws that will enable us to predict events *up to the limits* set by the uncertainty principle" (1988: 173, our emphasis). Hawking's theory is a good example of a science on the *borderline* between the modern and the postmodern, seeking to reconcile the general theory of relativity with quantum mechanics in a new "quantum theory of gravity." Despite his acceptance of the postmodern themes of contingency and uncertainty, Hawking continues the modern totalizing search for "a complete unified theory that will describe everything in the universe" (12).[13]

Indeed, many theorists use chaos and complexity theory to reconstruct modern science and its concepts of mechanism, determinism, law, causality, and prediction, while others call for a break with modern science and the development of a new postmodern science—as we see in the next section. Thus, some proponents of chaos and complexity theory use these discourses to develop and strengthen modern science itself, replacing Newtonian paradigms and models with ones tempered by relativity theory, quantum mechanics, and the more recent new sciences like chaos and complexity theory. Others, however, believe that a more radical break with modern science and paradigms is needed and call for new postmodern paradigms, although, as we shall see, the postmodern turn in science is as contested and variegated as it is in other fields.

TOWARD A POSTMODERN SCIENCE

> *We no longer live in the "modern" world.* The "modern" world is now a thing of the past. Our own natural science today is no longer "modern" science. Instead . . . it is rapidly engaged in becoming a "postmodern" science: the science of the "postmodern" world, of "postnationalist" politics and "postindustrial" society—the world that has not yet discovered how to define itself in terms of what it *is*, but only in terms of what it has *just-now-ceased to be.*
>
> —STEPHEN TOULMIN

As the implications and effects of the modern paradigm and worldview have become all too clear, many scientists are now breaking with the models of the past, frankly disowning their Baconian–Cartesian–Newtonian heritage, rejecting mechanism in favor of organism, seeing science as a construct in which theories, metaphors, instruments, and scientific practice help constitute its object, thus abandoning notions that science is a mirror of nature that supplies purely objective truth and calling for a postmodern turn in science based on principles of indeterminacy, discontinuity, chaos, complexity, and entropy. Some proponents of postmodern science attempt to restore a qualitative, normative, and even metaphysical and spiritual dimension to a scientific tradition whose *idées fixes* are cold objectivity, adherence to the "facts," reduction of living things to machine-like entities, and the mastery of nature. These advocates of postmodern science draw the conclusion that a new paradigm is necessary, one that is more philosophically sophisticated, scientifically complex, ethically sensitive, and ecologically sane (see the studies in Griffin, 1988a, 1988b, and Sassower, 1995).

From a diverse set of fields and sources, something like a postmodern paradigm in the sciences is emerging, but it has not yet been "normalized." Rather, it remains an alternative discourse, a counterhegemonic mode of thought that still has not eclipsed old modern assumptions. Moreover, there is no agreed-upon concept of postmodern science and various theorists present it differently: Some argue for a more holistic and ecological new postmodern science, while others merely use postmodern concepts to deconstruct and destroy, sophists of the contemporary moment (as we discuss later). On the whole, champions of postmodern discourse rarely analyze the continuities between the premises, practices, or values they are accepting and those they are rejecting or ask whether postmodern conceptions might not be better understood as a variant on modernity and modern theory (see Chapter 1). Though advocates of postmodern science are far from rigorous in these matters, they generally attempt to delineate what they understand to be modern, why they are rejecting it, and what they are proposing as alternatives.

For many theorists, the claim to a "postmodern science" does not mean that there are no fundamental continuities with "modern science." Much postmodern science retains the optimism of modern science, and even a sense of historical progress as exemplified in the call for a new "postmodernity." Unlike some other forms of postmodernism, many forms of postmodern science also retain the activist and emancipatory intent of modern science while adopting a normative stance that criticizes the modes in which this intent has been cast. Like its modern counterpart, many forms of postmodern science are committed to rational investigation and to norms of truth, while qualifying the kinds of truth and objectivity possible in science. Indeed, postmodern science often uses the norm of scientific evidence itself in pointing to the environmental catastrophes of modern science and industry. And much postmodern science continues to be oriented toward quantitative knowledge, experiment, prediction, and control.

Toulmin (1982b) characterizes postmodern science, of which he finds Gregory Bateson's work to be a model, as integrative and nondualist, overcoming rigid oppositions between both fact and value and humanity and nature. Like Griffin, Bohm, Oelschlaeger, and others, Toulmin affirms a postmodern science that renounces the positivist dichotomy between facts and values and between spectator and world, arguing:

> Instead of viewing the world of nature as onlookers from outside, we now have to understand how our own human life and activities operate as elements within the world of nature. So we must develop a more coordinated view of the world, embracing both the world of nature and the world of humanity—a view capable of integrating, not merely aggregating our scientific understanding, and capable of doing so *with practice in view.* (1982b: 256)[14]

The price to be paid for the change to an integrative framework, Toulmin insists, is that scientists can no longer enjoy the luxury of seeing themselves as amoral, detached specialists of knowledge and must take their place in the world of values and moral responsibilities, accepting the burden of the difficulties in reconciling the *vita contemplativa* with the *vita activa.* This price need *not* include the loss of the norms of truth, objectivity, prediction, and science itself, a conflation of science with myth, ideology, propaganda, or politics. Ecology, for example, which applies rigorous analytic methods toward an understanding of living systems in dynamic interaction, provides a model of "a postmodern consciousness that is neither reductionistic nor antirational" (Ferré, 1988: 94). As noted by Habermas (1968), the modern notion of objectivity masked the human interest in control and domination. The reconstructed, postmodern notion of objectivity allows science to overcome the myth of objectivism and to be guided by "the loving eye" rather than "the arrogant eye" (Frye, 1983).[15] Indeed, a respectful, empathetic relation to the natural world promises to bring more insight, understanding, and knowledge than the detached nonparticipating consciousness that seeks to control.

Thus, at a general level, there are some significant continuities between modern and postmodern science, but there are also fundamental shifts and reversals, involving tenets of modern science that postmodern science repudiates. Both modern and postmodern science utilize experimental and empirical methods of hypothesis, observation, experiment, and prediction; both are interested in detecting order, in control, and in discovering laws and regularities. Yet postmodern science turns more toward probability and statistical regularities and away from absolute certainty; it rejects notions of a fixed immutable order and absolute truth in favor of conceptions of evolving complexity and probability; it breaks away from mechanism and machine metaphors and affirms organism and biological models, and thus shifts from a self-contained and immutable universe to an open, self-organizing, dynamic cosmos that is constantly changing and evolving.

Postmodern science also puts a stronger emphasis on discontinuities in the universe than does modern science, drawing on catastrophe theory and recent studies that indicate that various natural catastrophes deeply influenced human and natural evolution, destroying entire regions of the earth and eliminating entire species like dinosaurs (Velikovsky, 1950; Thom, 1975; Zeeman, 1976, 1977). Indeed, following the theory of catastrophism, which opposed uniformitarianism and differed from the continuist emphases of Darwin, there is a current in geology and evolutionary theory that sees evolution taking place in spurts, through ruptures and catastrophes, thus making discontinuities more important (Gould, 1980: 179–185). Similarly, students of the solar system have hypothesized that cataclysms have been responsible for planetary evolution.[16]

But postmodern science is in flux, with intense paradigm battles going on between defenders of chaos theory and of complexity theory, between advocates of the old and of the new sciences, and between those who claim that we are at the "end of science" and those who believe that we are in a period of exciting development. As is evident by John Horgan's book *The End of Science* (1996), the postmodern sense of ending and exhaustion has even pervaded many of the sciences, such as evolutionary biology and physics. Horgan and other scientists ask us to seriously consider "the possibility—even the probability—that the great era of scientific discovery is over" (6). The "end of science" means not that scientists will no longer do research but, rather, that this research "may yield no more great revelations or revolution, but only some incremental, diminishing returns" (6).

This thesis, of course, is hugely controversial. Evolutionary biologists like Richard Dawkins, for example, believe that after Darwin and the "new synthesis" of biology and genetics, the basic maps of evolution are complete and no dramatic discoveries lay ahead. Gunther Stent, another strong advocate of this thesis, argues that the emergent theories of chaos and complexity will not bring any new major developments. Other scientists, like Roger Penrose, insist that we will always be in a process of discovery. In this vein, Stuart Kauffman states: "There's no finite number of ways to carve up the world. And it would appear that we are free to persistently invent novel ways to see the world that are fruitful. There is every reason to believe that there are new and beautiful laws to discover at a higher level" (http://www.hotwired.com/braintennis). Yet many scientists express ambiguity about the possibility that science may someday fill in the details of the Big Picture. Some scientists have even opted out of pure or applied science for what Horgan calls "ironic science," which, he claims, resembles literary criticism more than science in that it develops speculative, awe-inspiring, but nonfalsifiable interpretations of the world. Still other scientists and theorists, like Bohm, Prigogine, and Griffin, feel that the postmodern turn in science promises to bring new and great discoveries. It may indeed be the case that in some fields the *modern paradigm* is exhausted, but it takes considerable hubris to claim that science has completed its task, such that we know, in broad outlines, all that there is to know.

To summarize, the differences between modern and postmodern science can be represented in the following shift of emphases:

Modern/Postmodern

machine/organism
control nature/respect nature
alienation from nature/reintegration into nature
passive matter/active matter (autopoiesis)
determinism/indeterminancy
reversible time/irreversible time
immutable order/chaos
reductionism/complexity
certainty/probability
unmediated objectivity/hermeneutics
absolute space and time/relativist space–time continuum
monoperspectival/multiperspectival (complementarity)
value free/value responsible

Some proponents of postmodern science wish to underline the discontinuities between the new and the old theories, while others stress continuities.

Yet considering the entire spectrum of conceptual changes in contemporary science, the term "postmodern science" seems justified. Problems occur, however, when theorists focus on one or two supposedly definitive "postmodern" characteristics that could plausibly be assimilated to a "modern" framework, as when Sheldrake (1988) characterizes postmodern science in terms of a shift to an evolutionary framework, or when Toulmin (1982a, 1982b) characterizes it as a turn toward a hermeneutical framework. Both cases are compatible with a theory that denies randomness and indeterminacy, two other key criteria for a postmodern science, and therefore might arguably be termed "modern." Similarly, many theorists attempt to link postmodern science with ecology, without adequately distinguishing between a *modern ecology*, such as developed in mainstream environmental science (which fails to break with mechanistic and scientistic ideologies), or by a theorist such as Murray Bookchin (1995a, 1995b) who retains strong links to the humanist and Enlightenment traditions. Such modern enterprises can be contrasted to a *postmodern ecology*, such as deep ecology, which radically rejects humanism as "anthropocentrism" in favor of an Eastern-inspired biocentrism that places all life forms—whether human beings, bears, or bumblebees—on the same ontological plane of "intrinsic value" (see Devall and Sessions, 1986; Oelschlaeger, 1991; Sessions, 1992).

For orthodox scientists, however, postmodern concepts are inherently subversive and destructive, and they oppose them resolutely as the latest threat to scientific rationality, thereby carrying the battles of the culture wars into new territory and reopening the rifts between the "two cultures" of science and the humanities, which certain trends within postmodern thought attempt to overcome.

SCIENCE WARS AND THE POSTMODERN AGON

An intelligence that is not humane is the most dangerous thing in the world.
—ASHLEY MONTAGUE

Science is too important to leave to scientists.
—GEORGE LEVINE

Do not confuse the moon with the finger that points at it.
—ZEN PROVERB

Although "postmodern science" actually represents a diversity of positions, it is often identified as an irrationalist and anti-scientific position that emanates from radical postmodern social and cultural theory. Through critiques of modern forms of scientific knowledge and practice, postmodern science has entered the fray of the culture wars, in which mainstream values, notions of truth, objectivity, and rationality, and the hegemony of the Western intellectual traditions in universities have been challenged by leftists, feminists, multiculturalists, deconstructionists, postmodernists, and others. In particular, postmodern critiques forced a crisis in both the humanities and the sciences, one that precipitated "science wars" when defenders of orthodox science began to respond to social constructionist positions in a series of articles, books, and conferences during the 1990s.

Of course, there is nothing new about social constructionist arguments that science is not a pure quest for truth but, rather, is an enterprise conditioned and mediated by its social and cultural environment. Such arguments have been made in different stages in the development of the philosophy of science, particularly through the work of Kuhn, who inscribed scientific rationality in a larger scientific *worldview* suffused with values; they have been developed by philosophers such as Peirce, Dewey, Burtt, Langer, Bachelard, Quine, Hesse, and Hanson; they have roots in the hermeneutic theories of Heidegger, Gadamer, and Ricoeur, as well as Mannheim's sociology of knowledge and its popularization by Berger and Luckmann; they have been elaborated by feminists such as Keller, Harding, and Haraway, and by multicultural and postcolonial critics of Western science; and, importantly, they have emerged from the field of natural science itself, beginning with quantum mechanics and continuing through the postmodern theories of a broad range of scientists whose views we have engaged in this chapter.

The combined thrust of these assaults has created anxiety among many scientists. The postmodern critique is especially disturbing to mainstream science because they appear during a time when major social forces are putting scientists on the defensive, including increasing public skepticism about the role of science and decreasing government funding for scientific research since the end of the Cold War in the late 1980s.[17] The result, in the current era of downsizing, has been huge budget cuts for science, as well as for all other academic fields. Critics tend to question the Big Science associated with Cold War military spending and the erection of a gigantic military–industrial complex that threatens the survival of the planet. This combination of attacks on science has galvanized some members of the scientific establishment to seek scapegoats for their fears, and its academic critics are a safe target for the Holy Warriors of the Higher Truth.

Angered by postmodern assaults on classical notions of truth and objectivity, Paul R. Gross and Norman Levitt responded vitriolically in *Higher Superstition: The Academic Left and Its Quarrels with Science* (1994). Outraged that nonspecialists in science—cultural and social theorists like Stanley Aronowitz, Katherine Hayles, and Andrew Ross—would dare challenge scientific practice and norms, indignant that such critical theorists would politicize the eternal verities of dispassionate truth, and bemused that they would make technical errors in the process, Gross and Levitt set out to discredit the motley army of scientific critics as ignorant, misguided, irresponsible enemies of reason and progress, as uninformed ideologues of the "academic left."

Previously, they claim, either natural scientists were too timid to challenge philosophical or political critiques of their discipline because they had no expertise in the philosophy and sociology of science, or they were too tolerant or even were perversely fascinated by the critical perspectives of outsiders. But Gross and Levitt see themselves as part of the new post-wimpy crowd, tired of having sand kicked in their faces by postmodern bullies, filled with "skepticism and revulsion" at these critiques and ready to fight back. Like the 98-pound weakling on the beach, however, they find themselves ill prepared to take on the task.

Following their lead, Alan Sokal, a physicist at New York University, penned a parody of postmodern science (1996a) purposely riddled with errors and undigestible jargon that was accepted for publication by the unwary editors of *Social Text*, an event that appeared to vindicate Gross and Levitt's charge that the "academic left" is scientifically illiterate and incompetent. Even more than the verbose and bilious book of Gross and Levitt—which was featured at the 1995 New York Academy of Science conference on "The Flight from Reason" and which the conservative National Association of Scholars sent to science deans and department chairs across the country—Sokal's hoax was widely publicized and discussed, featured in the pages of *Lingua Franca, The New York Times, Fortune, The Chronicle of Higher Education*, and *The Nation*, while triggering a lively debate on the Internet among academics and demonstrating the passion still generated by postmodern debates.[18]

Just as the militia movement is a reaction to the decline of white male culture, the hysterical critiques of Gross and Levitt, and the antics of Sokal—the Gang of Three—are a reaction to the decline of positivistic conceptions of scientific objectivity. The lines drawn in the science war debates are not between right and left (Sokal describes himself as a leftist and feminist who taught mathematics in Nicaragua during the Sandinista rule, and claims that Levitt is a socialist and Gross a liberal [1996b, 1996c]). Rather, the science wars are being waged between positivists and postpositivists, between those defending scientific realism and the preeminent role of science in Western culture and those challenging the ahistorical claims to truth, the alleged theoretical purity, and the dogma of being a privileged discipline and form of knowledge.

Gross and Levitt's Higher Arrogance

> The aroma of sour grapes is in the air.
> —Paul R. Gross and Norman Levitt

Gross and Levitt wage their polemic against the opponents of science, claiming that the critiques represent "higher superstition" as opposed to "higher education" or "knowledge." Superstition is, of course, irrational belief, founded on fear and ignorance, precisely what Gross and Levitt claim informs contemporary critiques of science. But, as we shall argue, these warriors against superstition and for the higher truth fail to distinguish between rational and irrational critiques of science and technology, they crudely caricature their opponents' views, and they themselves are irrational and unreasoning in both their attacks on their "enemies" and their uncritical defense of outmoded conceptions of science.[19] Furthermore, they are profoundly ignorant concerning the positions they malign, and their polemic is undermined by a series of performative contradictions between their stated goals and actual writing, such as when they denounce the politicization of academia at the same time they call for the scientific policing of the humanities (1994: 215ff.).

Gross and Levitt claim that they worry about "a certain intellectual debility afflicting the contemporary university: one that will ultimately threaten it" (1994: 7). But their own uninformed and emotive screed exhibits a high state of debility and fosters divisions and animosities across the disciplines that can hardly be healthy for the university as an institution. They claim that an arrogantly obscurantist assault is being mounted in which, in the authors' words, "the proliferation of distortions and exaggerations about science, of tall tales and imprecations, threatens to poison the intellectual cohesion necessary for a university to work" (1994: 7). But, in fact, it is Gross and Levitt who are demonstrably arrogant, ignorant, sloppy, and irresponsible in their review of the positions and critics they assail. We will accordingly attempt to show that Gross and Levitt are themselves guilty of a large array of the deadly theoretical sins that they charge to scientific critics and that their "refutations" involve puerile ad hominems arguments; straw critiques, misrepresentations, and caricatures; out-of-context quotes; avoidance of the central issues and arguments; blatant ignorance of the subject matter; failure to cite adequate primary and secondary literature; lack of alternative positions; and apparent failure even to read the texts they discuss. Moreover, while they claim to be worried about threats to the integrity of the university and by divisions fostered by postmodern critiques, their own vicious polemics threaten to further divide a university under siege by a variety of reactionary social forces.

To begin, Gross and Levitt's attack on the so-called academic Left is seriously misdirected because it is the Right who has traditionally carried out totalizing critiques of science and technology (e.g., conservative anti-Enlightenment ideology and the work of Heidegger, Ellul, and so on). Moreover, it is right-wing forces today (e.g., fundamentalist religious groups, creationists, and anti-government and pro-market organizations who are against all government spending) who—in addition to certain New Age philosophies and a pervasive anti-intellectualism in U.S. culture—are leading the real existing offensive against science. But instead of taking on the actual forces undermining the status and health of science in contemporary society, Gross and

Levitt carry out a highly biased and distorted critique of a wide range of academic critics of science.

Blatantly misrepresenting the arguments and intentions of a wide and diverse range of scientific critics, they lump a disparate range of positions together as postmodernist, the "postmodern Left," or the "academic Left." Despite their awkward qualifications, Gross and Levitt's central term "academic Left" is vacuous and bereft of analytic value. They say the term does "not refer merely to academics with left-wing political views" (1994: 9) but, rather, aims "to designate those people whose doctrinal idiosyncracies sustain the misreadings of science" (9), which includes postmodernists, "radical feminists" (a technical term designating feminists who privilege gender over class but which Gross and Levitt misrepresent to include feminists in general), multiculturalists, and "radical environmentalists." There are obviously significant differences among these groups, but Gross and Levitt lump them together under one amorphous, misleading, ideologically loaded label (e.g., "academic and activist critics of science" would be better than the term "academic Left").[20]

The inaccuracy of their term "academic Left" is obvious if one considers the politics of many postmodernists, radical feminists, and environmentalists, which often have nothing to do with or are hostile toward traditional "left" politics. For instance, from a biocentric, often misanthropic perspective, Dave Foreman and others of the "radical environmental movement" are contemptuous of humanist values in general, to say nothing of left-wing or Marxist-style humanism. Foreman, in fact, left the group he cofounded, Earth First!, precisely because of the increasing influence of leftist social concerns that distracted from the pure wilderness focus he wanted to maintain and conflicted with his biocentric philosophy. But such subtleties escape Gross and Levitt, who attempt to reduce complexities to simple categories and who engage in ad hominem arguments and covert red-baiting. Their arrogant and authoritarian mentality is betrayed in one of many contradictions in their book: Claiming not to take pleasure in their acrimonious polemics, they ask the reader to believe that their "chief hope" in writing "is to convert friends (whose asseverations are for the moment our subject), or at least to persuade them to reflect" (1994: 2). Some 30 pages later, however, they offer a lengthy section on the "academic left" entitled "The Face of the Enemy"!

Most devastatingly, for all their condescension toward cultural theorists who misinterpret science, Gross and Levitt themselves misread the texts of Derrida, Foucault, and almost every major target of their critique. While they denigrate nonscientist critics for lack of solid knowledge of the scientific tradition, their assault on postmodern theory reveals that they themselves are totally ignorant of postmodern positions, which they distort and caricature. The main figures and concepts of postmodern theory are hardly discussed at all, even though this is a major focus of their critique. Derrida is disposed of in a page and a third (1994: 76–77), Foucault in two paragraphs (77), and Lyotard and Baudrillard in two sentences each (79–80)! Not surprisingly, in superficial readings largely derived from secondary references, they make blatant interpretive errors, claiming that neither Derrida nor Foucault believe in an external world and reading Foucault as reducing knowledge to power (75–78). Gross and Levitt apparently are unaware that Derrida himself rejected the idealist interpretation of his epigram "There is nothing outside of the text" (1981a: 51ff.). Nor do they have a clue that Foucault allowed room for scientific objectivity untainted by social interests and power or that he derided the attempt to *reduce* knowledge to power (see Best, 1995).

Gross and Levitt also mock Derrida's "deep epistemological pessimism" and claim that deconstructionism "holds that truly meaningful utterance is impossible, that language is ultimately impotent, as are the mental operations conditioned by linguistic habit" (1994: 76). Had Gross and Levitt actually read Derrida, they would have found the exact opposite to be the case, that language for Derrida is a particularly potent proliferator of meaning and that the mind and language are extremely creative and productive in their dissemination of meaning. They are also wrong in claiming that "few serious philosophers had much use for" Derrida's ideas (76). Indeed, almost everything that Gross and Levitt say about postmodern theory is flat out wrong, belying their claims to be responsible scholars rather than propagandists and hack polemicists. Given that they were writing a book on postmodern critiques of science, one would imagine that they would bother to read representative examples of the primary and secondary literature, but obviously they did not.

Indeed, their goal is not to understand the postmodern critiques of science and society but to attack the most blatant errors made by radical critics of science to discredit the entire academic and postmodern Left. It is not surprising that they identify postmodern theory per se with the most extreme versions of it, obscuring its variety and diversity and thus providing a caricature straw man that they proceed to demolish. Thus, as Gross and Levitt excoriate their friends/enemies for faulty and incomplete histories of modern science, as well as for derivative and erroneous understandings of esoteric scientific theory, they succumb to the very same problems in their interpretation of postmodern theory and postmodern critics, not the least of which is their display of a "remarkable arrogance" (1994: 106) toward their subject matter and colleagues. As they excoriate the Left for dogmatism, one would expect nondogmatic arguments from them, producing evidence for everything they try to "falsify," but they make rigid, rhetorical, and ad hominem arguments against everything that is not mainstream science, including phenomena like alternative medicine ("an ancient amalgam of quackery and self-delusion" [251]) and animal rights (part of the "intellectual junk-food of the 'New Age' movement" [199]).

In addition, they reduce the complex origins and uses of postmodern theory to the psychological deficits of its advocates, describing Foucault as "deeply neurotic" with "a self-despising personal life" (1994: 71). Their polemics rival Rush Limbaugh's for nastiness and snide dismissiveness; they can't just refer to "feminists" but have to say "fire-breathing feminist zealots" or "a gaggle of post-everything feminists" (37). Similar disdainful epithets are thrown around throughout the book against gays and lesbians, blacks, multiculturalists, and ecologists. In general, Gross and Levitt try to dismiss the "Left" critiques of science by reducing logic to psychology, claiming that the "Left," mourning the failure of the radical project in history, is seeking a scapegoat to vent its frustrations on, finding the perfect target in science, which for Gross and Levitt represents the highest achievement of Western civilization (27, *passim*). Of course, their "critique" is a blatant but unacknowledged reprise of Nietzsche's (1967b) theory of *ressentiment*. Nietzsche argued that *ressentiment* (a form of extreme resentment of the weak against the strong) drives the moral criticism not only of Christianity but also of anarchists and socialists. He reviled radical social critics as spiteful "lower types" whose jeremiads against a brutally exploitative 19th-century capitalism were symptomatic of their own personal inadequacies, jealousy, and pettiness rather than of legitimate problems of social injustice. By using the psychological category of resentment, Gross and Levitt seek to discredit the social and political insights of radical criticism, but they only obscure the complex social, historical, intellectual, and political context of the emergence and development of postmodern discourse (such as we are trying to clarify in this book).

If such psychological categories must be dredged up, they clearly apply to Gross and Levitt's own envy-driven descriptions of the "academic left's" prestige, allegedly high salaries, and social influence (e.g., 1994: 34, 103, 177, 237–238), as well as to Levitt's bitter complaints about earning only a "marginal" living as a mathematician (115). They systematically project their own failings (i.e., overweening arrogance, sharp biases, lack of understanding of the ideas that they are criticizing, etc.) onto their opponents. One might read *Higher Superstition*, therefore, as an acerbic, biased account of contemporary academic life, one that is so resentful of challenges to their outmoded Science-as-God worldview that it grossly (pun intended) distorts everything it examines as it revels in ad hominem/feminem arguments, exuding the bad aroma of Higher Arrogance on every page.

Ironically, in their imperious response to cogent criticisms of the methodological assumptions and practices of science, Gross and Levitt display the very linkage between knowledge and power that they otherwise try to erase by constantly evoking the superiority of science over all other academic disciplines; indeed, they argue that if the entire humanities faculty of MIT was eliminated the science and engineering faculties could make do, whereas the opposite is not the case (1994: 243). There is a turf war with money and authority at stake, and science has taken some hits with the cancellation of major government programs, like the construction of the Superconducting Supercollider (SSC), and accompanying cuts in university science programs. In this situation, Gross and Levitt's intervention can be read as an attempt to advance the interests of their own narrow coterie in budget battles within academia.

Their "critique"—when it is not offensive or uninformed—is always undialectical and purely negative, erupting with *Schadenfreude* whenever they detect something erroneous or uniformed in the texts of their "enemies." But they rarely attempt to grasp or explicate what might be salient or informed critiques of the sciences by those on the "academic left"; thus, their book is more of a biased polemic than the serious "objective" scholarly enterprise they champion. Indeed, despite the demagoguery of Gross and Levitt, most contemporary critics are not blaming science and technology per se for social ills, they are not attacking rationality in itself, and they are not denying the existence of an objective world independent of any observer. The critiques of science coming from Marxists, feminists, postmodernists, and others are characteristically *dialectical* in nature, granting the important benefits of scientific and technological developments but also critical of unethical forms of research (on both human and nonhuman animals) and the application of science to warfare, environmental disasters, and other ills of our time. But these dissidents from the modern faith assign blame where it belongs—on the people, policies, and institutions that abuse science and technology, as well as on forces like the profit motive and competition—rather than on science and technology alone, considered apart from social relations and interests.

A distinction therefore must be made between the critique of science per se and the critique of modern scientific ideologies and practices. It is the Right, accompanied by a few extremists on what Gross and Levitt call the "academic Left," that attacks science per se, whereas most contemporary critics of science criticize the modern scientific worldview and its ideology of scientism and positivism, or specific abuses of science and technology. These abuses include imperial claims that science is the sole source of truth and objectivity (as Gross and Levitt seem to imply), that its view of the world is the only correct one, and that other discourses (art, philosophy, social theory) are only so much "unalloyed twaddle," "hermeneutic hootchy-koo," and "magical thinking" (their terms).

Gross and Levitt do make a distinction between a "weak" and "strong" constructivism (1994: 42–43). A "weak" version admits to some social influences on science but without undermining substantive notions of truth and objectivity, as the "strong" thesis does by reducing science to mere ideology, power, parable, allegory, or convention. But they present a caricature of what they call the "strong constructivism" thesis, equating it with positions few social constructionists would adhere to. They complain that historians and sociologists of science "spin perverse theories" (43), but they themselves construct a caricature of the social constructivist position that they illustrate with extremely perverse and unrepresentative examples. They claim that "hundreds of left-wing social theorists dote on" what they call "Tooth Fairy Hypotheses" but offer no convincing documentation of these claims. Moreover, their ignore a vast range of positions that discuss the social construction of knowledge, ranging from that posited by Berger and Luckmann (1967) to that of the Frankfurt School (Kellner, 1989a).

Although Gross and Levitt attack their targets for not justifying their positions and for providing misleading accounts of science, they themselves rarely bother to explain difficult scientific theories and seldom explicate or defend their own positions, preferring polemic over pedagogy. Occasionally, however, they lay their cards on the table and reveal what they actually stand for, as in the following embarrassing and juvenile praise of positivism: "We are unabashed technocrats, unashamed of the instrumentalism behind such assertions. . . . Let us raise a glass to Bacon! . . . The more Baconian science we get, the easier it will be to believe that we have a fighting chance, if no more than that, on this lovely planet that spins its way through an unimaginably violent—and indifferent—space" (1994: 178). Here their unabashed and vulgar version of positivism is transparent, disclosing "how easily," to use their own words, "a redemptive vision [viz., science] can slip free of reason" (216). Indeed, what is "lovely" about smog, oil spills, tree stumps, sewage, waste dumps, smokestacks, litter, and clogged expressways, to say nothing about children with distended stomachs or cities smoldering in the rubble of war? And how does one rationally combine the terms "lovely" and "unimaginably violent" in the same breath? Or the terms "violent" and "indifferent," for that matter? Still worse, how can these intoxicated sops for Bacon praise instrumentalism while *also* admitting that "the instrumentalism available to our worst impulses have grown unimaginably lethal" (218)? Obviously, logic and consistency are missing in these scientific ideologues who endlessly praise such ideals but rarely follow them.

Thus, the would-be Emperors of the Higher Truth are shown to be naked in their dogma and ignorance. Defenders of an outmoded positivism, objectivism, and instrumentalism, they seem unaware of the developments in contemporary science and philosophy of science toward a postpositivist understanding of knowledge, mediated by postmodern critiques. Their views are also politically problematic. Although Gross and Levitt seek to disavow themselves from the political Right, their use of the term "left" plays into the right-wing assault on all critical modes of thought that have emerged since the 1960s that attack traditional paradigms, and they have largely been endorsed and promoted by rightists. As Ellen Schrecker (1996: 61) pointed out, the attacks on university-based critics of science tend to undermine critical academic discourse during a time when the Right "is fighting a broad-based campaign to demonize those sectors of the academic community that encourage critical thinking and offer an alternative perspective on the status quo." At a time when government budgeting has targeted all university programs, this division among academics, so skillfully fostered by Gross, Levitt, and Sokal, is most unfortunate.

Consequently, the Gang of Three's target is misplaced, limiting their critique to academics whereas the really dangerous critics of science are those politically active right-wing groups who oppose secularism, rationalism, and scientific theory *tout court*. Although Sokal (1996b)claims his hoax is aimed at saving the Left from pretension, jargon, and isolation from the public (1996b), in fact his intervention is congruent with that of Gross and Levitt, and all three serve to discredit radical critiques of science, as well as critical theory, cultural studies, and other progressive forms of thought, thereby playing the game of the Right.[22]

Sokal's Hoax

> Hertz points out that propositions in physics have neither the task nor the capacity of revealing the inherent essence of natural phenomena. He concludes that physical determinations are only pictures, on whose correspondence with natural objects we can make only the single assertion, viz., whether or not the *logically* derivable consequences of our pictures correspond with the empirically observed consequences of the phenomena for which we have designed our picture.
>
> —Werner Heisenberg

> All the laws of nature that are usually classed as fundamental can be foreseen wholly from epistemological considerations. They correspond to *a priori* knowledge, and are therefore *wholly subjective*.
>
> —Sir Arthur Eddington

Sokal begins his *Social Text* hoax with an attack "on the dogma imposed by the long post-Enlightenment hegemony over the Western intellectual outlook which . . . [holds] that there exists an external world, whose properties are independent of any individual human being and indeed of humanity as a whole" (1996a: 219). In his *Lingua Franca* exposé (1996b), he cites this passage and suggests that critics of science are denying that external reality exists outside of our social constructs. But his bugbear here is Bishop Berkeley and absolute idealism and not the social constructionist view of science. Only an extreme idealist would argue that it is the mind, or our ideas, that constitute the physical world. Rather, as Stanley Fish (1996) cogently argued in his rejoinder to the philosophically simplistic Sokal, it is not the physical world itself that is socially constructed but, rather, our concepts, theories, paradigms, and methods through which we investigate and describe the world. In an unpublished response to Fish (found on his Internet homepage: http://www.physics.nyu.edu/sokal.index.html), Sokal claims that Fish misrepresents his views and asserts that he himself was arguing that it is not the laws of nature that are socially constructed but only our theories about them. Against Sokal, one could argue that whereas it is true that the physical constitution of objects and their workings are not socially constructed, surely "laws of nature" are. "Laws of nature" refer to those constructs through which we describe physical events; they are subject to social consensus, contestation, revision, refinement, development, and even paradigm shifts—as everyone but the most simplistic and dogmatic positivist would agree.

Except for the isolated example of Hindess and Hirst (1975), which even they later recanted, we are aware of no "left" or "postmodern" theorists who deny the existence of the objective world. And "laws of nature" are themselves constructs that are subject to change and development. Newton's "law of gravity," for instance, postulated a gravitional force that caused the motion attributed to gravity, whereas Einstein, in his general relativity theory, argued that the motions that we attribute to gravitional effect are due not to force but to curvatures of space–time. In the future, yet other "laws of nature" may be constructed that will replace previous ones.

In a very important sense, science *is* a social construction: It is made by people working in groups, it is a *social* activity, and it is made according to certain models, methods, and principles, and thus it is a *construction*. Sokal therefore attempts to discredit a sound and cogent position with critique of a silly one that few others than absolute idealists would hold. Consequently, Sokal's suggestion that anyone who believes the laws of gravity to be a mere "social construction" are welcome to jump out of his 21st-floor apartment window misses the point (1996b: 62). The social constructionist would certainly splatter on the street, but his or her construct of reality contains this knowledge and prevents all but those who choose to do so from jumping. To be sure, if Sokal jumped out of his window in despair 100 years ago, today, or in the foreseeable future, the result would be the same, but it is possible that the physical description of what was occurring with his body and the physical explanation of falling bodies could change, and these thus are social constructions. Denying that laws of nature are social constructs presupposes a fixed, immutable, absolute truth, precisely the conception skillfully undermined by postmodern critiques. Against an objectivist conception of scientific law, the social constructionist argues that our scientific laws and explanations are constantly changing in accord with discoveries in science, shifts in intellectual fashions, and mutating paradigms of knowledge and inquiry, and thus that there is no objectivity, knowledge, or truth independent of our conceptual schemes, that all discourse is socially constructed and subject to change and modification.

Sokal thus confuses the social construction thesis with denial of "the real world," of facts, evidence, and objectivity. He fails to even make the obvious distinction made by Gross and Levitt between weaker and stronger versions of social constructionism and equates the notion that science is socially constructed with the denial of the existence of an external world, of facts, and of evidence, as if postmodern critics of modern science were absolute idealists or textualists who literally believed that there was nothing outside of the senses or the text. Yet it is obvious that science takes place in an institutional setting where there are rules of evidence, dominant paradigms, established research procedures, power struggles, and challenges to the existing regime, making science a contested terrain where relations of power and authority often determine what is "real" and what is accepted or rejected. When they first appeared, germ theory, genetic theory, evolutionary theory, and quantum theory were rejected by the scientific establishment and were accepted only after intense struggle. Scientists work within established discourses, institutions, and practices that are influenced by external factors. They may have ideals of objective truth, but social relations, discourses, and institutions are always influencing how their truths are received and transmitted. Rather than blindly attacking science, "social constructionists" typically are interested in analyzing the conceptual schemes, assumptions, and social forces that influence scientific method and practice, often in order to warp the humane and liberating potential they know science to have. Despite the dogmatic realist fantasies of the Gang of Three, science is never separate from culture and society.[23] Scientists, alas, are mere mortals. As such, they live in a world with other human beings, they absorb its wisdom and values, and they are vulnerable to various ideological biases and institutional pressures, as well as the temptations that come with research grants, rewards, and promotion.

Unlike some "strong constructionists," we are not suggesting that there is nothing unique about scientific method—as opposed to endeavors in literature, religion, or philosophy—nor that scientific objectivity is merely chimerical. In its appeal to falsification, verification, and evidence, through its use of deductive and inductive logic and the experimental method, in its practical uses and consequences, we believe there are important qualities about science that distinguish it from literature, myth, or mere rhetoric and ideology (although Kuhn and others point out that the sources of scientific discovery rarely match the logical archetypes of discovery represented in the lovely stories scientists tell of how they work). Yet, like any other discourse, science is embedded in a host of social institutions, it employs concepts, metaphors, narratives and other discursive forms, it begins from distinct presuppositions, and it is susceptible to influence and bias. As geneticist R. C. Lewontin describes:

> Science, like other productive activities, is a social institution completely integrated into and influenced by the structure of all our other social insti-

> tutions. The problems that science deals with, the ideas that it uses in investigating those problems, even the so-called scientific results that come out of scientific investigation, are all deeply influenced by predispositions that derive from the society in which we live. Scientists do not begin life as a scientist, after all, but as social beings immersed in a family, a state, a productive structure, and they view nature though a lens that has been molded by their social experience. (1992: 3)[24]

Only the most naive and self-deluded member of the scientific community could fail to acknowledge that observation is never theory free, that the biases of the interpreter can predetermine experimental results, often grossly so, as in the case of sociobiological racism and sexism, or that, most crudely, science is often a tool of corporate, government, and military interests.[25] Gross and Levitt admit that science has in the past sustained "the most ferociously antiegalitarian ideas—racist eugenics, 'Social Darwinism,' and the like," but they then claim such a science "has long been effaced, while the claims put forth to bolster the egalitarian view have endured, on the whole, rather well" (1994: 23). Such statements are incredibly naive, overlooking the publication and publicity of works, like Richard J. Herrnstein's and Charles Murray's *The Bell Curve: Intelligence and Class Structure in America Life*, that advance racism, sexism, and other forms of prejudice (see the critiques in Kinchloe et al., 1996, and Fraser, 1995).

Despite the direct implication of science—better, of many *scientists*—in racist, sexist, militarist, anthropocentric, and classist policies, we do not believe that astronomy is no different from astrology, that all theories are equally valid, or that "facts" do not exist. Whatever the extra-logical sources of scientific discovery, the processes of verification and confirmation can be rigorous and exact. Social constructionist arguments therefore simply cause us to question certain constructions, uses, and consequences of science and technology but do not entail the overthrow of scientific method as such (except in extreme constructionist versions that we cite and criticize later).

Thus, the Gang of Three give a caricatured and one-dimensional portrayal of postmodern critiques of science and completely misrepresent critical science studies. They fail to see that many critical positions come from figures within the natural sciences themselves, and they do not distinguish between the ways in which postmodern concepts and critiques can strengthen science, such as by reinscribing the norm of objectivity within its actual social, historical, and linguistic context rather than in the mythic appeal to theory-free observation. Ultimately, the Gang of Three are neo-positivists swimming against the tide of conceptual developments and paradigm shifts throughout the academic realm, unable to see that even their own fields of mathematics and physics have taken a decidedly postpositivistic turn that both puts in question the concepts of science and objectivity and offers alternative conceptions.[26] They therefore appear as scientistic ideologues, uncritical defenders of scientific doxa, who are blind to the fallacies and problematic nature of their own assumptions about culture, science, and reality, preferring to concoct conspiracy theories about creeping academic leftism and irrationalism.

Of course, as we are arguing, there are still raging debates concerning the nature of postpositivistic or postmodern science, and no new paradigm has emerged that has won general consent.[27] Rather, we are in a period between paradigms, between the modern and the postmodern, a period when there is necessarily confusion, conflict, panic, distortion, and hyperbole. A positive aspect of this interregnum is that all issues and orthodoxies are up for grabs and there is lively discussion of a wide range of important issues. One beneficial consequence of the Sokal affair is thus instituting public debate over science and scientific discourse, its social construction and critique, and the transformations within philosophy of science itself. Moreover, the science wars are significant because they demonstrate a postmodernization of discourse, pointing to the rapidity with which ideas and issues can circulate, the ways that even complex intellectual and political issues can be presented to a broader public, and the ways that the media and new information technologies mediate both intellectual and political debate. The debate engaged a wide public, and one of the most interesting aspects of this curious episode is the speed and extent to which the Sokal hoax circulated through the media and the culture at large.[28]

The Sokal hoax thus demonstrates the openness to debate of the new communications media and of a public sphere where individuals can intervene without the mediating gatekeepers of journals or other watchdogs. Any individual can enter the debate, demonstrating the viability of the new public spheres and cyberdemocracy made possible by the new technologies (see Kellner, 1995b). Yet precisely the publicness of intellectual discourse and the high political and cultural stakes involved indicate that participants must be careful in their language, must argue clearly and cogently, and must be wary of the wiles of their opponents. Individuals and groups with a message to communicate must also devise strategies to circulate their positions through the Internet, through print and broadcast media, and through other public spheres to gain maximum exposure. Sokal's prank represented a shrewd use of the media and could be admired as a successful provocation, or perhaps as a neo-dadaist cultural intervention.

More to the point, the hoax discloses contestations between modern and postmodern paradigms and the intensity of the passions involved, and to us it signifies that we are indeed living in a period between epochs, when dominant paradigms are under contestation and new paradigms are emerging. Sokal would not be worth even mentioning except that his "intervention" generated an unparalleled response in the mainstream media, on the Internet, and in a broader cultural sphere that is already producing announcements of books, conferences, and other forums to discuss the affair.[29] If this generates serious discussion of the science wars and the relative merits and limitations of the opposing positions, it is to the good, though so far it has engendered more heat than light.

Toward Transdisciplinary Discourse

Yet what is needed is transdisciplinary dialogue and not acrimonious polemics. Gross's and Levitt's book and the Sokal affair reveal that C. P. Snow's "two cultures" (1964 [1959]) have grown even further apart, with more pronounced animosity, conflict, and mutual ignorance than when he was writing. This is unfortunate given that some of the most interesting work today takes place in the borderlands between the sciences and the humanities, producing transdisciplinary bridge-building work of the sort that is distinctive of the best of postmodern thought. But Gross and Levitt go so far as to *attack* interdisciplinary studies as the home of fragmented parcels of "Theorese" (1994: 75), a slam that hardly does justice to the many important contributions of these programs throughout the country. Although, as they claim, the proliferation of various "studies" programs may have balkanized theoretical discourse, one role of interdisciplinary programs is precisely to combat this fragmentation and to combine various theories in a multiperspectival critical vision. Such perspectives, however, are demonized as relativistic by Gross and Levitt in favor of Newton's old "single vision," rightly challenged by William Blake, as well as by later forms of science, but resurrected by the Gang of Three and their scientistic colleagues who believe that science is the royal road to truth.

In general, Gross, Levitt, and Sokal have not contributed to mutual understanding between the two cultures. Rather, their intervention serves to cut off communication, arrogantly ascribing to themselves and their fellow scientists alone the right to speak intelligently about science, concerned only to show that certain nonscientists are ignorant fools, unable to understand science at all, or are biased ideologues, seeking to attack everything they resent. What is needed, by contrast, is dialogue that respects the other, that accepts differences, that attempts to discover what is positive in one's interlocutor's views, and that is *constructively* critical.

But the Gang of Three neither promote dialogue, nor do they further contemporary understanding of science itself. They scorn critics of science for their lack of understanding of basic concepts of contemporary science, but they do little to enlighten their readers. Instead, they expend merely negative energy attacking alleged distortions. But if they really wanted to promote the interests of science like Carl Sagan or Stephen Jay Gould, they might explain complex scientific concepts to the public, they might enlighten the public concerning the virtues of science and scientific method, and thus do something positive for science. In the long run, they are probably hurting the cause they want to promote because they perpetuate the belief that science is too difficult for any but specialists and reproduce the impression that specialists are arrogant and imperial, disdaining to explain their field to the masses and viciously attacking all critics of their religion. Thus, they ultimately perpetuate the image of a closed fraternity of science, controlled by high priests who go on witchhunts against any heretics or detractors from the faith.

While Gross, Levitt, and Sokal should be aware of the challenge to modern positivist paradigms from within science, they do not adequately engage them or register the seriousness of the critiques and challenges. They ultimately carry out an ostrich critique, with their heads buried in a pile of references to the worst excesses of postmodern theory and the academic Left, oblivious to the serious challenges to the neo-positivist concept of science and the ferment in the field that is putting in question the received wisdom that the Gang of Three uncritically accept. Consequently, although they may be right that many people making critiques of science often get their facts wrong, make mistakes, and have axes to grind, they fail completely to distinguish the stronger from the weaker critiques and ignore the wide range of theorists—from postmodernists to feminists to postcolonialist theorists to scientists themselves—questioning their neo-positivist faith and offering new and potentially more liberating and productive notions of science.[30]

Crucially, Sokal and his neo-positivist comrades seem unaware that there are a tremendous diversity of positions within the postmodern turn in science and that the postmodern has many faces, not all of which are hostile or destructive to scientific norms. Indeed, there is a major division in postmodern science between "reconstructive" and "deconstructive" approaches, which reflect different ways of appropriating postmodern ideas. As expressed by "reconstructive" postmodernists like Bohm, Griffin, Haraway, Harding, and Prigogine, substantive criteria are employed to judge the adequacy of scientific knowledge, while truth and objectivity remain guiding norms. Modern and postmodern approaches are combined, for example, in the dialectic of order and disorder, determinacy and indeterminacy, and structure and complexity in recent new scientific discourses.

For "deconstructive" or radical postmodernists like Feyerabend (1978a, 1978b), however, objectivity is dismantled altogether, and scientific results are made dependent upon vagaries such as a satisfactory sex life. In fact, much as do Horkheimer and Adorno (1972) in their critique of the Enlightenment, Feyerabend deconstructs the opposition between science and myth to argue that "science is much closer to myth than a scientific philosophy is prepared to admit" (1978a: 15). Rejecting the "law-and-order" approach of modern scientific epistemology which polices the right path to knowledge, Feyerabend anticipates Lyotard's (1984) claim that progress in science is characterized not by reaching the "truth," once and for all in a progressive accumulation of knowledge, but, rather, through a proliferation of theories. And the only principle for Feyerabend likely to generate a multiplicity of theories, to bring progress in knowledge, is the "anarchist" principle that "anything goes." Hence, "anarchism, while perhaps not the most attractive *political* philosophy, is certainly excellent medicine for *epistemology*, and for the *philosophy of science*" (1978a: 17).[31]

Clearly, this is an extreme position, but it hardly epitomizes "postmodern science" per se. In general, postmodern science seems aptly characterized by Griffin (1988a, 1988b) as a "constructive" or "revisionary" rather than "deconstructive" or "eliminative" postmodernism. Though postmodern science is critical of modern science, it does not in most cases seek to destroy classical and modern notions such as self, purpose, meaning, reality, or truth. As Gross, Levitt, and other neo-positivists fail to see, many versions of postmodern science refuse to terminate the critique of realism and determinism in relativism and nihilism, and postmodern criticism is directed against *scientism*, not science; *technocracy*, not technology; and *rationalism* (or "logocentrism"), not reason. As Griffin defines it, postmodern science seeks to loosen the boundary between scientific and "nonscientific knowledge" in order to incorporate other realms of knowledge and value in the sciences, involving "a new unity of scientific, ethical, aesthetic, and religions intuitions" (1988a: x) and a "creative synthesis" of premodern, modern, and postmodern ideas.

In harmony with Toulmin's argument that science must link theory with practice, Griffin claims postmodern science should work to transcend the pernicious ideologies of modernity, such as individualism, anthropocentrism, nationalism, patriarchy, economism, consumerism, and militarism, in order to provide support "for the ecology, peace, feminist, and other emancipatory movements of our time, while stressing that the inclusive emancipation must be from modernity itself" (1988a: xi). The new values, Griffin argues, must be incorporated in the struggle for a new society *with the support of postmodern science.* Bohm and other practitioners of postmodern science are calling not only for a new science but for a new society that they call "postmodern." They seek to merge science with social theory and insight, and they believe that postmodern science offers important insights for the reconstruction of society, such as are available in the science of ecology, which provides the empirical and normative foundations for a sustainable society, or in the autopoietic conception of life as spontaneous, creative, and evolving.

Similarly, "postcolonial science studies" emerged during the 1980s to reconstruct modern science in a richer form or to develop alternatives to Western science. Composed of scientists, social theorists, and activists of both European and Third World descent, postcolonial science has challenged the modern notion of a "universal science" rooted in pure observation to reveal its social and political influences and interests and its Eurocentric biases. There are a significant variety and diversity of positions within postcolonial science, some of which draw on postmodern critiques of modern science and some of which polemicize against these critiques, arguing that developing countries need more science and technology rather than Western critical discourse, to deal with problems of development. Some postcolonial critics seek to deconstruct the binary oppositions that inform false notions of universality and objectivity—such as are grounded in the distinctions between rationality–irrationality, historical–natural, and civilized–primitive— in order to construct "stronger" standards of truth and objectivity that are socially and historically contextualized (see Harding, 1993, 1994). Others develop more sociological critiques of the role of modern science and technology in the historical trajectory of imperialism and contemporary global domination of the developing world by neo-capitalist superpowers and transnational corporations.

Postcolonial science critics tend to argue that modern science developed not simply on a path toward progress in pure knowledge, but also in response to the military, economic, and political needs of European expansion and imperialism. Many critics point out the irony that as European powers were suppressing "savage" or "inferior" non-Western cultures, they also appropriated these cultures' scientific achievements as their own. Indeed, postcolonial science critics deconstruct the opposition between Western and non-Western knowledges by uncovering the contributions of China, India, Africa, Asia, Islam, preColumbian America, and other cultures to "Greek culture" or "European science" (see, e.g., Needham, 1969; Diop, 1974; Goonatilaha, 1984; Bernal, 1987; Weatherford, 1988). Just as environmentalists seek to preserve endangered species, many postcolonial critics attempt to preserve the still-valuable forms of "endangered knowledges" of cultures being absorbed by global capitalism and to transform them into ones that can function effectively for their societies in today's global political economy.

Although it is all too easy, à la the Gang of Three, to ridicule some of the more extreme and jargon-ridden expositions of postmodern theory and science (an anathema we share), Sokal's form of satire and the uninformed diatribes of Gross and Levitt occlude important transdisciplinary discussions concerning the critiques of modern science and serious attempts to delineate a new postmodern paradigm. Admittedly, as we have been arguing throughout our studies, postmodern discourse lends itself to all kinds of uses and abuses and is able to front for a variety of agendas. One cannot say that postmodern discourse per se is either pro- or anti-science, only that there are postmodern positions that are used by different people in various ways. Whatever the outcome of the science war debates, it is clear that the postmodern turn in various fields remains highly contested and that there is little possibility of any consensus in the foreseeable future over its meaning and direction.

NOTES

1. The first use of the term "postmodern science" appears to have occurred in 1964 in Floyd Matson's book *The Broken Image: Man, Science, and Society,* although Huston Smith (1982) used the term "postmodern physics" in 1962, assuming a broader notion of postmodern science. Frederick Ferré deployed the concept in 1976, in his *Shaping the Future: Resources for the Postmodern World,* six years before Stephen Toulmin published "The Construal of Reality: Criticism in Modern and Postmodern Science" (1982a) and *The Return to Cosmology: Postmodern Science and the Theology of Nature* (1982b). During this same period, Prigogine and Stengers (1984) developed new paradigms, as did Griffin and his colleagues (1988a, 1988b), Sheldrake (1990), Oelschlaeger (1991), Sassower (1995), and a wide range of theorists associated with chaos and complexity theory. In the 1980s, a "postcolonial science studies" emerged, using postcolonial and multicultural concepts to reconstruct the ahistorical basis of modern science, while drawing from the contributions of various cultures to scientific knowledge and challenging the hegemony of Western science. Thus, the concept of postmodern science has firmly entrenched itself in scientific discourse and continues to proliferate, as we show in this chapter. We use the term "postmodern science" broadly to refer not only to actual scientists and physicists employing postmodern ideas, such as David Bohm, Charles Birch, and Rupert Sheldrake, but also to theorists of science who deploy postmodern discourse.

2. Positivism holds that only science and the scientific method provide positive knowledge of the world, as opposed to metaphysical speculation or morally and politically biased opinion and ideology. The term was deployed by Auguste Comte and 19th-century social and natural scientists and was systematized and defended by the Vienna Circle and logical positivists in the post-World War I period, becoming a dominant ideology by midcentury. For its critics, ranging from phenomenologists to neo-Marxists to feminists, it was construed as a narrow and false concept of science in the 20th century. Today, post-positivist concepts of the philosophy of science are widely accepted, although as we will see in our discussion of the science wars, there has been a recrudescence of positivist orthodoxy in some quarters, in large part as a response to the postmodern critique of modern science.

3. Unlike most of the other theorists we consider, Prigogine and Stengers do not explicitly call their theories "postmodern science," although they delineate a "new science," which they distinguish from modern science, and they share key themes with other proponents of postmodern science.

4. Scientists have verified that the universe is limitless. According to Hawking (1988), analysis of light waves through use of the Doppler effect, such as was initially done by Edwin Hubble in the 1920s, shows that the universe is still expanding, since the frequencies (the number per second) of light waves from other galaxies are low (redder) on the electromagnetic scale, meaning that they are moving away from us.

Moreover, Hawking claims, the farther a galaxy is from us, the faster it is moving away. Newton and even Einstein believed that the universe is static, set firmly in place by gravitational influence; yet evidence suggests that the universe expands at a rate that supersedes the force of gravity. For Hawking, "the discovery that the universe is expanding was one of the great intellectual revolutions of the twentieth century" (1988: 30).

5. For studies showing the strong influence of the concept of entropy in cybernetics, systems research, and information theory and its application to a wide range of natural and social sciences, see Buckley (1968); Bateson (1972, 1979); Wilden (1980); Campbell (1982); and Capra (1996). Obviously, concepts like entropy will be deployed differently in the fields of thermodynamics, information theory, and ecology, though its use shows that there are important similarities between these fields as well as differences.

6. As argued by Paul Erlich et al., however, biological evolution does not contradict the law of entropy because it displaces energy loss from one region to another:

> The catch [to the erroneous view that biological evolution refutes the second law of thermodynamics] is that Earth is not an isolated system; the process of evolution has been powered by the sun, and the decrease in entropy on Earth represented by the growing structure of the biosphere is more than counterbalanced by the increase in the entropy of the sun. (1993: 73)

Similarly, Georgescu-Roegen finds entropy to be an incontrovertible law for all life:

> The truth is that every living organism strives to maintain its own entropy constant. To the extent to which it achieves this, it does so by sucking low entropy from the environment to compensate for the increase in entropy to which, like every material structure, the organism is continuously subject. But the entropy of the entire system—consisting of the organism and its environment—must increase. (1993a: 79)

7. Evidence for this hypothesis began accumulating since the early 19th century, and the hypothesis was widely accepted by the end of the century. For discussions of complementarity and how light and objects can be interpreted as both particle and wave, see Matson (1966: 132–133) and Berthold-Georg Englert et al., "The Duality in Matter and Light," *Scientific American* (December 1994: 86–92).

8. Yet Russell's set paradox also caused an uproar in the mathematics world by identifying a set (a class of objects) whose members were both in it and not in it: the set of all sets that are not members of themselves. Mathematicians and logicians have tried to dispel all paradoxes and contradictions to maintain the formal clarity of binary logic, but new paradoxes keep emerging. For a good discussion of these issues, see Kosko (1993). Bart Kosko is a pioneer in "fuzzy logic," a direct offshoot of the work of Heisenberg and Gödel. Fuzzy theorists hold that everything is a matter of degree, a gray zone between black and white. Hence, they reject binary logic as overly simplistic; they substitute multivalence for bivalence and the dialectical principle of "A *and* not A" for Aristotle's celebrated law of contradiction, whereby something is *either* A *or* not A.

9. The development of chaos theory was made possible by the technology of computers, such as the now-retired Digital Orrery, since they can perform unimaginably complex calculations that are too difficult and too time consuming for the human mind. Though supercomputers seem to have become the realization of the

Laplacian demon who could successfully grasp the nature of the distant past and distant future, often proving stunningly accurate in their predictions, their calculations can run aground when complex movements and massive spans of time are involved (see Peterson, 1993). Ironically, the more the human mind advances in its knowledge of the physical universe, the more it becomes aware of its limitations. Yet, to twist the paradox one more turn, it is precisely through an understanding of chaos that scientists and mathematicians are increasingly able to improve their predictions of the behavior of matter and complex systems.

10. As Prigogine and Stengers state in more detail:

> At the microscopic level, the laws of classical mechanics have been replaced by those of quantum mechanics. Likewise, at the level of the universe, relativistic physics has displaced Newtonian physics. Classical physics nevertheless remains the natural reference point. Moreover, in the sense that we have defined it—that is, as the description of deterministic, reversible, static trajectories—Newtonian dynamics still may be said to form the core of physics. (1984: 68)

11. According to Johnson (1996), chaos theory is divided into two distinct areas of study: the "*order-behind-chaos* school" popularized by James Gleick's book *Chaos* (1987) and the "*order-out-of-chaos* school" associated with Prigogine and Stengers and the Santa Fe Institute. The latter branch, Johnson observes, is known as "complexity theory" because of its focus on complex systems operating at the interface of order and chaos. For a good account of complexity theory, see Coveney and Highfield (1995). According to Horgan (1996: 194ff.), there are over 31 definitions of complexity theory and intense debates over its similarities to and differences from chaos theory. Horgan proposes the term "chaoplexity" to highlight to the overlappings in the concepts (1996: 191ff.).

12. This dialectical approach differs from that of Jacques Monod, who leaps from deterministic to indeterministic extremes, arguing that "chance alone is at the source of every innovation, of all creation in the biosphere" (1974: 110). For Birch (1988), Monod is a "mechanistic evolutionist" who simply reverses the logic of mechanism.

13. Hawking admits that we can never be completely certain we have the right theory, but he adds that if the theory is mathematically consistent and always provides predictions that agree with observations, "we could be *reasonably* confident that it was the right one" (1988: 167, our emphasis). He therefore rejects the skepticist claim that there could be no theory of the universe if it is governed totally by chance events, as well as a Feyerabend-like claim that there could be numerous theories of the universe, none of which are more true than the other and which do not cohere into a general picture.

14. In a similar vein, Bohm suggests:

> Because we are enfolded inseparately in the world, with no ultimate division between matter and consciousness, *meaning and value are as much integral aspects of the world as they are of us*. If science is carried out with an amoral attitude, the world will ultimately respond to science in a destructive way. Postmodern science must therefore overcome the separation between truth and virtue, value and fact, ethics and practical necessity. (1988: 67–68)

In the same mode, Prigogine and Stengers argue:

> We can no longer accept the old a priori distinction between scientific and ethical values. This was possible at a time when the external world and our internal world appeared to conflict, to be nearly orthogonal. Today we know that time is a construction and therefore carries an ethical responsibility. (1984: 312)

15. Ferré summarizes the necessary change aptly: "Modern science, in portraying the world as a Machine, alienated modern consciousness from the purposive, the responsible, and the whole; the task of postmodern science is to let us keep the modern tools of analysis sharp in their proper role as tools and to send us back into the Garden to work with respect and caution" (1988: 96). For other attempts to redefine objectivity within the boundaries of a more humane vision of the world, see Fromm (1947) and the proposals for a "new science" in Marcuse (1964).

16. Most recently, there has been an argument concerning what sort of cataclysm shaped the face of Venus (see *The New York Times*, July 16, 1996: B5) and Mars in the aftermath of the July 1997 *Pathfinder* voyage.

17. On the social context behind the current science wars, see some of the articles in the *Social Text* Science Wars issue (Numbers 46–47 [1996]), especially the articles by Stanley Aronowitz, Emily Martin, Dorothy Nelkin, and Andrew Ross.

18. It is clear that Sokal got many of his examples from Gross and Levitt, and he is dishonest in utilizing their sources and arguments without properly attributing their contribution to his hoax. In his "satire" he describes their critiques as "right-wing" and "a vicious right-wing attack" (1996a: Notes 41, 52). But in his published pieces in *Lingua Franca* (1996b, 1996d), he does not indicate his relationship to their work—although in a May 15, 1996, *All Things Considered* interview on National Public Radio and a *New York Times* interview (May 18, 1996: All), Sokal admits that Gross and Levitt's jeremiad was his inspiration. Moreover, in an Internet text that he circulated as a "proposed [but unpublished] article for *The Nation*," Sokal admits that Levitt "has become a friend" and describes their work as an "important book" (Sokal, 1996c). Further, there is evidence that others, whom he has also failed to credit in his published accounts of the fraud, collaborated with Sokal. Ruth Rosen in a *Los Angeles Times* article (May 23, 1996) indicates that she contributed to the article, Barbara Epstein was identified in an *In These Times* article as a "contributor" (May 27, 1996: 23), and we have talked to others who claim that they too contributed, indicating that Sokal had circulated his satire to a circle of friends and colleagues who helped him produce it. Indeed, it appears that Sokal was really not intelligent or conversant enough with the sources he quotes to write the satire himself. One has to understand the object of one's satire to effectively mock it, and Sokal has not really demonstrated that he understands the critique of science and positivism that he is attacking; instead, by his published responses to his hoax he indicates that he is himself a victim of an obsolete positivist ideology of science. Yet since Sokal's own writings on his stunt lack substantive content, it is difficult to discern what his own positions are, or if he even has positions.

19. Hart (1996) demonstrates how Gross and Levitt systematically distort and misrepresent the views of Haraway, Harding, Hayles, Latour, and Shapin and Shaffer. We will show how they totally misrepresent and fail to understand postmodern theo-

ry. In view of the shoddy scholarship of their work, we are astonished at the positive reception this book has received within the scientific community and broader public. As Hart has shown, the reviews of *Higher Superstition* have generally been uncritically celebratory, commending them for a "fine job," for "covering an enormous amount of literature with . . . care," and for exposing ignorance in the enemy camp of the humanities. We offer an alternative evaluation of their work.

20. It becomes clear by page 38 of their book that their real enemy is not the "academic left" but academic *postmodernists*, or radical social constructionists of various stripes. Indeed, they often substitute the term "academic postmodernism" for "academic left" or use the term "the postmodern academic left" (e.g., Gross and Levitt, 1994: 104). The concept of "academic left," by contrast, confuses the issue that their real target is Nietzsche, not Marx (indeed, Marx began the radical tradition of dialectical critique and support of the sciences and, if anything, was too uncritical). They see Nietzsche's doctrine of perspectivism to be "the central tenet of . . . the academic left" (38), but this is not a widely accepted position of traditional Marxists or of others, like radical environmentalists and animal rights activists, who are also blended into the "academic left" soup.

21. See Hart (1996) for many examples. Attacking an earlier study of postmodern science written by Best (1989), Gross and Levitt attempt to smear him as a blind champion of postmodernism, despite his explicit criticisms and distance from many postmodern positions. And though they cited Best's collaboration with Kellner (1991), they obviously have not read it, or they would have discovered a defense of the project of the Enlightenment—also found in Best (1995). Though they feel the need to correct Best for his error in interpreting Newton's mathematics, they reveal themselves to be incapable of even correctly citing the works and authors they list in their bibliography, misreferencing Kellner in their bibliography as "David" rather than "Douglas" and referring to the journal in which Best's article appeared as *Science and Culture* rather than *Science as Culture* (1994: 95). These blunders clearly disclose Gross and Levitt's haphazard "scholarship." Moreover, they identify Best as an acolyte of Jeremy Rifkin, and then proceed to appropriate his critique of Rifkin as their own (172). If they were to read the works they cite, they would know that Best's "heroes" are not Foucault, Derrida, Lyotard, and Baudrillard (95) but, rather, that he identifies with neo-Marxists and activists in new social movements. Nor has Best ever claimed or suggested that he is anti-science or relativistic in the sense that Gross and Levitt assert. As is clear in his work, as well as in his collaborations with Kellner, there is a defense of reconstructed notions of truth and objectivity and dialectical views on science common to the tradition of the Frankfurt School. Reading Gross and Levitt's critique of Best's views is thus like watching TV news coverage of an event one attended in person: It is totally distorted. But their misreading of Best is only an instance of their misreading of *everyone* they discuss: For example, they generate a straw critique of Sandra Harding based on only one cited paragraph and accuse Andrew Ross of parroting New Age thought when in fact he provides critical analyses. Consequently, anyone in the scientific world or elsewhere seeking a reliable reading of postmodern theories and contemporary analyses of science should definitely *not* consult Gross and Levitt.

22. In *The Flight from Science and Reason* (Gross et al., 1996), a collection of papers from a 1995 conference with that title, funded by the ultraconservative John M.

Olin Foundation, Gross and Levitt bring more lost neo-positivist sheep into their flock, all bleating in unison the same plaintive wail against the academic and postmodern left and other alleged enemies of science; the book is marked by repetitive genuflecting toward the great tome of Gross and Levitt, shrill hysteria regarding the attacks on science and reason, and largely undocumented and exaggerated claims concerning the flight from reason and science. Like the McCarthyists of the previous generation to whom they are intellectually and in some cases ideologically affiliated, conservative defenders of the bastion of Western civilization take isolated instances of excessive verbal radicalism as symptomatic of general subversion and treason to Western ideals. Yet like McCarthyism, the level of exaggeration and fantasy of these screeching polemics is such that their enemies are imaginary and their battles against barbarism largely quixotic. We should also note that although the Right had a predictable field day using the Sokal affair to attack the academic Left in publications like *The Wall Street Journal*, *Fortune*, and *The Washington Times*, in which they called upon state legislators to monitor the "nonsense" going on in the universities, and to denigrate the humanities and cultural studies per se, Sokal also received positive publicity in progressive publications like *The Nation*, *In These Times*, and *Z Magazine*. The latter critiques point to a split within the Left between its more theoretically inclined and university-based cadres and those outside the university or opposed to the new developments in theory. Whereas we agree that the Left should deploy clear and comprehensible discourse, should pursue rigorous standards of scholarship and truth, and should avoid mystifying obscurantism, we believe that the more advanced theoretical discourses can promote progressive causes, can be translated into intelligible language, and should not be rejected out of hand. We also believe that the critiques of science advanced by *Social Text*, other progressive journals such as *Science as Culture*, and a variety of individuals deploying critical theories are sound and should be embraced by progressives and all concerned citizens. Discrediting critiques of science, cultural studies, and the humanities does no one any good except the most retrograde and reactionary social forces, and those on the Left who celebrated the Sokal hoax should rethink their position.

23. Hence the title that Robert Young and Les Levidow gave to their excellent journal *Science as Culture*; many of Young's papers are collected in his Web site (http://www.shef.ac.uk./uni/academic/N-Q/psysc/staff/rmyoung/papers/index.html). The Gang of Three seems to be unaware that a vast literature critical of science has emerged from Britain and a variety of other places and is not just a fantasy of the American academic Left. They also never engage the cogent critical positions of the Frankfurt School, who were major players in the last round of the "positivism debate," which took place in the 1960s; see the discussion in Kellner (1989a).

24. Nor is Lewontin the only scientist who accepts social constructionist arguments; one finds throughout Prigogine and Stenger's book (1984) luminaries like Erwin Schrödinger, who observed that "there is a tendency to forget that all science is bound up with human culture in general, and that scientific findings, even those which at the moment appear the most advanced and esoteric and difficult to grasp, are meaningless outside their cultural context" (quoted in Prigogine and Stengers, 1984: 18).

25. For examples of the impact of corporate control of university-based scientific research, see Daniel Zalewski, "Ties That Bind" (*Lingua Franca*, June–July 1997:

51–60). As an example of the extra-scientific motivations and biases that often underlie scientific "objectivity," Roger Urlich, a former animal experimenter deeply regretful of the "years of torture" he inflicted on animals, revealed the following to *Monitor*, a magazine published by the American Psychological Association:

> Initially my research was prompted by the desire to understand and help solve the problem of human aggression, but I later discovered that the results of my work did not seem to justify its continuance. Instead I began to wonder if perhaps financial rewards, professional prestige, the opportunity to travel, etc. were the maintaining factors, and if we of the scientific community (supported by our bureaucratic and legislative system) were actually a part of the problem. (quoted in Singer, 1990: 70)

For some very disturbing evidence of the way many "scientists" actually treat animals in laboratory settings (including the infamous Head Injury Clinic at the University of Pennsylvania infiltrated by the Animal Liberation Front), see Blum (1994) and the video *Hidden Crimes*, produced by Suppress.

26. For yet another example of neo-positivism, see Weinberg's (1996) response to the Sokal hoax. Weinberg claims that the results of physics—until scientists grasp the origin of the universe or final laws of nature—have no relevance for culture, politics, or even philosophy. Thus, Heisenberg's conclusions on how quantum mechanics alters realist epistemology are dismissed as mere "philosophical wanderings," while Prigogine's theorizing is renounced as "hype." Weinberg is quite comfortable in his assertion that scientific thought exists in a "one-to-one correspondence with aspects of objective reality" and that science and culture exist as two utterly opposed worlds (11ff.). He therefore resents the battering of cultural critiques against the steel doors of science and feels, with Gross, Levitt, and Sokal, that orthodox science must protect its fortress against "the irrational tendencies that still beset humanity" (11).

27. Parusnikova (1992), for instance, identifies two main strands of postmodern theory, one of which he claims is compatible with science and the other not. The first "tendency" of postmodern theory is the Lyotardian idea that the world is fragmented into a plurality of discourses, each local and autonomous. This view not only captures the multiplicity of scientific discourses and allows for unconventional approaches and greater flexibility, but it also bolsters the standard self-image of science as an imaginative and anti-authoritarian discipline. The second tendency is the Derridean approach that seeks to deconstruct any claims to clear and present meaning in order to expose incoherences and contradictions and to frustrate any positive claims to truth. Parusnikova is right that this aspect of Derrida's philosphy is antithetical to the spirit and practice of science, but he fails to see the significance of the challenge Lyotard poses as well. By localizing all truth claims to specific discourses or communities, Lyotard does not necessarily disrupt norms of truth, evidence, and experimentation; but he goes right to the heart of modern science by attacking its universalizing ambitions and the positivist search for a complete description of reality, a description that would unify the natural and social sciences in one method and logic. Science is endangered by this kind of postmodern critique only if by "science" we mean "positivism." One begins to understand why Gross, Levitt, and Sokal are so reactive: Their own gods and most cherished beliefs are under attack.

28. Within days of the publication of the hoax, articles appeared in *The New York Times*, *Newsweek*, *The Washington Post*, and many other periodicals that usually do not pay attention to academic debates and cultural wars. National broadcast me-

dia also took note of the affair (e.g., National Public Radio, CNN, *The Rush Limbaugh Show*), as did professional journals. Most significantly, the Internet was ablaze with arguments, flames, and "interventions" on the affair. Several discussion lists that we subscribe to were dominated by the "Sokal affair." Within a week after Sokal's initial publications in *Lingua Franca* and *Social Text* in May 1996, a *Lingua Franca* editor said that he had compiled over 1,000 Internet messages on the affair, and right after the affair, the "alt.postmodernism" Internet list had over 1,000 entries devoted to the incident. There is a Web site that documents the continuing unfolding of the saga into 1997, in addition to Sokal's home page on the Internet, which contains a selection of texts and entries that debate his prank (see http://weber.u.washington.edu/~jwalsh/sokal and http://www.physics.nyu.edu/sokal.index.html). We should also note that the National Association of Scholars has a Web site and mailing list (http://www.nas.org/nassnl/contents.htm) that monitors discussions of the science wars and that posts quotations by the "academic Left," often taken out of context for easy attack by anyone wishing to join in the fray.

29. Indeed, his hoax seems to have produced a career for the previously unknown Sokal, who had been denied tenure at the NYU Courant Institute of Mathematical Sciences, and was, in effect, demoted to its Physics Department. His career in a tailspin, Sokal engaged in a clever piece of neo-dadaist satire that caught the moment and made him a much-cited source of anti-postmodern discourse and a much-sought-after lecturer and conference speaker. Consequently, he was invited in fall 1996 to keynote conferences at several U.S. universities, in addition to participate in public debates throughout the country. As of early 1997, over 20 public forums have taken place on the Sokal affair, including well-publicized affairs at Duke, NYU, and Michigan. In December and January 1996–1997, a series of articles were published in *Le Monde* and *Die Zeit*, and were discussed in such English journals as *Nature*. Thus, the Sokal hoax has made the previously unknown Sokal a global postmodern celebrity.

30. For an excellent summary of these challenges to modern science and alternative conceptions, see Harding (1994) and her reader (1993), which collects various constructionist, feminist, postcolonial, and other critiques.

31. As he explains, Feyerabend believes political anarchism "cares little for human lives and human happiness (except for the lives and the happiness of those who belong to some special group); and it contains precisely the kind of Puritanical dedication and seriousness which I detest" (1978a: 21, Note 12). In this gross caricature, Feyerabend conflates anarchism with Leninism and Stalinism. He goes on to declare himself a "flippant Dadaist" because in addition to not being able to "hurt a fly" (apparently he is ignorant of the history of violent conflicts between dadaists and surrealists in France), a dadaist abandons all pretense to seriousness: "A Dadaist is convinced that a worthwhile life will arise only when we start taking things *lightly* and when we remove from our speech the profound but already putrid [!] meanings it has accumulated over the centuries ('search for truth'; 'defense of justice'; 'passionate concern'; etc., etc.)" (21, Note 12). Hence, Feyerabend shows himself to be a radical postmodernist not only in his theory of knowledge but also in his theory of morals, a nihilist who mocks political and normative concerns and commitment to any cause whatsoever, save his own goal of dismantling any and all standards for truth.

CHAPTER SIX

Between the Modern and the Postmodern

Paradigm Shifts in Theory and Politics

> A profound shift is taking place in the metaphysical assumptions underlying the modern world . . . one can be sure that the new world that emerges will be so discrepant with the present as to fully justify the appellation *postmodern world.*
>
> —Willis W. Harman

> Chaos theory is an idea too rich, a metaphor too powerful not to spread "rhizomatically" into other fields like crabgrass on a suburban lawn.
>
> —John Barth

The scientific developments we just described have significant similarities to recent changes in the arts and social theory, leading us to believe there is a postmodern paradigm shift taking place in multiple fields of knowledge and the arts. It appears that the epistemological, metaphysical, and ethical assumptions about the nature of the world are rapidly changing in all fields, creating new configuration of thought, what Kuhn calls a "paradigm shift." In his most general sense of the term, a "paradigm" is an "entire constellation of beliefs, values, techniques, and so on shared by the members of a given community" (1970: 175). We have argued in the previous chapters that we are currently undergoing a major paradigm shift within the culture at large, parallel to the shift from premodern to modern societies and from medieval to modern theory. The debates, controversies, and diverse discourses of the postmodern are evidence of the reality of and stakes involved in the shift. The proliferation of postmodern discourse in popular and academic domains—and the polemics against it—indicate that there is significant cultural capital at stake in the postmodern turn and thus also interests, reputations, and concrete material investments.

Typically, one paradigm is replaced by another when a discipline reaches a crisis state that calls into question the explanatory adequacy of the existing paradigm, such that emergent problems are no longer seen only as "anomalies" and ad hoc "solutions" are no longer convincing. The shift to a new paradigm is a noncumulative, discontinuous development whereby novelty rules and tacit assumptions, theories, and techniques emerge that are incommensurably different from what preceded. The fresh paradigm dictates new kinds of research and defines new problems. In the early stages of a paradigm, there is no firm consensus over the nature of the new knowledge. As Kuhn points out, theorists can "agree in their *identification* of a paradigm without agreeing on, or even attempting to produce, a full *interpretation* or *rationalization* of it" (1970: 44).

We believe that this is the present situation with the emerging postmodern paradigm. The crisis in the mechanistic and deterministic paradigm of modern thought has been building for some time, of course, being registered in fields ranging from the sciences, phenomenology, ethnomethodology, critical theory, and feminism to postmodern theory. But there is as of yet, and there might never be, a consensus on the "rules" of the postmodern, let alone on its definition. A complex discourse like the "postmodern" is not an essence to be defined but is, rather, a discursive construct containing the sort of "family resemblances," described earlier by Wittgenstein (1958), of similar themes and methodologies operating in various disciplines. Nor is there widespread agreement as to what elements of modern thought remain valid and the extent to which modern theories should be superseded. The discourse of the postmodern remains a new "research program" (Kuhn) that reworks key aspects of modern thought and culture in a plurality of ways, that generates new theories and concepts, and whose outcome is indeterminate.

The question immediately arises as to how to know one is witnessing a paradigm shift, however new or inchoate, rather than identifying superficial similarities, random coincidences, ephemeral fashions, borrowed concepts, and trivial analogies. How can one know, in other words, that the changes in question stem from a common cultural context and add up to a shift in the *logic* of thought, to a new worldview that unifies various disciplines, to a *post*modern paradigm that promises to be as far-reaching and comprehensive as was the deterministic and mechanistic paradigm of modern thought in the 18th and 19th centuries? As Foucault observed, it is impossible to completely describe a new paradigm until after it has fully taken shape. Yet even Foucault, who professed not to know what the word "postmodern" meant, stated in the 1960s that "something new is about to begin, something we glimpse only as a thin line of light low on the horizon . . . [that] may well open the way to a future thought" (1973: 384, 386).

Since we claim that the postmodern paradigm is still only an *emerging* paradigm and is not yet "normalized," "dominant," or entrenched, it is unlikely that we can predict the exact forms and effects of this "future thought."

Ferré, for example, argues that "postmodern concepts and values are becoming pervasive, if not by any means yet dominant, in the various sciences" (1988: 93). Indeed, the weight of evidence suggests that a generalized postmodern paradigm is taking shape before our eyes. One has to look, first, to the sheer quantity of similarities across various disciplines, ranging from architecture and anthropology to economics and environmental ethics, from philosophy and physics to science and sociology, to theater and theology. In our studies, we traced the broad outlines of paradigm shifts from the modern to the postmodern in the fields of theory, art, and science. In this concluding chapter, we accordingly define the contours and features of this postmodern paradigm that these domains share in common.

Second, to grasp the importance of the paradigm shift from the modern to the postmodern, one must consider the extent and depth to which new postmodern ideas have restructured the field of knowledge in various disciplines. Even the fields of business organization and management have been reconceptualized, and in practice are being reconstituted, around the postmodern paradigm (see, e.g., the studies of postmodern management in Boje and Dennehy, 1994; Boje et al., 1996; and Hirshhorn, 1997). Yet despite the extent and depth of the variety of shifts from the modern to the postmodern, and the different ways that postmodern discourse is used to organize a wide range of new assumptions, terms, and methods, there are some core family resemblances of ideas and shared positions. Hence, although there is a bewildering variety of concepts identified as "postmodern" in various fields of knowledge and the arts, we can identify four main thematic similarities that break with distinct modern concepts and themes.

1. Postmodernists reject unifying, totalizing, and universal schemes in favor of new emphases on difference, plurality, fragmentation, and complexity. Thus, Derrida espouses the philosophy of *différance*, Lyotard champions the *différand*, Rorty calls for a multiplicity of cultural voices to join in the great conversation, Jencks embraces an architecture of multiple codes and styles, Feyerabend calls for the proliferation of scientific concepts, multiculturalists embrace the diversity of subject positions and identities, while postmodern science theorists generate new hypotheses and competing paradigms and some uphold the wisdom and scientific contributions of premodern and non-Western worldviews. Similarly, the emphasis on discontinuity in postmodern architecture parallels Foucault's archaeological approach to history, the jump-cut style of Godard's films, the disjointed aesthetic of music videos, and multilayered and nonlinear hypertexts. The theme of discontinuity also has application in the quantum mechanics theory of the discontinuous nature of subatomic particles, as well as in the psychological emphases of Lacan, Jameson, and Deleuze and Guattari on the schizophrenic nature of postmodern subjectivity. The theme of the end of the subject and of the author and

the theme of originality and authenticity resonate in areas ranging from philosophy and literary criticism to photopiracy and the sampling techniques of rap music.

The theme of "the end of grand narratives" in the postmodern condition (Lyotard, 1984) plays out differently in the various sectors of contemporary culture. In philosophy and social theory, it stands for the renunciation of grand systems of philosophy, of totalizing theory that attempts to capture the dynamics of the whole, in favor of local narratives, minor knowledges, ironic science, or more modest "poor thought" (Vattimo, 1988). It also involves rejection of narratives of progress in favor of discourses of conflict, crisis, or decay. In science, the refusal of foundations and a unified theory involves a turn toward fractal knowledges, toward seeking statistical probabilites in local regions, rather than searching for theories of the universe as a whole. In art, the postmodern involves abandoning the modernist search for the great work, for monumental oeuvres that would transform art and life, and turning to more ironic, sportive modes that simply appropriate, quote, and play with tradition.

2. Postmodernists renounce closed structure, fixed meaning, and rigid order in favor of play, indeterminacy, incompleteness, uncertainty, ambiguity, contingency, and chaos. Here we find parallels among the new theories of science that reject older theories of order, causality, mechanism, and prediction; the poststructuralist theories (of Roland Barthes, Derrida, et al.) of semantic indeterminacy and instability; the metafictional break with narrative structure in the work of Burroughs, Robbe-Grillet, Barth, and Pynchon; the emphasis on entropy in "steady-state" economics; and the free-form jazz of Ornette Coleman, the chance music of John Cage, and the experimental dance form of Merce Cunningham. Similarly, the concept of entropy has seeped into the postmodern zeitgeist, linking thermodynamics and physics with the life sciences and ecology, with information theory and communication, and with postmodern literature and social theory, since entropy plays a key role in writers such as Pynchon and Baudrillard.

The emphasis on indeterminacy, uncertainty, and contingency has led to the rejection of essentialism in many fields. The quantum theory that light is both wave and particle anticipates the postmodern rejection of identity logic that informs some modern theories of subjectivity, society, history, textuality, and meaning, all of which have been defined in Western thought as having a preexisting, unchanging essence. Just as postmodern science emphasizes a complexity of the natural world that far exceeds the simplistic causal models of the Newtonian paradigm, so postmodern social theory highlights the intricacy, decenteredness, and open-endedness of language, texts, discursive formations, cultural identities, and so on, claiming that the totalizing and essentializing biases of modern social theorists obscures semantic and cultural heterogeneity. And as modern claims to truth and universal themes are rejected as pretentious and overly serious, so new postmodern emphases on play emerge, ranging from poststructuralist and metafictional play with language to Feyerabend's dadaist play with notions of truth and justice to Baudrillard's play with the detritus of modernity. The ludic turn in postmodern theory involves, in some cases, an evacuation of normative standards, a position that many oppositional postmodernists resist.

Indeed, a major division runs through the postmodern turn between a ludic and oppositional postmodernism. In theory and art, ludic postmodernism involves renunciation of the seriousness of modern theory and modernist art in favor of a more playful attitude with cultural forms that is more experimental, ironic, and fragmentary. Yet oppositional postmodern art actively attempts to deconstruct conservative and traditional cultural forms and uses postmodern techniques (quotation, implosion, fragmentation, the popular) to create an art of resistance that criticizes existing society. Likewise, an oppositional postmodern theory and politics appropriates the techniques, style, and ideas of postmodern theory to transform theory and politics in order to engage the novel dynamics and problems of contemporary society.

3. Postmodernists abandon naive realism and representational epistemology, as well as unmediated objectivity and truth, in favor of perspectivism, anti-foundationalism, hermeneutics, intertextuality, simulation, and relativism. This development is visible in the perspectival outlook of Nietzsche and Foucault; in the standpoint theory of feminism and identity politics; in Maxwell's theory of electromagnetism, Einstein's theory of relativity, and Bohr's theory of complementarity; in the pragmatism of Dewey and Rorty; the textualism of Derrida; in the hyperrealism of Baudrillard; in the semiotic relativism of Jencks; and in the epistemological "anarchism" of Feyerabend. As Campbell summarizes these changes:

> No final, wrapped up, all-inclusive theory of reality will ever be perfected. The nature of language, the forms of logic, the duality of matter beneath the surface we observe, the power of rules to generate new structures, the limits to knowledge, the special character of complex as opposed to simple systems, all point to this conclusion. In this respect, science and art, philosophy and politics, history and psychology, meet on common ground, so that the barriers between the cultures break down under the recognition that all are incomplete and always will be; that no single discipline or school of thought has a monopoly on the truth. The truth itself has become more difficult to define as a result of the last half-century of discoveries in what used to be known as the exact sciences, making them richer but not necessarily more exact, and disturbing them to their foundations. (1982: 111)

Yet another distinction must be made within postmodern theory between those extreme postmodern theorists who reject truth, objectivity, and meaning altogether in favor of an ultraskepticism and relativism, and those who want to reconstruct modern epistemological concepts to provide new normative foundations for philosophy, social theory, and critique. Such reconstructive postmodern theorists temper skepticism with respect for mapping social structures, relations, and tendencies, attempting to develop theories adequate to conceptualize the developmental tendencies and emergent phenomena and trends of the present age. This difference in epistemology discloses once again that the discourse of the postmodern is a contested terrain between moderate and extreme postmodernists. Thus, the same division found in postmodern science between more reconstructive and deconstructive postmodern theory is present in a variety of theoretical domains.

4. A new emphasis is evident on deconstructing boundaries within and among different disciplines in the postmodern turn. Hence, Derrida attacks the distinction between philosophy and literature, Foucault works across various disciplinary boundaries such as history and philosophy, postmodern metafiction implodes the borders between history and fiction, and pop art challenges the barriers between art and everyday life. Distinctions between academic disciplines are being broken down even in the sciences, notorious for their specialization and provincialism. As exemplified by Griffin, Toulmin, Harding, ecologists, and critical theorists, advocates of postmodern science seek to eliminate the fact–value and theory–practice dichotomies in order to bring science into the broader horizons of ethics and social policy. The postmodern cultural avant-garde dissolves barriers between literature, theory, and criticism, producing new textual strategies and modes of writing. Postmodern artists like Laurie Anderson enact complex multimedia performances, computer art bridges the gulf between art and technology, and world music appropriates cross-cultural influences in a now-global music environment. Much of the best philosophy and social theory takes places in a transdisciplinary space (i.e., the work of many working within the paradigms of the Frankfurt School, postmodern and poststructuralist theory, cultural studies, feminism, multiculturalism, and the like). It is precisely those who are unable to break out of the paradigms and assumptions of their discipline, like Gross, Levitt, and Sokal, who are regressive forces in the present.

BORDER CROSSINGS

The paradigm shift from disciplinary to transdisciplinary approaches has spilled over into manifold realms of theory, the arts, and the sciences. Cross-disciplinary border crossing is evident in the field of literary criticism, where a number of essays and books have explored the connections between quantum mechanics, chaos theory, and postmodern fiction and literary theory (Hayles, 1984, 1990, 1991; Kuberski, 1994; Livingston, 1997). As Steven Johnson writes of this boundary-busting phenomenon: "Overcoming the hostility between the lab and the library, literary critics have sneaked over the disciplinary border into the remote fields of chaos and complexity theory, where they have found new tools for making sense of texts" (1996: 42). In light of the influence of chaos theory, novels and literary texts are being redefined as "complex systems" (Joyce's *Ulysses*, for example, is read by Franco Moretti [1996] as a balance of orderly and chaotic "behavior" in its narrative structure and semantic complexity).

As Peter Coveney and Roger Highfield (1995) show, the ideas of chaos and complexity theory have influenced the plays of Hugh Whitemore and Tom Stoppard, the poetry of Paul Muldoon, and even the work of various architects (see, e.g., Jencks, 1995). In many cases, these influences are quite deliberately assimilated, as in the case of Tom Stoppard's play *Arcadia* or Michael Crichton's *Jurassic Park*. "Complex" images such as fractals and Lorenz's butterfly have also richly stimulated the arts, where the use of computers and genetic programming techniques spawns fascinating creations. For example, William Latham's and Karl Sims's "evolutionary art" erases the line between art and technology and blurs distinctions between art and life through computer methods used to "breed" evolving designs. These artists employ "aesthetic selective pressure" that chooses which images and shapes will "survive" the simulations (see Coveney and Highfield, 1995: 341–343).

There is, of course, a long-standing tradition of the arts borrowing from the sciences, so it should not be too surprising to witness the appropriation of chaos and complexity theories in contemporary plays, novels, and literary criticism. Literary naturalists of the 19th century incorporated Darwin's theory of natural selection into their work, just as James Joyce incorporated some of Einstein's ideas of relativity and curved space into *Finnegan's Wake*, and Pynchon employed ideas of entropy and indeterminacy. Further, the genre of science fiction is a direct modeling of fiction on scientific ideas and imagination. Scientists often balk at the extension and dilution of scientific concepts into literary or philosophical metaphors, especially when they think they detect a political agenda at work (see Gross and Levitt, 1994). Sometimes, indeed, the metaphors are stretched and forced, as when Arthur Kroker (1992) speaks of "fractal subjects" or literary theorists analyze novels as complex systems. But transposing metaphors from one discipline to another can stimulate new thinking and generate new insights, and the importing of concepts from contemporary science into other fields has promoted important forms of transdisciplinary work.

Often, theorists and historians privilege science as the epicenter of conceptual changes that come to influence other disciplines. In many cases this is true; one need only consider the impact the modern scientific revolution had on the Enlightenment and on modern philosophers as divergent as Kant, Marx, and the logical positivists or the influence of ecology on social policy and ethics. In other cases, however, the converse is true, as developments in pragmatist philosophy have strongly shaped later concepts of science and, as argued by Leonard Shlain (1991), the arts have anticipated changes and developments that were later foregrounded in science, as he claims Manet, Monet, and Cézanne intuited the revolution in physics initiated by Einstein.[1] Ultimately, the "roots" of the postmodern seem to be rhizomatic, criss-crossing and spreading from numerous loci while coalescing into a new paradigm.

Yet if there is an Ariadne's thread that winds through all postmodern developments, it is the linguistic turn that shattered realist assumptions about language, knowledge, and representation. The linguistic turn is an explicit realization that (1) the primary way human beings know and participate in their world is through language and (2) different linguistic maps bring different senses of reality and claims to truth. The linguistic turn, therefore, is the eruption into human consciousness of the perspectival, contextual, and contingent nature of all truth claims. Language does not re-present reality, rather it shapes and constructs it, refracting the light of the world through its unique phonetic and conceptual prisms. As discussed in philosophy by pragmatists such as Peirce, Dewey, and the later Wittgenstein, truth claims of any kind are embedded in distinct linguistic conventions and communities. As played out by philosophers of science (see Kuhn, 1970; Hesse, 1980), "truth" exists relative to a community of scientific practitioners who create various procedures to guide research and judge epistemic validity. Since there are alternative scientific communities, there are also competing descriptions of truth and strategies of inquiry.

The recent emphases in so many fields—from art and literary criticism to philosophy and science—on language, information, communication, cybernetics, semiosis, dissemination, and the like, underscore the shift from the deterministic and mechanistic paradigm of modern thought to a postmodern paradigm that sees the universe and human agency as active, dynamic, spontaneous, and creative. Life is not mere matter in motion but a self-organizing, self-regulating, communicating (in the broadest sense) cybernetic force; life involves not only an exchange of matter and energy, but also an exchange of information. Human beings and the texts they create are endlessly rich in signification and meaning, producing polysemic forms of ambiguity demanding a multiplicity of interpretations.[2] And the emergence of postindustrial society involves a shift "from metallurgy to semiurgy" (Baudrillard), from the production of goods to the production of information, and from a traditional industrial blue-collar working class to white collar specialists in computer technologies.

Numerous theorists have argued that a postmodern paradigm shift is occurring. From the standpoint of literary criticism, for example, Hans Bertens (1995) concluded that ontological uncertainty and indeterminacy characterized the postmodern shift. Social theorist Zygmunt Bauman (1992) interprets the postmodern as a renunciation of certainty, universality, and other features of modern theory in favor of uncertainty, contingency, and ambiguity. Bauman suggests that the postmodern philosophical positions follow from relativity theory, quantum mechanics, and the undecidability principles of mathematics, suggesting connections between major conceptual shifts in the fields of philosophy and science. His picture, and that of Bertens and others, remains incomplete, however, until we (1) add the influences of systems theory, information theory, and chaos and complexity theory, all of which problematize determinist schemes of explanation; (2) connect these elements with similar changes occurring with literature, the arts as a whole, politics, and culture and everyday life in the broadest sense (see Chapters 3 and 4); (3) accentuate other key factors in the postmodern paradigm that involve the will to implode boundaries, transdisciplinary and multiperspectival thinking, eclecticism and emphases on radical difference, and the hermeneutic attack on realism and foundationalism; and (4) relate these changes to new social developments (such as the impact of computer technologies and the global restructuring of capitalism) as interpreted by critical social theory.

The postmodern paradigm shift suggests that the gulf between the "two cultures" of science and the humanities is in some ways being bridged. The shared emphasis that truth is conditioned by language and culture in general and the incorporation of temporality and historical methods into science, as well as into literature and philosophy, dissolve rigid methodological boundaries among various fields while still allowing science to be characterized as a unique enterprise with its own distinctive methods and results. The transdisciplinary turn does not imply that scientists know their Plato or Shakespeare better today, any more than it implies that people rooted in the humanities know their quantum physics or chaos theory; rather, it implies that a conscious use of postmodern concepts and methods may provide some common ground of discussion and that science, technology, and culture are becoming increasingly imbricated with one another. Of course, as the recent science wars show, postmodern concepts and critiques are passionately debated and contested and often dogmatically renounced, and a future "postmodern consensus," or a shared postmodern paradigm, is unlikely. At least for now, conflict between modern and postmodern paradigms, and over the postmodern itself, is intensifying, and the extent and passion of this conflict suggests that the postmodern debates will be with us for some time.

The erosion of the boundaries between the two cultures is a key insight of cyberpunk fiction, with its vision of the implosion of science, technology, and human beings in today's technoculture, as well as its use of the concepts of entropy, chaos, and ecology. As Bruce Sterling explains:

> Traditionally, there has been a yawning cultural gap between the sciences and the humanities: a gulf between literary culture, the formal world of art and politics, and the culture of science, the world of engineering and industry.

> But the gap is crumbling in unexpected fashion. Technical culture has gotten out of hand. The advances of the sciences are so deeply radical, so disturbing, upsetting, and revolutionary, that they can no longer be contained. They are surging into culture at large; they are invasive; they are everywhere. (1986: xii)

Symptomatically, extreme postmodern theorists like Baudrillard and Kroker draw heavily from scientific findings and scientific metaphors. Baudrillard, for instance, makes frequent use of terms such as "black hole," "implosion," and "entropy," while Kroker and his associates proliferate cybernetic metaphors, which are also central to the work of Gregory Bateson (1972, 1979), Capra (1982, 1996) and those involved in systems research (see Buckley, 1968). But though postmodern science moves from mechanism to organism, some versions of postmodern theory expunge nature and organism altogether from their conceptual framework, depicting the erasure of nature and the human by technology in visions of the high-tech dys/u/topias of ascendent capitalism.

Though the descriptions of postmodernists like Baudrillard are extreme and unwarranted and one can question the validity of applying metaphors such as "black holes" to human society, it is a mistake to dismiss such work in toto. Instead, we can read hyperbolic postmodern theory as science fiction, more valid as a "literary" than a "theoretical" mapping (the borders are in question) that, though not literally true or accurate, portrays *tendencies* of our current world and frightening partial truths, painting a grisly picture of things to come or currently happening (see Kellner, 1995a, Chap. 8). The major shortcoming of dystopic postmodernism is that it fails to theorize important countertendencies, contradictions, and forms of resistance. Thus, while this work may not always be true or accurate, it nonetheless vividly dramatizes some terrifying possibilities of the present and perhaps shocks us into critical awareness of our contemporary world; paraphrasing the 1980s TV show *Max Headroom*, it depicts a world only "twenty minutes into the future." Indeed, as one follows the rapidly changing, fractal contours of high-tech society, Baudrillardian postmodern theory often emits a Doppler-like red glow, as the situations it describes often move right by us, from the future into the past.

Both modern theory and postmodern theory are in agreement that contemporary society and culture is wracked with fragmentation, conflicts, contradictions, and disorder. Modern theory and politics wants to discover resolutions to these conflicts, to (re)create harmony and order. Some versions of postmodern theory and politics, by contrast, live within the fragmentation and disorder, affirming positive possibilities within the whole, devising more modest survival strageties for life in the fragments, or attempting to solve piecemeal problems. Yet we confront a highly conflicted contemporary situation marked by growing implosion of boundaries and social domains, accompanied by increasing fragmentation, strife, and divisiveness. For instance, at the same time that aspects of gender, race, and class divisions are imploding in a multicultural, consumer society, inequalities and conflicts between these groups are growing. Baudrillard (1983a, 1983b, 1993) argues that a radical implosion is occuring in the postmodern scene between the economic, political, and cultural, and some sociologists are claiming that implosion and dedifferentiation is *the* criterion of the postmodern (see Lash, 1990). Yet while there are clearly tendencies toward implosion and the erasure of the boundaries and divisions of modernity, there are also growing divisions and conflicts, as are manifest in the "culture wars" among liberals, conservatives, and radicals, and the "theory wars" between proponents of various strands of modern and postmodern theory. In addition, increasing racial tensions, with the rise of Neo-Nazis throughout the world and the militia movement in the United States, as well as the greatest levels of global economic inequality in the last half-century, with over half of the world's population earning less than $2 a day (*The New York Times*, July 15, 1996: A3), promise increased conflict and turmoil in the future. Moreover, although it is becoming increasingly clear that the crises of the environment, the economy, and the world's energy resources are interconnected, the planet is dividing into ever-more sovereign nations and a coherent global strategy to overcome these crises is still lacking.

One significant debate that we take up in the next section involves controversy over the significance of nature and ecology in today's increasingly technological world.

ECOLOGY AND THE DEFICITS OF POSTMODERN SCIENCE AND SOCIAL THEORY

> A growing sense is now evidenced that we can and should leave modernity behind—in fact, that we *must* if we are to avoid destroying ourselves and most of the life on our planet.
>
> —David Griffin

> We *cannot* go on as we are; we must have something really new and creative. This step cannot be merely a reaction to the breakdown of the modern world order, but must arise out of a fresh insight.
>
> —David Bohm

> Nature speaks with a thousand voices, and we have only begun to listen.
>
> —Ilya Prigogine and Isabell Stengers

> Society seems intent . . . on making a noteworthy contribution to entropy.
>
> —T. W. Adorno

Although postmodern science and social theory have strong similarities, there are also important differences between them. While both are critical of modern rationality, postmodern science tends to be far more optimistic about the value of science, technology, and rationality than is most postmodern social theory. Much postmodern science believes that science and technology can be rationally managed and applied in order to harmonize with nature rather than destroying it; it claims that we can overcome the destructive aspects of modern epistemology and embark on a more emancipatory trajectory that some theorists see as the transition to a new postmodern global order (see Griffin, 1988a, 1988b; Harding, 1993). Unlike Baudrillard and other theorists who claim that we are already in a new postmodernity, some postmodern theorists argue that we are all too trapped in modernity and that the path to postmodernity is a positive trajectory yet to be achieved. Yet others claim that postmodern consciousness is merely a stage in the awakening of critical consciousness from the damaging myths of modernity, a moment of nihilistic paralysis, and we need to go "beyond the post-modern mind" (Smith, 1982) into a more positive mode of thought that retains progressive modern norms in reconstructed form.

Moreover, where quantum mechanics and chaos theory adopt a relational logic that understands entities or events to exist within a larger whole or relational field, postmodern social theory frequently employs a reified logic that isolates phenomena and eschews contextual and systemic analysis, which it equates with the supposedly terroristic totalization of dialectics. Postmodern science adopts a holistic logic that understands nature as comprised of self-organizing systems and the individual person as a unified mind–body whose physical and mental health is closely interdependent; this is opposed to some postmodern social theory that stresses the fragmented and schizophrenic nature of contemporary consciousness and experience. Thus, though some postmodernists reject the concept of the subject and decenter the individual to language, desire, or social processes, the focus in postmodern science is on the individual as an active, creative, and ethical subject.

In breaking with the assumptions of the classical theory of representation, both postmodern science and social theory move toward a pragmatic theory of truth and knowledge, but more extreme forms of postmodern theory radicalize the break with representation initiated by Nietzsche, the pragmatists, and quantum mechanics. The idea that physical reality cannot be objectively or precisely determined has generated within some versions of postmodern theory nihilism and despair about the possibility of mapping an increasingly complex and abstract social world. The problematization of the subject–object dichotomy in quantum mechanics is radicalized in Baudrillard's implosive theory, which describes an object world that has totally engulfed the subject in its proliferating networks, as well as in Feyerabend's exultant epistemological anarchism.

In postmodern social theory, entropy has proven to be an indispensable metaphor to describe the degradation of our contemporary social and cultural existence. The scientific emphasis on the inevitable disintegration of natural systems has its analogue in social theory in the apocalyptic sense of the end of modernity, the postmodern "aesthetics of disappearance" of the grand referents and narratives of modernity, and the fin-de-millennium sense of despair, exhaustion, and nihilism. Unlike postmodern science, however, the social theory concept of entropy is stripped of any positive, ethical, or ecological values and disconnected from any opposing notions of autopoiesis or self-evolution.

For Lyotard, the "postmodern condition" arises when there is a widespread "incredulity toward metanarratives" (1984: xxiii). Although the move is not taken by the major postmodern social theorists, it is but a short step to an ecological perspective in which postmodern "incredulity" realizes the bankruptcy of modern views of progress and understands that the modern age of cheap, nonrenewable resources is ending. The entropic notion that there are real limits to the uses and amount of (available) energy in the universe directly contradicts Enlightenment, linear-progressive visions of history and the modern belief in a cornucopic world. As population theorist William Catton (1980) argues, the last four centuries of exuberant growth and progress were made possible by the nonrepeatable achievements of the discovery of a second hemisphere and the unimpeded exploitation of fossil fuels, and we now live in a "postexuberant" era of resource depletion.

The shift from the modern belief in inexhaustible resources—which remains the dominant view of the advanced industrial world and continues to seduce the developing world—to the postmodern realization of scarcity and finitude allows for a new ethic of conservation, a new appreciation of ecology, a critique of consumerism, and a new vision of "sustainable" societies and consumption habits that are ecologically sound. Thus, in economics, the entropy principle has created a paradigm shift from the notion of a growth economy to a "steady-state" economy (see the essays in Daly and Townsend, 1993). In this context, the principle of entropy recognizes that economic activity is merely an extension of the environment, rather than vice versa, and hence must never be allowed to outstrip an ecosystem's ability to recycle waste and renew available resources.

For ex-World Bank economist Herman E. Daly,

> the major blind spots of growth economists are their lack of appreciation of the second law of thermodynamics and its manifold implications and their failure to recognize the magnitude and fragility of the life support services provided by the very natural systems whose proper functioning is being disrupted by the ever-larger entropic flow of materials and energy required by economic growth. (Daly and Townsend, 1993: 52–53)

Similarly, M. King Hubbert believes that the modern world has evolved into "a culture so heavily dependent upon the continuance of exponential growth for its stability that it is incapable of reckoning with problems of nongrowth"

(1993: 125). For Jeremy Rifkin, the modern world faces the "diminishing returns of technology" and lives in an entropic situation in which "monumental collapse could come at any time" (1989: 285). The only way out of this impasse, Rifkin argues, is to reject all reformist schemes and the deluded dream of a technological fix and to reconstruct society through a new "postmodern worldview" organized around the principle of entropy: "The entropic way of thinking provides a context for a postmodern form of consciousness that is far more sophisticated and challenging than the forms of knowing we have relied on in the past" (18).

Like entropy theory, chaos theory teaches important ecological lessons. Whereas the old mind-set held that small changes in a system produced equally small results, chaos theory warns us that such initial changes could have enormous consequences. Like the concept of entropy, chaos theory "alert[s] us to the possibility of a sensitive and fragile world, necessarily changing the way we think about it and (ought to) treat it" (Kamminga, 1990: 58). The implications of chaos theory are that nothing is certain, only probable, and that complex processes cannot be completely known, accurately predicted, or fully controlled. Our attempts to "control" the natural world therefore frequently backfire through ignorance of certain "initial conditions" of complex ecological relations. We build levies and dams to control the flow of a river, but this interferes with subtle, unknown ecological dynamics and we inadvertently cause chronic flooding to areas like Egypt and the U.S. Midwest. We try to exterminate insect populations with deadly chemicals such as DDT, but we only sicken and kill ourselves and create insect species resistant to chemical poisons.

Our illusions that we have vanquished diseases like polio have recently been shattered by the new resistance of deadly diseases to antibiotics that have been overused to treat mild ailments and that pour into our bodies through heavy consumption of meat and dairy products. In addition to the return of old diseases, in the last few decades we have witnessed the emergence of altogether new diseases like Ebola, Lassa fever, AIDS, and Legionnaires' disease. According to many medical experts, we now live in a "post-antibiotic" era (Garrett, 1994) that belies the modern illusion of control of the body through drug technologies. There is evidence that the invasion of life with synthetic chemicals has even produced "hormone impostors" and "endocrine disrupters," synthetic chemicals triggering a vast array of problems in both animal and human populations, ranging from hormonally triggered cancers to neurological disorders to reduction in sperm counts—perhaps the revenge of nature against the human species (see Colborn et al., 1996). Perhaps most frighteningly, genetic engineers, like Dr. Frankenstein, have embarked on a reckless tempering with the creation of life without considering the potential long-range consequences of their actions.

Hence, postmodern developments have important implications not only for the "high" or theoretical sciences but also for the practical use of technologies and for other sciences such as medicine and ecology. The emergence of an alternative health care community, from which significant numbers of Americans now seek treatment for problems that bedevil conventional medicine, as well as the combined use of standard and alternative therapies in hospitals and operating rooms and the huge popularity of New Age philosophies, demonstrates the emergence of a postCartesian, postmodern paradigm in medical science and health care, of a holistic approach that treats the mind–body continuum as a natural ecology.

Prigogine and Stengers, Griffin, Sheldrake, Oeschlaeger, Rifkin, and numerous others privilege ecology as central to postmodern science and see one of the key tasks of the new science to be overcoming the crippling opposition between human beings and the natural world, which is fundamental to the entire Western tradition. Prigogine and Stengers, for example, believe that postmodern science can take a nonexploitative relationship to nature and other human beings. This involves a "reenchantment of nature" that sees nature as more than merely a dead machine, that seeks a rapprochement between human beings and the natural world, and that refuses to anathematize reason as repressive or merely instrumental: "We no longer have to settle for the previous dilemma of choosing between a science that reduces man to being a stranger in a disenchanted world and antiscientific, irrational protests" (Prigogine and Stengers, 1984: 36). They call for a "new dialogue" with nature that will bring us closer to the "universal message" of science (7).

Prigogine and Stengers do not clarify what "reenchantment" means, what the nature of this "universal message" is, or how this is to be accomplished. They do not discuss the social changes required in order to change science itself. Their claim that we are moving toward a "new naturalism" obscures the extent to which we are also plunging into an increasingly artificial, chemically saturated, technically mediated, mass media and cybernetic environment, such as that described by some postmodern theorists. The extreme optimism of their work with regards to tendencies within science mystifies the extent to which science is still thoroughly implicated in repressive social and military policies.

Rifkin, conversely, discusses institutional changes but only in the vaguest way, and his position is marred by mystical tendencies that wax poetically about love, the "timeless" realm of the spirit, and the "natural goodness of [the] cosmic process" (1989: 291). In Rifkin's New Age ecometaphysics, nature is inherently good, and all forms of social evil result from our basic alienation from its "laws." The ultimate solution to our current social and personal crisis is to surrender to natural processes: "The only hope for the survival of the species is for the human race to abandon its aggression against the planet and seek to accommodate itself to the natural order" (287).

Such thinking recycles the quietism inherent in many forms of Eastern philosophy, which encourages us to merge with the undifferentiated whole, a potentially dangerous idea that could stifle individuality and critical reason. Moreover, phrases such as "harmony with nature" are vacuous and dangerous. If nature is evolving, running down, chaotic, or indifferent, how does one seek harmony with it and what good would result? If nature inflicts a virus, famine, or flood on the human world, should we do nothing but allow nature to "take its course"? While the basic idea here is sound—that nature has its own autonomy, which we must respect and adhere to insofar as the social world is inextricably embedded within natural ecosystems—not all human interventions in the natural world have negative effects, and, increasingly, the continued survival of many species and ecosystems today depends on rational and democratic human intervention (see Light and Higgs, 1997, on ecological restoration debates).

In conflating natural economy with political economy, Rifkin and others obscure the class relations and social inequities that emerge from the social production process. There can be no "harmony with nature" as long as capitalism and its alienating effects mediate relations between nature and human beings. Our connection to the natural world has to be understood as socially and historically mediated, and the *category* of nature itself must be seen as a historical construct, which is not to say that nature itself is nothing but a "text."[3] Postmodern social theorists on the whole are too sophisticated and historical in their outlook to fall prey to the kind of metaphysics and ecobabble that mars some postmodern science, but many neglect ecology entirely. Key postmodern theorists analyze language, discourse, and signification as ideal worlds of meaning divorced from social and natural environments. Derrida, Lyotard, Foucault, Rorty, and others focus strictly on critiques of Western philosophy while failing to engage its core anthropocentric outlook. Baudrillard (1988a) declares that we are imprisoned in a world of images, texts, and signs but finds no natural degradation in his semiotic utopia: he speeds through the U.S. West with the top down and whiskey bottle in hand, commenting on Americana, but he fails to observe the rapid disappearance of wilderness and wildlife.

Postmodern theorists are helpful in showing how we are deeply enmeshed in technological, social, and semiotic systems, but, unlike the Frankfurt School and postmodern science, they fail to analyze our embeddedness in the natural world. Ironically, some of the key influences on postmodern theory—Horkheimer and Adorno's *Dialectic of Enlightenment* and Heidegger's critique of technology and the "enframing" attitude of Western culture—were directly concerned with the domination of nature and new relations to nature but were translated by many postmodern theorists into strictly social critiques. Thus, whereas Herbert Marcuse called for a "new science" that would include new values and new relations to nature, and the overcoming of the rift between humans and nature (see Marcuse, 1964, and the discussion in Kellner, 1984a), most postmodern social theorists neglect nature and the need for ecological perspectives and fail to develop a critical theory of how society is degrading the natural environment.[4]

While the postmodern critique of humanism could be easily applied to an understanding of the psychological and social dynamics informing anthropocentrism, this move has not been made by any leading postmodern social theorist. No major postmodern philosopher or social theorist, in other words, has explored the connections between the deification of the Western rational self, the alienation from the natural world, and the modern project of the domination of nature. Such critiques, in contrast, have been the focus of deep ecologists, who operate in a postmodern and posthumanist framework and who find crucial the later Heidegger's critique that "technological society" is inherently a mode of dominating nature and "Being." Deep ecology, however, is typically mystical, and its deification of nature usually leads to neglect of the socioeconomic forces that are destroying nature. The situation is different, however, in the rapidly developing fields of environmental theory and ethics, where postmodern concepts and methods have spawned a new wave of environmental thinking that is highly critical of modern capitalism, modern science, and the modernist assumptions of much environmental thought (see Oelschlaeger, 1991). However, there has also been a critical reaction against postmodern concepts, not only by Gross, Levitt, and Sokal but also by environmentalists concerned for its relativistic implications; these critics argue that postmodern claims that all things are linguistic or cultural "constructions" potentially make Disneyland as valid an "environment" as the Brazilian rainforest and undercut normative attempts to condemn the degradation of the earth.[5]

The project of Bacon and Descartes has been played out, and we now live with the consequences of human arrogance and the war against the natural world. The environmental crisis is the combined result of a long-standing Western anthropocentric tradition, the modern mechanistic worldview predicated on violent opposition to nature, and the capitalist economic system that requires endless growth, expansion, and accumulation. The transcendence of an insane, destructive society for a sound ecological world demands radical social and personal changes, a revolution in the way we organize our political and economic life and relate to the natural world.

As Griffin, Toulmin, Prigogine, and others are urging, science, which has done so much to shape the dualistic and destructive worldview of modernity, can now play a key role in changing it. With many scientists having abandoned the fact–value dichotomy, science is in a position to contribute to a rethinking of the fundamental policies and practices that shape our current world and to embark on more a emancipatory path. In the movement toward the postmodern, a pseudo-objective "science" informed by an interest in controlling and exploiting is giving way to a "metascience" whose full normative context includes reference to ethics, values, social justice, and ecology.

For Prigogine and Stengers, the human race is now in a crucial period of transition, which makes it "more important than ever to keep open the channels between science and society" (1984: xxx). Yet if the goals of postmodern science are to remain anything but utopian fantasies or academic theorizing, and if ecological theories are to avoid a reactionary cast of Malthusian misanthropy, lifeboat ethics, and mysticism, they have to be articulated with a critical social theory with radical political intent. More than ever, science today, as a key force of production, is controlled by business and government and is tied to the profit imperatives of transnational capital. Research for AIDS and other vital human projects is subordinated to studies benefiting corporations and the military or is controlled by entrenched interests (see Duesberg, 1996). A critical analysis of science is necessary to understand the mechanisms of domination in society, the ways these affect science itself, the political implications of scientific research, and the sociopolitical determinants on the behavior of natural systems (the greenhouse effect, for example, is not simply a natural phenomenon but the result of specific social and economic policies).

In its emphasis on the pervasive effects of technological systems, its linking of knowledge to power, its political analyses of discourse and micropowers such as one finds in schools or psychiatric institutions, and its multiperspectival outlook, postmodern social theory can contribute to such a critical theory. Yet some versions of postmodern social theory suffer from defects that result from the rejection of important tenets of modern social theory, such as the attempt to grasp systemic relations in society, to ground social critique in normative assumptions, or to transform society along emancipatory lines. Any future postmodern science and ecological theory of real value must merge with a critical social theory that employs the best insights of both modern and postmodern theory.

POSTMODERN POLITICS AND THE BATTLE FOR THE FUTURE

The old conceptual and political systems are breaking up, and new forms of order are emerging out of the chaos. The twilight zone of the modern–postmodern yields new dangers, such as the potential loss of modern humanism, the Enlightenment, and radical social traditions, as well as new possibilities, such as emerge from new technologies, new identities, and new political struggles. The old theories, concepts, and modes of thought and analysis will only go so far in theorizing, analyzing, and mapping the new constellations, thus requiring new strategies, discourses, and practices. Accordingly, in addition to the transformations in theory from the modern to the postmodern that we have discussed in this book, there are calls for a new postmodern politics to overcome the limitations of modern politics.

Thus, the postmodern paradigm would not be complete without discussing changes in the nature of politics, and, indeed, the contemporary terrain shows a mutation in political thought and practice that parallels and is informed by changes in theory. As with postmodern theory, there is no single "postmodern politics" but, rather, a conflicting set of positions that emerge from the ambiguities of social change and multiple theoretical perspectives on the postmodern. The different categories of postmodern politics are not merely conceptual distinctions but are actual political tendencies played out in the public sphere, in the universities, in the workplace, and in everyday life.

Generally characterized, the project of modern politics was to define and implement universal goals like freedom, equality, and justice, in an attempt to transform institutional structures of domination, and thus to intervene into the public sphere. Modern politics often involved a politics of alliance and solidarity whereby oppressed groups joined together to struggle for common interests. The American Revolution declared the universal rights of "all people" to be "self-evident truths" revealed by "the light of reason." The French Revolution championed the universal "Rights of Man" on the basis of *liberté, égalité, fraternité,* and shortly thereafter, in 1792, Mary Wollstonecraft published her treatise *A Vindication of the Rights of Woman* (1975). Trying to realize these universal appeals beyond the limiting context of bourgeois class relations, Marx urged that the "Workers of the World Unite!" to create an international politics of solidarity designed to overthrow bourgeois property forms. Despite war, poverty, hunger, economic depression, and fierce forms of oppression and suffering, modern politics was optimistic in its outlook; indeed, it was often religious in its teleological faith that the progressive logic of history would soon be realized. Thus, modern politics was informed by strong normative values and utopian visions of a world of universal freedom, equality, and harmony.

A postmodern politics, by contrast, began to take shape during the 1960s, when numerous new political groups and struggles emerged, and is strongly informed by the vicissitudes of liberation movements in France, the United States, and elsewhere, as well as by emerging postmodern theories (see Chapter 1). The utopian visions of modern politics proved, in this context, difficult to sustain and were either rejected in favor of cynicism, nihilism, and, in some cases, a turn to the right or were dramatically recast and scaled down to "modest" proportions. The modern emphasis on collective struggle, solidarity, and alliance politics gave way to extreme fragmentation, such that the "movement" of the 1960s splintered into various competing struggles for rights and liberties. The previous emphasis on transforming the public sphere and institutions of domination gave way to new emphases on culture, personal identity, and everyday life, as macropolitics were replaced by the micropolitics of local transformation and subjectivity.

In the aftermath of the 1960s, novel and conflicting conceptions of politics emerged. A postmodern politics, for example, would include the antipolitics of Baudrillard and his followers, who exhibit a cynical, despairing rejection of the possibility of emancipatory social transformation. Indeed, according to this position we are stranded at the end of history, paralyzed and frozen, as the masses collapse into inertia and indifference and as simulacra and technology triumph over human agency. Thus, from Baudrillard's perspective, all we can do is "accommodate ourselves to the time left to us" (1988b: 44).

Another form of postmodern politics also rejects utopian visions of liberation, global politics, and attempts at large-scale social transformation but eschews the nihilism of Baudrillard in favor of an emphasis on piecemeal reforms and local strategies. This is the position of Foucault, Lyotard, and Rorty, all of whom reject a global politics of systemic change in favor of a politics of modifications at the local level designed to enhance individual freedom. Foucault and Lyotard reject utopian thought and the category of "totality" as terroristic while searching for new "styles" of life "as different as possible from each other" (Foucault) and for a proliferation of "language games" in "agonistic" opposition to one another (Lyotard). Rorty merely—and meekly—seeks "new descriptions" of reality that pluralize the voices in the social "conversation," as he replaces normative critique with "irony" and restricts philosophy to a limited role in private life. This form of postmodern politics, consequently, is but a refurbished liberal reformism that fails to break with the logic of bourgeois individualism and subverts attempts to construct bold visions of a new reality to be shaped by a radical alliance politics.

Conversely, a third form of postmodern politics, advanced by Laclau and Mouffe, stakes out a position between the modern and the postmodern in order to use postmodern critiques of essentialism, reductionism, and foundationalism to reconstruct Enlightenment values and socialist politics through a logic of contingency and plurality (see Laclau and Mouffe, 1985). Rejecting the Marxist reduction of radical politics to a class struggle that privileges the working class, Laclau and Mouffe embrace the "new social movements" of the 1970s and 1980s as multiple sources of radical change that can bring about radical democracy.

According to Mouffe, Enlightenment universalism was instrumental in the emergence of democratic discourse, but

> it has become an obstacle in the path of understanding those new forms of politics, characteristic of our societies today, which demand to be approached from a nonessentialist perspective. Hence, the necessity of using the theoretical tools elaborated by the different currents of what can be called the postmodern in philosophy and of appropriating their critique of rationalism and subjectivism. (1988: 33)

Universal values—for example, the concept that everyone has certain rights—are not entirely abandoned, but they enter into a "new kind of articulation" with particular values and a logic of irreducible difference. Yet for this postmodern politics, the rejection of essentialism and lack of solid "foundations" does not entail the abandonment of radical politics. As Laclau puts it:

> Abandonment of the myth of foundations does not lead to nihilism, just as uncertainty as to how an enemy will attack does not lead to passivity. It leads, rather, to a proliferation of discursive interventions and arguments that are necessary, because there is no extradiscursive reality that discourse might simply reflect. Inasmuch as argument and discourse constitute the social, their open-ended character becomes the source of a greater activism and a more radical libertarianism. Humankind, having always bowed to external forces—God, Nature, the necessary laws of History—can now, at the threshold of postmodernity, consider itself for the first time the creator and constructor of its own history. The dissolution of the myth of foundations—and the concomitant dissolution of the category "subject"—further radicalizes the emancipatory possibilities offered by the Enlightenment and Marxism. (1988: 79–80)

The shift to a postmodern logic, in other words, leads to "an awareness of the complex strategic–discursive operations implied by [the] defense" of Enlightenment values (Laclau, 1988: 72). Thus, for Laclau and Mouffe postmodern philosophy and social theory do not entail a rejection of key political commitments to modernity itself. For them, nothing in the radical political project is lost with the rejection of foundationalism, and everything is gained through the liberating effects of a new logic of difference and contingency. In Mouffe's words, "far from seeing the development of postmodern philosophy as a threat, radical democracy welcomes it as an indispensable instrument in the accomplishment of its goals" (1988: 44). To speak ironically, we could say that the postmodern critique puts the modern project on even firmer "grounds" than Enlightenment rationality, insofar as its values are not simply dogmatically stated but are given pragmatic and rational grounds of justification. Hence, their approach is very similar to that of Habermas, who sees the Enlightenment as an "unfinished project" and seeks communicative grounds of normative justification (see Best, 1995), with the key difference that Laclau and Mouffe believe that postmodern theory has radical democratic potential whereas Habermas believes that it weakens Enlightenment values and aids irrationalist, conservative traditions.

Finally, there is a fourth form of postmodern politics, perhaps the dominant form of politics today, known as "identity politics," which often has radical aspirations but which usually falls short of advancing systemic change and new forms of radical struggle. Identity politics has its origins in the "new social movements" of the 1970s and 1980s and, ultimately, the struggles of the 1960s (see Chapter 1). The "movement" of the 1960s both pursued a coalition and alliance politics *and* challenged the dominant powers on multiple levels—gender, race, the hierarchical structure of the universities, colonial domination, U.S. imperialism in Vietnam, the alienated nature of work, sexual repression, and the oppressive organization of everyday life. In the 1970s, however, the "movement" fragmented into the "new social movements," which included feminist, black liberation, gay and lesbian, and peace and environmental groups, each fighting for its own interests (e.g., blacks saw the emerging environmental movement in the late 1960s as a bourgeois diversion from civil rights struggles, and environmentalists emphasized wilderness issues while ignoring problems of urban pollution). By the 1980s and 1990s, as the balkanization process continued, the "new social movements" had become transformed into "identity politics," the very name suggesting a turn away from general social, political, and economic issues and toward concerns with culture and personal identity.

Identity politics bears the influence of postmodern theory, which is evident in its critique of modern reductionism, abstract universalism, and essentialism, as well as in its use of multiperspectival strategies that legitimate multiple political voices. Foucault's genealogical politics, for example, is explicitly designed to liberate suppressed voices and struggles in history from the dominant narratives that reduce them to silence (see Foucault, 1980). In identity politics, individuals define themselves primarily as belonging to a given group, marked as "oppressed" and therefore as outside the dominant white, male, heterosexual, capitalist culture. These identities revolve around a "subject position," a key identity marker defined by one's gender, race, class, sexual preference, and so on and through which an individual is made subordinate to the dominant culture. Although class is certainly a major form of identity, identity politics typically is defined in opposition to class politics.

But although postmodern theory usually attacks essentialism, there is a mode of essentialism in many forms of identity politics, which privileges gender, race, sexual preference, or some other marker as *the* constituent of identity. Moreover, by fetishizing a single all-defining personal identity (woman, black, gay, etc.), identity politics also ignores the insights of postmodern theory that identities are multiple and socially constructed and that they need to be reconstructed in an emancipatory, autonomous, and self-affirming fashion. In other words, some versions of identity politics fetishize given constituents of identity, as if one of our multiple identity markers were our deep and true self, around which all of our life and politics revolve.

In some forms, identity politics also dovetails with liberal interest group politics that seeks to advance the interests of a single specific group, typically in opposition not only to the dominant groups but also to other marginalized and oppressed groups. Thus, in contrast to the universal and collective emphases of modern politics, a postmodern identity politics tends to be insular and something of a special interest group, perhaps itself a postmodern phenomenon. Hence, whereas modern politics focused on universalistic goals like gaining civil liberties, reducing inequalities, or transforming structures and institutions of domination, postmodern identity politics singles out the specific interests of a group and constructs identities through identification with the group and its struggles.

Of course, critics of modern politics have indicated from the beginning that the universalistic claims of modern theorists and politicians were cloaks for advancing the particular interests of ruling groups, mainly white male property owners. The cardinal rights advanced by the bourgeois revolutions in America and in France were those of property rights, which granted supreme economic and political power to white male capitalists in flagrant contradiction to their democratic rhetoric. Yet the new universalist ideology of modern politics unleashed a power that the ruling classes could not restrain; it inspired and legitimated the struggles of the very groups it was used to suppress, including those advocating identity politics today, who denounce universalist appeals as inherently ideological and oppressive.

Yet classical Marxism also advanced a reductionist and essentialist view of politics that is repudiated by postmodern politics. Marx theorized laborers as a "universal class" that by emancipating itself would emancipate all other oppressed groups. According to Marx's scheme, subjectivity is constituted as a class identity and all social antagonisms devolve around production as the essence of the social. Later Marxists continued with this policy, subsuming other key social issues to the "woman question," the "race question," the "national question," and so on, failing to see how gender, race, nationality, and other forms of identity were crucial and often more directly relevant for many different groups of people, just as nationalism provided a far more powerful identity than did international workers' solidarity for various European workers during World War I.

Yet Marxist politics was not seriously challenged until the 1960s, with the explosion of new struggles and identities that contested advanced capitalist society in its totality. Identity politics as it is defined today departs—explicitly or implicitly—from a critique of Marxist politics. The break from the essentialist and reductionist logic informing Marxist conceptions of class struggle has had some liberating effects in the political field. It allowed for new conceptions of micropolitics, pluralist democracy, and a politicization of the multiple ways in which the subject is constituted across numerous institutional sites and in everyday life.

One of the key insights of this period—theorized by Foucault—was that power is everywhere, not only in the factories but in schools, prisons, hospitals, and all other institutions. This insight is both depressing, since it acknowledges that power saturates all social spaces and relations, and exhilarating, because it allows for and demands new forms of struggle. Hence, multiple forms of resistance open up along every line of identity that is controlled or normalized. The movements of the period challenged capitalism, state power and bureaucracy, and the repressive organization of everyday life in the midst of consumer society, along with various modes of ideologically constituted identities.

Postmodern politics, following capital and state intervention processes themselves, represents a politicization of all spheres of social and personal existence, which were previously ignored or rejected as proper political spaces by modern and Marxist approaches. With postmodern politics, every sphere of social life becomes subject to questioning and contestation, and the sites of struggle multiply. With the pluralistic approach, power is more vulnerable to attack, and hence Foucault emphasized the contingency and frailty of power relations. Where the Leninist would argue that pluralized struggle only dissipates the centralized forces needed to combat capital and the state, a politically radical postmodernist would respond that the new struggles attack the weak links of the system and spread resistance everywhere, thereby allowing for the general attack that Leninists think is necessary for overthrowing capitalism.

Hence, the 1960s brought a shift from a macropolitics that focused on changing the structure of the economy and state to a micropolitics that aims to overturn power and hierarchy in specific institutions and to liberate emotional, libidinal, and creative energies repressed by the reality principle of bourgeois society. An important aspect of micropolitics, as evident in the work of Lyotard, Foucault, Deleuze, and Guattari, is a politics of subjectivity that theorizes the conditions under which the modern subject has emerged as an effect of power, what Foucault calls the "subjectification" of individuals. This entails primarily a struggle against the "microfascism" latent in everyone, to be combated by breaking out of, in the terms of Deleuze and Guattari, the "molar" pole of desire (such as informs all normalized subjectivities) and finding the "molecular" lines of escape. For Foucault, the politics of subjectivity involves a "politics as ethics" that creates new subjects on the Greek model of an "aesthetics of existence" (see Best and Kellner, 1991; Best, 1995).

Postmodern models of politics are trying to redefine the "political" based on changes in society, technology, economics, and everyday life. A postmodern *cultural politics*, building on the insights of Gramsci, the surrealists, Lefebvre, and the Situationists, thematizes culture as a crucial terrain of power and struggle. To the extent that social reproduction is now largely achieved at the levels of culture and everyday life, where the individual is a target of total administration, questions of subjectivity, ideology, culture, aesthetics, and utopian thought take on a new importance. The instrumentalist, pragmatic, or rationalist conception of political struggle, which attempts to shape "political consciousness," class or otherwise, and to mobilize critical insight into a political movement that transcends questions of culture, is insufficient because it begs the question of how a political movement is possible in the first place, given the degree of subjective identification with dominant modes of thought and behavior throughout society. As thinkers like Reich and Adorno saw, fascism has roots not only in the crisis of monopoly capital but also in the repression of the instinctual structure and the emergence of an "authoritarian personality."

Thus, if people live immersed in a culture colonized by capitalism, a culture of spectacles that binds affect and mobilizes pleasures to its sights, sounds, and experiences, then the struggle for culture, subjectivity, and identity is no longer secondary to the struggle for society, and both cultural politics and identity politics are crucial for breaking from the dominant ideologies and creating new forms of life and consciousness. Given the need to produce new subjectivities, political education, rational persuasion, and moral appeals remain of the greatest importance, but they can be very weak opponents to the seductive pleasures of MTV, blockbuster films, the Internet, fashion and advertising, and commodity consumption of all kinds. In Marcuse's words: "No persuasion, no theory, no reasoning can break this prison [of subjectivity], unless the fixed, petrified *sensibility* of the individuals is *'dissolved,' opened to a new dimension in history*, until the oppressive familiarity with the given object world is broken—broken in a *second alienation*: that from the alienated society" (1972: 71–72).

It is culture that molds the sensibilities, and thus, a radical cultural politics attempts to undo the enculturation of the dominant culture by providing new ways of seeing, feeling, thinking, talking, and being. Progressives today must not simply fall back on the old valorization of critical realism and its narrow cognitive models, as valuable as didactic and pedagogical art might be. What is ultimately needed are new affective structures and modes of experience that can act as catalysts, and the possibility of broader social and political transformations. Here, the political function of critical art becomes, negatively, a defamiliarization from the dominant mode of *experiencing* reality, what Marcuse has termed an alienation from alienation. Such has been the practice of Brecht's epic theater, Antonin Artaud's theater of cruelty, or Godard's anti-narrative films, all of which sought to question and displace the dominant mode of experiencing reality rather than to reproduce it through staid aesthetic conventions. Positively, a cultural politics has the task of "aesthetic education," the reshaping of human needs, desires, senses, and imagination through the construction of images, spectacles, and narratives that prefigure different ways of seeing and living.

Situationist art, for example, practiced both functions, the negative through its deconstruction of advertisements and other images (*détournement*), and the positive through experiences with the "constructed situation," a practice earlier advanced by the surrealists in their various exercises and games (such as "the exquisite corpse") designed to liberate unconscious creative forces. Paradoxically, today we find the atrophy of the senses in their hypertrophic extension throughout the spectacle and its images and commodity empires. Against Lukács, we emphasize the importance of formal innovation and avant-gardism in the arts, where such new techniques and modes of vision can help people break with repressive identifications with both the utilitarian (instrumental reason) and affective (sign value) modes of experience constituted by advanced capitalism. A new society will never be attainable until it is experienced as a need, for new modes of community, work, experience, social interaction, and relations to the natural world that can never be satisfied within capitalism and therefore cannot be co-opted by economic reforms.

As Bahro (1978) saw, capitalism generates needs and desires, which it ultimately cannot satisfy, for freedom, justice, self-realization, and a good life, and a radical cultural politics will both depict how the current mode of social organization restricts, limits, and deforms desire, freedom, and justice and project visions of how these aspirations could be realized. Both the radical negations of certain forms of critical modernism (i.e., those of Kafka, Beckett, the German expressionists, etc.) and the utopian dimension of art stressed by theorists such as Bloch and Marcuse are thus more relevant than ever today when radical critique is needed to free individuals from forms of oppression of which they are often unaware and when a better way of life for all is technically possible.

While today we need the expansion of localized cultural practices, they attain their real significance only within *the struggle for the transformation of society as a whole.* Without this systemic emphasis, cultural politics and identity politics remain confined to the margins of society and are in danger of degenerating into narcissism, hedonism, aestheticism, or personal therapy, where they pose no danger and are immediately co-opted by the culture industries. In such cases, *the political is merely the personal*, and the original intentions of the 1960s goal of broadening the political field are inverted and perverted. Just as economic and political demands have their referent in subjectivity and everyday life, so these cultural and existential issues find their ultimate meaning in the demand for a new society and mode of production.

Identity politics have been liberatory in breaking away from the abstract universalism of the Enlightenment and the reductionist class politics of Marxism, but they tend to be insular and fragmenting, focusing solely on the experiences and political issues of a given group, even splintering further into distinct subgroups, such as the disparate factions within the feminist community. Identity politics are often structured around simplistic binary oppositions—such as Us versus Them and Good versus Bad—that pit people against one another, making alliances, consensus, and compromise difficult or impossible. This has been the case, for example, with tendencies within radical feminism and ecofeminism that reproduce essentialism by stigmatizing men and "male rationality" while exalting women as the bearers of peaceful and loving values and as being "closer to nature" (see Biehl, 1991). The black liberation movement of the 1960s and the early politics of Malcolm X were exclusionist and racist, literally demonizing white people as an evil and inferior race. Similarly, the sexual politics of some gay and lesbian groups tend to focus exclusively on their own interests, while the mainstream environmental movement is notorious for resisting alliances with people of color and grassroots movements (see Dowie, 1995).

Even though each group needs to asserts its identity as aggressively as possible, postmodern identity politics should avoid falling into seriality and sheer fragmentation. These struggles, though independent of one another, should be articulated within counterhegemonic alliances and attack power formations on both the micro- and macro-levels. Not all universalistic appeals are ideological in the sense criticized by Marx; there *are* common grounds of experience, common concerns, and common forms of oppression that different groups share and that should be articulated—concerns such as the degradation of the environment and common forms of oppression that stem from capitalist exploitation and alienated labor.

To overcome alienation and oppression, the implementation of participatory democracy is proposed by a variety of tendencies within postmodern theory. In modern democratic theory, the notion of representative democracy superseded in liberal capitalist societies the stronger forms of participatory democracy advocated by the Greeks and by modern theorists like Rousseau, Bakunin, and Marx. The postmodern political turn, then, involves a radicalization of the theme of participatory democracy that is advocated in a variety of fields and domains of social life. Within the mode of theory, the democratic turn involves a shift toward more multiperspectival theorizing that respects a variety of sometimes conflicting perspectives, rather than, as in modern theory, seeking the one perspective of objective truth or absolute knowledge. In opposition to discourses of the unity of absolute truth, postmodern micropolitics stresses difference, plurality, conflict, and respect for the other.

In science, the postmodern turn involves increased emphasis on the scientific community and the various ways that consensus is reached, competing hypotheses are tested, and knowledge is gained through dissensus and the exploration of contrasting positions, as well as coming to agreement over facts and theories. While modern science often remains an elitist and domineering enterprise, multicultural science recognizes the contributions to knowledge of diverse cultures and renounces the arrogance of believing that only the Western way of knowing is valid and that all other forms of knowledge are inferior and defective.

In art, postmodern democracy includes increased collaborative work in multimedia, renouncing the myth of the great artist and even decentering the theory of the author, seeing that all art involves a form of collaboration and cultural dialogue. In postmodern culture, there is emphasis as well on public arts, on public access television, community radio, Internet activism, and on developing more interactive forms of culture that include popular participation. Indeed, the postmodern turn involves seeing how the audience is part of the collaborative process, that art involves participation of the audience in the creation of meaning and aesthetic significance, thus overcoming the divisions between the author, work, and audience, reified by some versions of modernist aesthetics. The emphasis on the motif of the popular unites postmodern developments in theory, the arts, science, and politics. In various fields, there is renunciation of the elitism and specialization endemic to the modern paradigm in favor of discourse and works that are more accessible to popular audiences. Of course, this is not always the case, and postmodern theoretical discourse is often as obscure and inaccessible—if not more so—as some modern discourse. Yet emphasis on the popular, on democratic participation, and on effective communication in the public sphere provides a counterforce to postmodern obscurantism.

In addition, postmodern culture tends to be more inclusive rather than exclusive, celebrating plurality, difference, and the acceptance of otherness. To be sure, some forms of identity politics are separatist and privilege the standpoint and interests of other groups in an exclusivist fashion, but the participatory democratic strain of the more progressive aspects of the postmodern mitigate against such exclusivity and separatist politics. Attacks on hierarchy and domination in postmodern theory thus provide the basis for a more egalitarian and democratic vision in diverse areas of human life.

Yet it would be a mistake to draw too sharp a distinction between the modern and postmodern paradigms and to vilify the modern as the site of all that is repressive and retrograde, and the postmodern as the mode of progressiveness and emancipation. There are regressive and progressive aspects in both the modern and postmodern traditions, and our claim is that we are currently suspended between two historical epochs—the modern and the postmodern—each of which has its own theoretical articulations and discourses, narratives, forms of art and cultural expression, scientific paradigms, politics, and forms of everyday life. The problem for those of us trying to theorize this great transformation, this rapid move into a new space, is to think together the modern and the postmodern, to see the interaction of both in the contemporary moment and to deploy the resources of both modern and postmodern theory to illuminate, analyze, and critique this space.

Our contemporary situation thus finds us between the modern and the postmodern, the old and the new, the traditional and the contemporary, the global and the local, the universal and the particular, and any number of other competing matrices. Such a complex situation produces feelings of vertigo, anxiety, and panic, and contemporary theory, art, politics, and everyday life exhibit signs of all of these symptoms. To deal with these conflicts, we need to develop new syntheses of modern and postmodern theory and politics in order to negotiate the complexities of our current era.

Indeed, both modern and postmodern positions have strengths and limitations, and we should seek a creative combination of the best elements of each. Thus, we need to combine modern notions of solidarity, alliances, consensus, universal rights, macropolitics and institutional struggle with postmodern notions of difference, plurality, multiperspectivalism, identity, and micropolitics. The task today is to construct what Hegel called a "differentiated unity," in which the various threads of historical development come together in a rich and mediated way. The abstract unity of the Enlightenment, as expressed in the discourse of rights or human nature, produced a false unity that masked and suppressed differences and privileged certain groups at the expense of others. The postmodern turn, conversely, has produced warring fragments of difference, exploding any possible context for human community. This was perhaps a necessary development in order to construct differences, but it is now equally necessary to reconstruct a new social whole, a progressive community in consensus over basic values and goals, one that is richly mediated with differences that are articulated without being annulled.

Thus, one of the main dramas of our time will be our choice of the road to travel into the future, the road that leads, to paraphrase Martin Luther King, Jr. (1969), to community, or the one that verges toward chaos. Similarly, will we take the course that leads to war or the one that brings peace? The one that establishes social justice or the one that allows ever-grosser forms of inequality and poverty? Will we stay on the same modern path of irrational growth and development, of the further expansion of a global capitalist economy (the world of NAFTA and GATT) that has taken us to the brink of social and environmental collapse, or will we create a sustainable society that lives in balance with the natural world? Will we chart a whole new postmodern path, with all its attendant snares and dangers, blind to the progressive heritage of the past? Or will we stake out an alternative route, radicalizing the traditions of modern Enlightenment and democracy, guided by the vision of a future that is just, egalitarian, participatory, ecological, healthy, happy, and sane?

Whatever choice we make, the multiple roads to the postmodern have led us into a postmodern adventure based on new social processes, theories, aesthetic practices, science, and politics. The postmodern turn has challenged and uprooted entrenched norms, practices, theories, and ways of seeing, resulting in fragmentation, instability, indeterminancy, and uncertainty but also in new challenges, excitement, and possibilities to develop new modes of thought and action. The postmodern adventure, unfolding in the twilight between the modern and the postmodern, involves a mapping of this new space–time continuum, situating us at the current historical crossroads where we can explore our options and suggest some new directions for the beleaguered forms of life on this planet.[6]

NOTES

1. On Shlain's thesis,

> revolutionary art can be understood as the preverbal stage of a civilization first contending with a major change in the perception of the world. . . . [Art is] not only . . . an aesthetic that can be pleasing to the eye, but . . . a Distant Early Warning System of the collective thinking of a society. Visionary art alerts the other members that a conceptual shift is about to occur in the thought system used to perceive the world." (1991: 18)

Throughout his book, Shlain elaborates on this insight, earlier advanced by people such as art critic John Ruskin ("There is in art a clairvoyance for which we have not yet found a name, still less an explanation") and Marshall McLuhan ("If men were able to be convinced that art is precise advanced knowledge of how to cope with the psychic and social consequences of the next technology, would they all become artists?") (quoted in Shlain, 1991: 18–19).

2. Poststructuralist theories of language, however, are paradoxical because they typically proclaim the "death of the author" as an active agent while granting full power to language as an autonomous entity, subsuming intentional consciousness under the structures and codes of language.

3. See Lukács (1971) and Feenberg (1991) for discussion of nature as a social construct. The American Indians, for example, did not see the undeveloped parts of the natural world as "wilderness"; rather, they saw every aspect of nature as suffused with a living spirit. The American settlers, by contrast, experienced the same "wilderness" as a hostile and desolate world that had to be subdued to erect the forces of civilization. Similarly, what many take to be a "natural" landscape, such as a prairie, may in fact be the result of extensive human intervention with fire, cattle, and clear-cutting.

4. An exception is Félix Guattari, whose book *Les trois écologies* (1983), has not been translated or widely read in postmodern circles. See also Zimmerman (1994).

5. For critics like Michael Soule, postmodern theory is the "ideological assault" on living nature that reinforces the physical assault carried out by global capitalism. For his and other essays critical of postmodern concepts in environmental theory, see Soule and Lease (1995).

6. This is precisely the project we will develop in our forthcoming book, *The Postmodern Adventure,* in which we will provide some genealogical studies of the transition from modernity to postmodernity and examine the trajectories and vicissitudes of global capitalism, the warfare state, the proliferation of new technologies, technoscience, the challenges to youth and the emergence of new youth cultures, and the playing out of identity politics in the O. J. Simpson trials, the militia movement, and various forms of terrorism. We will thus examine some of the defining phenomena of our time, deploying the resources of critical social theory and cultural studies.

Bibliography

The reference list assembled here contains the books and key scholarly articles that we have drawn on and cited in our studies. Newspaper and magazine articles, as well as Internet sources, are cited in the notes at the end of each chapter.

Adorno, Theodor W. (1973). *Negative Dialectics.* London: Routledge and Kegan Paul.

Alexander, Jeffrey (1995) *Fin de Siècle Social Theory: Relativism, Reduction, and the Problem of Reason.* London and New York: Verso.

Antonio, Robert J. (1995) "Nietzsche and Classical Social Theory." *American Journal of Sociology* 101(1): 1–43.

Antonio, Robert J., and Kellner, Douglas (1992) "Metatheorizing Historical Rupture: Classical Theory and Modernity," in *Metatheorizing,* ed. George Ritzer. Newbury Park, CA: Sage.

____________ (1994) "Postmodern Social Theory: Contributions and Limitations," in *Postmodernism and Social Inquiry,* eds. David R. Dickens and Andrea Fontana. New York: Guilford Press.

Argyros, Alexander (1992) *A Blessed Rage for Order: Deconstruction, Evolution, and Chaos.* Ann Arbor: University of Michigan Press.

Aronowitz, Stanley, and de Fazio, William (1994) *The Jobless Future.* Minneapolis: University of Minnesota Press.

Ashton, Dore (1973) *The New York School.* New York: Viking Press.

Bacon, Francis (1960) *The New Organon and Other Writings.* New York: Library of Liberal Arts.

Bahro, Rudolph (1978) *The Alternative in Eastern Europe.* London: New Left Books.

Barrett, William (1958) *Irrational Man: A Study in Existential Philosophy.* New York: Anchor Books.

Barth, John (1988) "The Literature of Replenishment," in *Essentials of the Theory of Fiction,* eds. Michael J. Hoffman and Patrick D. Murphy. Durham, NC, and London: Duke University Press.

Barthes, Roland (1975) *The Pleasures of the Text.* New York: Hill and Wang.

Bateson, Gregory (1972) *Steps to an Ecology of Mind.* New York: Ballantine Books.

____________ (1979) *Mind and Nature: A Necessary Unity.* New York: Dutton.

Baudelaire, Charles (1970) *Paris Spleen.* New York: New Directions.

Baudrillard, Jean (1968) *Le système des objets.* Paris: Denoel-Gonthier.

____________ (1970) *La société de consommation.* Paris: Gallimard.

____________ (1975) *The Mirror of Production.* St. Louis, MO: Telos Press.

____________ (1981) *For a Critique of the Political Economy of the Sign.* St. Louis, MO: Telos Press.

____________ (1983a) *Simulations.* New York: Semiotext(e).

____________ (1983b) *In the Shadow of the Silent Majorities.* New York: Semiotext(e).

____________ (1983c) "The Ecstasy of Communication," in *The Anti-Aesthetic,* ed. Hal Foster. Port Townsend, WA: Bay Press.

____________ (1984) "Interview: Game with Vestiges," *On the Beach* 5 (Winter): 19–25.

____________ (1987) *Forget Foucault.* New York: Semiotext(e).

____________ (1988a) *America.* London: Verso.

____________ (1988b) "The Year 2000 Has Already Happened," in *Body Invaders: Panic Sex in America,* eds. Arthur and Marilouise Kroker. Montreal: New World Perspectives.

____________ (1990) *Cool Memories.* London: Verso.

____________ (1993) *Symbolic Exchange and Death.* London: Sage.

____________ (1994) *Simulacra and Simulation.* Ann Arbor: University of Michigan Press.

Bauman, Zygmunt (1991) *Modernity and Ambivalence.* Cambridge, Eng.: Polity Press.

____________ (1992) *Intimations of Postmodernity.* New York and London: Routledge.

Baynes, Kenneth, et al. (1987) *After Philosophy: End or Transformation?* Cambridge, MA: MIT Press.

Beauvoir, Simone de (1952) *The Second Sex.* New York: Knopf.

Bell, Daniel (1976) *The Coming of Post-Industrial Society.* New York: Basic Books.

Benedikt, Michael (1987) *For an Architecture of Reality.* New York: Lumen Books.

____________ (1991a) *Deconstructing the Kimball: An Essay on Meaning and Architecture.* New York: Lumen Books.

____________, ed. (1991b) *Cyberspace: First Steps.* Cambridge, MA: MIT Press.

Benevolo, Leonardo (1977) *History of Modern Architecture.* Cambridge, MA: MIT Press.

Benjamin, Walter (1969) *Illuminations.* New York: Schocken Books.

____________ (1973) *Charles Baudelaire: A Lyric Poet in the Era of High Capitalism.* London: New Left Books.

Benstock, Shari (1986) *Women of the Left Bank: Paris, 1900–1940.* Austin: University of Texas Press.

Benton, Ted (1984) *The Rise and Fall of Structural Marxism.* London: Macmillan.

Berger, Peter L., and Luckmann, Thomas (1967) *The Social Construction of Reality.* Garden City, NY: Anchor Books.

Bergman, Peter (1987) *Nietzsche.* Bloomington: Indiana University Press.

Berman, Art (1994) *Preface to Modernism.* Urbana and Chicago: University of Illinois Press.

Berman, Marshall (1982) *All That Is Solid Melts in the Air.* New York: Simon and Schuster.

Berman, Morris (1981) *The Reenchantment of the World.* New York: Bantam.

Bernal, Martin (1987) *Black Athena: The Afroasiatic Roots of Classical Civilization: Vol 1.* New Brunswick, NJ: Rutgers University Press.

Bernstein, Richard (1994) *The Dictatorship of Virtue.* New York: Knopf.
Bernstein, Richard J. (1991) *The New Constellation.* Cambridge, Eng.: Polity Press.
Bertens, Hans (1995) *The Idea of the Postmodern.* New York and London: Routledge.
Best, Steven (1989) "Jameson, Post-Structuralism, and the Critique of Totality," in *Postmodernism/Jameson/Critique,* ed. Douglas Kellner. Washington, DC: Maisonneuve Press.
____________ (1995) *The Politics of Historical Vision: Marx, Foucault, Habermas.* New York: Guilford Press.
Best, Steven, and Kellner, Douglas (1988) "Watching Television: The Limits of Postmodernism," *Science as Culture* 4: 44–70.
____________ (1990) "Modernity, Mass Society, and the Media: Reflections on *The Corsair Affair,*" in *International Kierkegaard Commentary: The Corsair Affair,* ed. Robert Perkins. Macon, GA: Mercer University Press.
____________ (1991) *Postmodern Theory: Critical Interrogations.* London and New York: Macmillan and Guilford Press.
____________ (forthcoming) *The Postmodern Adventure.* New York: Guilford Press.
Bhabha, Homi K. (1994) *The Location of Culture.* London and New York: Verso.
Biehl, Janet (1991) *Rethinking Ecofeminist Politics.* Boston: South End Press.
Birch, Charles (1988) "A Postmodern Challenge to Biology," in *The Reenchantment of Science: Postmodern Proposals,* ed. David Ray Griffin. Albany: State University of New York Press.
Blake, Peter (1974) *Form Follows Fiasco: Why Modern Architecture Hasn't Worked.* Boston: Little, Brown.
Blake, Peter (1993) *No Place Like Utopia: Modern Architecture and the Company We Keep.* New York: Knopf.
____________ (1996) *The Master Builders: Le Corbusier, Mies van Der Rohe, Frank Lloyd Wright.* New York: Norton.
Bloch, Ernst (1986) *The Principle of Hope.* Cambridge, MA: MIT Press.
Bloom, Allan (1987) *The Closing of the American Mind.* New York: Simon and Schuster.
Bluestone, Barry, and Harrison, Bennett (1982) *The Deindustrialization of America.* New York: Basic Books.
Blum, Deborah (1994) *The Monkey Wars.* New York: Oxford University Press.
Bogard, William (1996) *The Simulation of Surveillance.* New York and Cambridge, Eng.: Cambridge University Press.
Bognar, Botond (1995) *The Japan Guide.* New York: Princeton Architectural Press.
Boggs, Carl (1976) *Gramsci's Marxism.* London: Pluto Press.
Bohm, David (1957) *Causality and Chance in Modern Physics.* Philadelphia: University of Pennsylvania Press.
____________ (1988) "Postmodern Science and a Postmodern World," in *The Reenchantment of Science: Postmodern Proposals,* ed. David Ray Griffin. Albany: State University of New York Press.
Bohr, Niels (1958) *Atomic Theory and Human Knowledge.* New York: Wiley.
Boje, David M., and Dennehy, Robert F. (1994) *Managing in the Postmodern World.* Dubuque, IA: Kendall/Hunt.
Boje, David M., Gephart, Robert P., and Thatchenkery, Joseph (1996) *Postmodern Management and Organization Theory.* Thousand Oaks, CA, and London: Sage.

Bookchin, Murray (1995a) *Re-Enchanting Humanity.* London: Cassell Press.

____________ (1995b) *The Philosophy of Social Ecology: Essays on Dialectical Naturalism,* revised edition. Montreal: Black Rose Books.

Bordo, Susan R. (1987) *The Flight of Objectivity: Essays on Cartesianism and Culture.* Albany: State University of New York Press.

Bradbury, Malcolm, and McFarlane, James (1976) *Modernism.* London and New York: Penguin Books.

Bronner, Stephen Eric, and Kellner, Douglas, eds. (1989) *Passion and Rebellion: The Expressionist Heritage.* New York: Columbia University Press.

Brooks, H. Allen, ed. (1987) *Le Corbusier.* Princeton: Princeton University Press.

Broude, Norma, and Garrard, Mary D., eds. (1982) *Feminism and Art History: Questioning the Litany.* New York: Harper and Row.

Buck-Morss, Susan (1989) *The Dialectics of Seeing.* Cambridge, MA: MIT Press.

Buckley, Walter (1968) *Modern Systems Research for the Behavioral Scientist: A Sourcebook.* Chicago: Aldine.

Burger, Peter (1984) *The Historical Avant-Garde.* Minneapolis: University of Minnesota Press.

Burgin, Victor (1986) *Between.* Oxford: Blackwell.

Burnside, Scott, and Cairns, Alan (1995) *Deadly Innocence.* New York: Warner Books.

Burtt, E. A. (1932) *The Metaphysical Foundations of Modern Science.* New York: Doubleday.

Bury, J. D. (1973) "The Science of History," in *The Varieties of History: From Voltaire to the Present,* ed. Fritz Stern. New York: Vintage Books.

Butler, Christopher (1980) *After the Wake: An Essay on the Contemporary Avant-Garde.* Oxford: Oxford University Press.

Cage, John (1961) *Silence.* Middletown, CT: Wesleyan University Press.

Calinescu, Matei (1987) *Five Faces of Modernity.* Durham, NC: Duke University Press.

Callinicos, Alex (1990) *Against Postmodernism.* New York: St. Martin's.

Campbell, Jeremy (1982) *Grammatical Man: Information, Entropy, Language, and Life.* New York: Simon and Schuster.

Capra, Fritjof (1982) *The Turning Point: Science, Society, and the Rising Culture.* New York: Bantam Books.

____________ (1996) *The Web of Life: A New Scientific Understanding of Living Systems.* New York: Anchor Books.

Cassirer, Ernst (1951) *The Philosophy of the Enlightenment.* Princeton: Princeton University Press.

Catton, William R., Jr. (1980) *Overshoot: The Ecological Basis of Revolutionary Change.* Urbana: University of Illinois Press.

Caute, David (1988) *The Year of the Barricades: A Journey through 1968.* New York: Harper and Row.

Chomsky, Noam (1975) *Reflections on Language.* New York: Pantheon Books.

Clarke, Simon, et al. (1980) *One-Dimensional Marxism.* London: Allison and Busby.

Cohn-Bendit, Daniel, et al. (1968) *The French Student Revolt: The Leaders Speak.* New York: Hill and Wang.

Colborn, Theo, Dumanoski, Dianne, and Myers, John Peterson (1996) *Our Stolen Future.* New York: Penguin Books.

Comte, Auguste (1850) *Social Physics: From the Positive Philosophy of Auguste Comte.* New York: Blanchard.

Connor, Steven (1989) *Postmodernist Culture.* Oxford: Blackwell.

Cooper, Douglas (1970) *The Cubist Epoch.* London: Phaidon.

Correa, Charles (1987) "Chandigarh: The View from Benares," in *Le Corbusier,* ed. H. Allen Brooks. Princeton: Princeton University Press.

Coveney, Peter, and Highfield, Roger (1995) *Frontiers of Complexity: The Search for Order in a Chaotic World.* New York: Ballantine Books.

Craft, Catherine Anne (1996) *Constellations of Past and Present: (Neo)-Dada, the Avant-Garde, and the New York Art World, 1951–1965.* Ph.D. dissertation, University of Texas.

Crimp, Douglas (1980) "The Photographic Activity of Postmodernism," *October* 15: 91–101.

____________ (1983) "On the Museum's Ruins," in *The Anti-Aesthetic,* ed. Hal Foster. Port Townsend, WA: Bay Press.

Cvetkovich, Ann, and Kellner, Douglas, eds. (1997) *Articulating the Global and the Local: Globalization and Cultural Studies.* Boulder, CO: Westview Press.

Daly, Herman E., and Townsend, Kenneth N., eds. (1993) *Valuing the Earth: Economics, Ecology, Ethics.* Cambridge, MA: MIT Press.

Davies, Paul (1978) *The Runaway Universe.* New York: Harper and Row.

Davis, Douglas (1977) *Artculture: Essays on the Post-Modern.* New York: Harper and Row.

____________ (1980) "Post-Everything," *Art in America* (February): 11–13.

Debord, Guy (1970) *The Society of the Spectacle.* Detroit, MI: Black and Red.

____________ (1979) "Preface," in *The Society of the Spectacle,* fourth Italian edition. London: Chronos.

____________ (1990) *Comments on the Society of the Spectacle.* London and New York: Verso.

Debord, Guy, and Sanguinetti, Gianfranco (1994) *The Veritable Split in the International.* London: B. M. Piranha.

de Certeau, Michel (1984) *The Practice of Everyday Life.* Berkeley and Los Angeles: University of California Press.

Deleuze, Gilles, and Guattari, Félix (1977) *Anti-Oedipus.* New York: Viking Press.

____________ (1987) *A Thousand Plateaus.* Minneapolis: University of Minnesota Press.

Denzin, Norman K. (1991) *Images of Postmodern Society.* London: Sage.

Derrida, Jacques (1976) *Of Grammatology.* Baltimore: Johns Hopkins University Press.

____________ (1981a) *Positions.* Chicago: University of Chicago Press.

____________ (1981b) *Dissemination.* Chicago: University of Chicago Press.

____________ (1982) *Margins of Philosophy.* Chicago: University of Chicago Press.

____________ (1994) *Specters of Marx.* London and New York: Routledge.

Descartes, René (1960) *Discourse on Method.* New York: Macmillan.

Devall, Bill, and Sessions, George (1986) *Deep Ecology: Living as If Nature Mattered.* Layton, UT: Gibbs Smith.

Dewey, John (1979) *Theory and Practice.* New York: Paragon Books.

Dickens, David R., and Fontana, Andrea, eds. (1994) *Postmodernism and Social Inquiry.* New York: Guilford Press.

Diop, Cheikh Anta (1974) *The African Origin of Civilization: Myth or Reality?*, trans. M. Cook. Westport, CT: L. Hill.

Dostoyevsky, Fyodor (1974) *Notes from Underground.* New York: Bantam Books.

Dowie, Mark (1995) *Losing Ground: American Environmentalism at the Close of the Twentieth Century.* Cambridge, MA: MIT Press.

Duesberg, Peter H. (1996) *Inventing the AIDS Virus.* Washington, DC: Regnery.

Dutton, Thomas A., and Mann, Lian Hurst, eds. (1996) *Reconstructing Architecture.* Minneapolis: University of Minnesota Press.

Dyke, C. (1990) "Strange Attraction, Curious Liaison: Clio Meets Chaos," *The Philosophical Forum* 21 (4, Summer): 369–392.

Ebert, Teresa (1996) *Ludic Feminism and After.* Ann Arbor: University of Michigan Press.

Epstein, Barbara (1996) "The Postmodernism Debate," *Z Magazine* (October): 57–59.

Erlich, Paul R. (1968) *The Population Bomb.* New York: Ballantine Books.

Erlich, Paul R., Erlich, Anne H., and Holdren, John P. (1993) "Availability, Entropy, and the Laws of Thermodynamics," in *Valuing the Earth: Economics, Ecology, Ethics,* eds. Herman E. Daly and Kenneth N. Townsend. Cambridge, MA: MIT Press.

Esterbrook, Gregg (1995) *A Moment on Earth: The Coming Age of Environmental Optimism.* New York: Viking Press.

Evenson, Norma (1987) "Yesterday's City of Tomorrow Today," in *Le Corbusier,* ed. H. Allen Brooks. Princeton, NJ: Princeton University Press.

Ewen, Stuart, and Ewen, Elizabeth (1983) *Channels of Desire.* New York: McGraw-Hill.

Featherstone, Mike (1989) "Postmodernism, Cultural Change, and Social Practice," in *Postmodernism/Jameson/Critique,* ed. Douglas Kellner. Washington, DC: Maisonneuve.

____________ (1991) *Consumer Culture and Postmodernism.* London: Sage.

Feenberg, Andrew (1991) *Critical Theory of Technology.* Oxford: Oxford University Press.

Ferré, Frederick (1976) *Shaping the Future: Resources for the Postmodern World.* New York: Harper and Row.

____________ (1988) "Religious World Modeling and Postmodern Science," in *The Reenchantment of Science: Postmodern Proposals,* ed. David Ray Griffin. Albany: State University of New York Press.

Feyerabend, Paul (1978a) *Against Method: Outline of an Anarchistic Theory of Knowledge.* London and New York: Verso.

____________ (1978b) *Science in a Free Society.* London and New York: Verso.

Fiedler, Leslie (1971) *The Collected Essays of Leslie Fiedler: Vol. II.* New York: Stern and Day.

Fish, Stanley (1996) "Professor Sokal's Bad Joke," *The New York Times* (May 21): A23.

Foster, Hal (1983) "Introduction," in *The Anti-Aesthetic,* ed. Hal Foster. Port Townsend, WA: Bay Press.

____________ (1985) *Recodings.* Seattle, WA: Bay Press.

Foucault, Michel (1965) *Madness and Civilization.* New York: Random House.

____________ (1972a) *The Archaeology of Knowledge.* New York: Pantheon.

____________ (1972b) *Power/Knowledge.* New York: Pantheon.

____________ (1973) *The Order of Things.* New York: Vintage Books.

___________ (1975) *Discipline and Punish.* New York: Pantheon.

___________ (1977) *Language, Counter-Memory, Practice.* Ithaca, NY: Cornell University Press.

___________ (1980) *The History of Sexuality: Vol. 1.* New York: Vintage Books.

Frampton, Kenneth (1983) "Toward a Critical Regionalism," in *The Anti-Aesthetic,* ed. Hal Foster. Port Townsend, WA: Bay Press.

Frankel, Boris (1987) *The Post-Industrial Utopians.* Cambridge, Eng.: Polity Press.

Fraser, Steven, ed. (1995) *The Bell Curve Wars: Race, Intelligence, and the Future of America.* New York: Basic Books.

Frisby, David (1986) *Fragments of Modernity.* Oxford: Blackwell.

Fromm, Erich (1947) *Man for Himself.* New York: Holt, Rinehart and Winston.

Frow, John (1991) *What Was Postmodernism?* Sydney, Australia: Local Consumption Publications.

Frye, Marilyn (1983) "In and Out of Harm's Way: Arrogance and Love," in *The Politics of Reality.* Trumansburg, NY: Crossing Press.

Fukuyama, Francis (1992) *The End of History and the Last Man.* New York: Free Press.

Gadamer, Hans-Georg (1975) *Truth and Method.* New York: Seabury.

Galison, Peter, and Stump, David, eds. (1995) *The Disunity of Science.* Stanford, CA: Stanford University Press.

Gane, Mike (1991) *Baudrillard: Critical and Fatal Theory.* New York and London: Routledge.

___________, ed. (1993) *Baudrillard Live: Selected Interviews.* New York and London: Routledge.

___________ (1995) "Radical Theory: Baudrillard and Vulnerability," *Theory, Culture and Society* 12(4): 109–125.

Garrett, Laurie (1994) *The Coming Plague: Newly Emerging Diseases in a World Out of Balance.* London and New York: Penguin Books.

Gates, Bill (1995) *The Road Ahead.* New York: Viking Press.

Georgescu-Roegen, Nicholas (1993a) "The Entropy Law and the Economic Problem," in *Valuing the Earth: Economics, Ecology, Ethics,* eds. Herman E. Daly and Kenneth N. Townsend. Cambridge, MA: MIT Press.

___________ (1993b) "Selections from 'Energy and Economic Myths,'" in *Valuing the Earth: Economics, Ecology, Ethics,* eds. Herman E. Daly and Kenneth N. Townsend. Cambridge, MA: MIT Press.

Geldzahler, Henry (1969) *New York Painting and Sculpture: 1940–1970.* New York: Dutton.

Ghirardo, Diane (1984–1985) "Past or Post Modern in Architectural Fashion," *Telos* 62 (Winter): 187–195.

Gitlin, Todd (1987) *The Sixties: Years of Hope, Days of Rage.* New York: Bantam.

___________ (1995) *The Twilight of Common Dreams.* New York: Henry Holt.

Glass, Philip (1987) *Music by Philip Glass.* New York: Harper and Row.

Gleick, James (1987) *Chaos.* London and New York: Viking Books.

Goldberger, Paul (1983) *On the Rise: Architecture and Design in a Postmodern Age.* London and New York: Penguin Books.

Goldman, Robert (1992) *Reading Ads Socially.* New York and London: Routledge.

Goldman, Robert, and Papson, Stephen (1996) *Sign Wars: The Cluttered Landscape of Advertising.* New York: Guilford Press.

Goonatilaha, Susantheia (1984) *Aborted Discovery: Science and Credibility in the Third World.* London: Zed Books.

Gorz, Andre (1982) *Farewell to the Proletariat.* London and Boston: Pluto Press and South End Press.

Gottdiener, Mark (1995) *Postmodern Semiotics.* Oxford: Blackwell.

Gould, Stephen Jay (1980) *The Panda's Thumb: More Reflections in Natural History.* New York: Norton.

Graff, Gerald (1985) "Preface," in Charles Newman, *The Postmodern Aura: The Act of Fiction in an Age of Inflation.* Evanston, IL: Northwestern University Press.

Gramsci, Antonio (1971) *Prison Notebooks.* New York: International Publishers.

Greenberg, Clement (1961) *Art and Culture.* Boston: Beacon Press.

Grewal, Inderpal, and Kaplan, Caren, eds. (1994) *Scattered Hegemonies: Postmodernity and Transnational Feminist Practices.* Minneapolis: University of Minnesota Press.

Griffin, David Ray, ed. (1988a) *The Reenchantment of Science: Postmodern Proposals.* Albany: State University of New York Press.

____________, ed. (1988b) *Spirituality and Science: Postmodern Visions.* Albany: State University of New York Press.

Gropius, Walter (1965) *The New Architecture and the Bauhaus.* Cambridge, MA: MIT Press.

Gross, Paul R., and Levitt, Norman (1994) *Higher Superstition: The Academic Left and Its Quarrels with Science.* Baltimore: Johns Hopkins Press.

Gross, Paul R., Levitt, Norman, and Lewis, Martin W., eds. (1996) *The Flight from Science and Reason.* New York: New York Academy of Sciences.

Grossberg, Lawrence (1987) "The In-Difference of Television," *Screen* 28 (2): 28–46.

____________ (1993) "The Media Economy of Rock Culture: Cinema, Postmodernity and Authenticity," in *The Music Video Reader,* eds. Simon Frith, Andrew Goodwin, and Lawrence Grossberg. London and New York: Routledge.

Guattari, Félix (1983) *Les trois écologies.* Paris: Editions Galilée.

Guilbaut, Serge (1983) *How New York Stole the Idea of Modern Art.* Chicago: University of Chicago Press.

Habermas, Jürgen (1968) *Toward a Rational Society.* Boston: Beacon Press.

____________ (1983) "Modernity: An Incomplete Project," in *The Anti-Aesthetic,* ed. Hal Foster. Port Townsend, WA: Bay Press.

____________ (1987) *The Philosophical Discourse of Modernity.* Cambridge, MA: MIT Press.

____________ (1989) *The Structural Transformation of the Public Sphere.* Cambridge, MA: MIT Press.

Hall, Stuart (1991) "Old and New Identities, Old and New Ethnicities," in *Culture, Globalization and the World-System: Contemporary Conditions for the Representations of Identity,* ed. Anthony D. King. Binghamton: State University of New York Art Department.

Hamamoto, Darrell (1996) "Introduction," in *New American Destinies: A Reader in Contemporary Asian and Latino Immigration,* eds. Darrell Hamamoto and Rudolfo Torres. London and New York: Routledge.

Harding, Sandra (1986) *The Science Question in Feminism.* Ithaca, NY: Cornell University Press.

____________ (1991) *Whose Science? Whose Knowledge? Thinking from Women's Lives.* Ithaca, NY: Cornell University Press.

____________, ed. (1993) *The "Racial" Economy of Science.* Bloomington: Indiana University Press.

____________ (1994) "Is Science Multicultural? Challenges, Resources, Opportunities, Uncertainties," *Configurations* 2: 301–330.

Hart, Roger (1996) "The Flight from Truth and Reason: An Essay Review of *Higher Superstition,*" in *Science Wars,* ed. Andrew Ross. Durham, NC: Duke University Press.

Harvey, David (1989) *The Condition of Postmodernity.* Oxford: Blackwell.

Hassan, Ihab (1987) *The Postmodern Turn: Essays in Postmodern Theory and Culture.* Columbus, OH: Ohio State University Press.

Haug, W. F. (1986) *Critique of Commodity Aesthetics.* Minneapolis: University of Minnesota Press.

Hawking, Stephen W. (1988) *A Brief History of Time: From the Big Bang to Black Holes.* New York: Bantam.

Hayden, Dolores (1984) *Redesigning the American Dream.* Ontario: General Publishing.

____________ (1995) *The Power of Place.* Cambridge, MA: MIT Press.

Hayles, N. Katherine (1984) *The Cosmic Web: Scientific Field Models and Literary Strategies in the Twentieth Century.* Ithaca, NY: Cornell University Press.

____________ (1990) *Chaos Bound: Orderly Disorder in Contemporary Literature and Science.* Ithaca, NY: Cornell University Press.

____________, ed. (1991) *Chaos and Order: Complex Dynamics in Literature and Science.* Chicago: University of Chicago Press.

Heidegger, Martin (1962) *Being and Time.* New York: Harper and Row.

____________ (1977) *The Question Concerning Technology.* New York: Harper and Row.

Heisenberg, Werner (1958) *Physics and Philosophy.* New York: Harper and Row.

____________ (1971) *Physics and Beyond.* New York: Harper and Row.

Hesse, Mary (1980) *Revolutions and Reconstructions in the Philosophy of Science.* Bloomington: Indiana University Press.

Hindess, Barry, and Hirst, Paul (1975) *Pre-Capitalist Modes of Production.* London: Routledge and Kegan Paul.

Hirshhorn, Larry (1997) *Reworking Authority: Leading and Following in the Post-Modern Organization.* Cambridge, MA: MIT Press.

Hitchcock, Henry Russell, and Johnson, Philip (1966) *The International Style.* New York: Norton.

Hobbes, Thomas (1947) *Leviathan.* Oxford and New York: Oxford University Press.

Hollinger, Robert (1994) *Postmodern Social Theory.* Thousands Oaks, CA: Sage.

Hong, Howard, and Hong, Edna, eds. (1982) *The Corsair Affair.* Princeton: Princeton University Press.

Horgan, John (1996) *The End of Science: Facing the Limits of Knowledge in the Twilight of the Scientific Age.* Reading, MA: Helix Books.

Horkheimer, Max, and Adorno, Theodor. (1972) *Dialectic of Enlightment.* New York: Continuum.

Howe, Irving (1970) "Mass Society and Postmodern Fiction," in *Decline of the New,* ed. Irving Howe. New York: Horizon.

Hubbert, M. King (1993) "Exponential Growth as a Transient Phenomenon in Human History," in *Valuing the Earth: Economics, Ecology, Ethics*, eds. Herman E. Daly and Kenneth N. Townsend. Cambridge, MA: MIT Press.

Hutcheon, Linda (1988) *A Poetics of Postmodernism*. New York and London: Routledge.

____________ (1989) *The Politics of Postmodernism*. New York and London: Routledge.

Huyssen, Andreas (1986) *After the Great Divide*. Bloomington: Indiana University Press.

Jacobs, Jane (1961) *The Death and Life of Great American Cities*. New York: Random House.

Jacobus, John (1966) *Twentieth-Century Architecture: The Middle Years 1940–65*. London: Thames and Hudson.

Jameson, Fredric (1971) *Marxism and Form*. Princeton: Princeton University Press.

____________ (1984) "Postmodernism: The Cultural Logic of Late Capitalism," *New Left Review* 146: 53–93.

____________ (1988) "Cognitive Mapping," in *Marxism and the Interpretation of Culture*, eds. Larry Grossberg and Cary Nelson. Chicago and Urbana: University of Illinois Press.

____________ (1991) *Postmodernism, or, the Cultural Logic of Late Capitalism*. Durham, NC, and London: Duke University Press.

Jantsch, Erich (1980) *The Self-Organizing Universe: Scientific and Human Implications of the Emerging Paradigm of Evolution*. Oxford: Pergamon Press.

Jardine, Alice A. (1985) *Gynesis: Configurations of Women and Modernity*. Ithaca, NY: Cornell University Press.

Jencks, Charles (1977) *The Language of Post-Modern Architecture*. New York: Rizzoli.

____________ (1988). *Architecture Today*. New York: Harry N. Abrams.

____________ (1995) *The Architecture of the Jumping Universe*. London: Academy Editions.

Johnson, Steven (1996) "Strange Attraction," *Lingua Franca* (March/April): 42–50.

Kamminga, Harmke (1990) "What Is This Thing Called Chaos?," *New Left Review* 181 (May/June): 49–59.

Kaplan, E. Ann (1987) *Rocking Around the Clock: Music Television, Postmodernism, and Consumer Culture*. London and New York: Methuen.

Kaplan, E. Ann, and Sprinker, Michael (1993) *The Althusserian Legacy*. London and New York: Verso.

Katsiaficas, George (1987) *The Imagination of the New Left: A Global Analysis of 1968*. Boston: South End Press.

Kauffman, Stuart A. (1991) "Antichaos and Adaptation," *Scientific American* (August): 78–84.

____________ (1995) *At Home in the Universe: The Search for the Laws of Self-Organization and Complexity*. New York: Oxford University Press.

Kearney, Richard (1988) *The Wake of Imagination*. Minneapolis: University of Minnesota Press.

Keith, Michael, and Pile, Steve (1993) *Place and the Politics of Identity*. New York and London: Routledge.

Kellner, Douglas (1973) *Heidegger's Concept of Authenticity*. Ph.D. dissertation, Columbia University.

____________ (1977) *Karl Korsch: Revolutionary Theory.* Austin: University of Texas Press.

____________ (1984a) *Herbert Marcuse and the Crisis of Marxism.* London and Berkeley and Los Angeles: Macmillan and University of California Press.

____________ (1984b) "Authenticity and Heidegger's Challenge to Ethical Theory," in *Thinking about Being: Aspects of Heidegger's Thought,* eds. Robert W. Shahan and J. N. Mohanty. Norman: University of Oklahoma Press.

____________ (1989a) *Critical Theory, Marxism and Modernity.* Cambridge, Eng., and Baltimore: Polity Press and Johns Hopkins University Press.

____________ (1989b) *Jean Baudrillard: From Marxism to Postmodernism and Beyond.* Cambridge, Eng., and Palo Alto, CA: Polity Press and Stanford University Press.

____________, ed. (1989c) *Postmodernism/Jameson/Critique.* Washington, DC: Maisonneuve Press.

____________ (1990) *Television and the Crisis of Democracy.* Boulder, CO: Westview Press.

____________ (1991) "Nietzsche and Modernity: Reflections on *Twilight of the Idols,*" *International Studies in Philosophy* 23(2): 3–17.

____________ (1992) *The Persian Gulf TV War.* Boulder, CO: Westview Press.

____________, ed. (1994) *Jean Baudrillard: A Critical Reader.* Oxford: Blackwell.

____________ (1995a) *Media Culture.* New York and London: Routledge.

____________ (1995b) "Intellectuals and New Technologies," *Media, Culture, and Society* 17: 201–217.

Kermode, Frank (1967) *The Sense of an Ending: Studies in the Theory of Fiction.* New York: Oxford University Press.

Kierkegaard, Søren (1978) *Two Ages: The Age of Revolution and the Present Age.* Princeton: Princeton University Press.

____________ (1982) *The Corsair Affair.* Princeton: Princeton University Press.

____________ (1987) *Either/Or.* Princeton: Princeton University Press.

____________ (1988) *Stages on Life's Way.* Princeton: Princeton University Press.

Kinchloe, Joel L., Steinberg, Shirley R., and Gresson, Aron B. III (1996) *Measured Lies: The Bell Curve Examined.* New York: St. Martin's Press.

King, Martin Luther, Jr. (1969) *Where Do We Go from Here?: Chaos or Community?* New York: Bantam.

King, Ross (1996) *Emancipating Space: Geography, Architecture, and Urban Design.* New York: Guilford Press.

Klotz, Heinrich (1988) *The History of Postmodern Architecture.* Cambridge, MA: MIT Press.

Knabb, Ken (1981) *Situationist International Anthology.* Berkeley: Bureau of Public Secrets.

Koelb, Clayton, ed. (1990) *Nietzsche as Postmodernist: Essays Pro and Contra.* Albany: State University of New York Press.

Kolb, David (1990) *Postmodern Sophistications.* Chicago: University of Chicago Press.

Korsch, Karl (1977) *Karl Korsch: Revolutionary Theory,* ed. Douglas Kellner. Austin: University of Texas Press.

Kosko, Bart (1993) *Fuzzy Thinking: The New Science of Fuzzy Logic.* New York: Hyperion.

Krauss, Rosalind (1983) "Sculpture in the Expanded Field," in *The Anti-Aesthetic*, ed. Hal Foster. Port Townsend, WA: Bay Press.

____________ (1985) *The Originality of the Avant-Garde and Other Modernist Myths.* Cambridge, MA: MIT Press.

Kroker, Arthur (1992) *The Possessed Individual.* New York: St. Martin's.

____________ (1993) *Spasm.* New York: St. Martin's.

Kroker, Arthur, and Cook, David (1986) *The Postmodern Scene.* New York: St. Martin's.

Kroker, Arthur; Kroker, Marilouise; and Cook, David (1989) *Panic Encyclopedia.* New York: St. Martin's.

Kroker, Arthur, and Weinstein, Michael (1994) *Data Crash.* New York: St. Martin's.

Kuberski, Philip (1994) *Chaosmos: Literature, Science, and Theory.* Albany: State University of New York Press.

Kuhn, Thomas S. (1970) *The Structure of Scientific Revolutions,* second edition. Chicago: University of Chicago Press.

Laclau, Ernesto (1988) "Politics and the Limits of Modernity," in *Universal Abandon,* ed. Andrew Ross. Minneapolis: University of Minnesota Press.

Laclau, Ernesto, and Mouffe, Chantal (1985) *Hegemony and Socialist Strategy: Toward a Radical Democratic Politics.* London and New York: Verso.

Laplace, Pierre (1951) *A Philosophical Essay on Probabilities.* New York: Dover Publications.

Lash, Scott (1990) *A Sociology of Postmodernism.* New York and London: Routledge.

Lash, Scott, and Urry, John (1987) *The End of Organized Capitalism.* Cambridge, Eng.: Polity Press.

____________ (1994) *Economies of Signs and Space.* London: Sage.

Layzer, David (1975) "The Arrow of Time," *Scientific American* 223 (6): 56–69.

Lefebvre, Henri (1969) *The Explosion: Marxism and the French Upheaval.* New York: Monthly Review Press.

____________ (1984) *Everyday Life in the Modern World.* New Brunswick, NJ: Transaction Books.

Leiss, William (1974) *The Domination of Nature.* Boston: Beacon Press.

Leopold, Aldo (1989) *A Sand County Almanac.* Oxford and New York: Oxford University Press.

Levin, Harry (1966) "What Was Modernism," in *Refractions,* ed. Harry Levin. Oxford and New York: Oxford University Press.

Levin, Kim (1988) *Beyond Modernism.* New York: Harper and Row.

Levins, Richard, and Lewontin, Richard (1985) *The Dialectical Biologist.* Cambridge, MA: Harvard University Press.

Lewis, Lisa (1990) *Gender Politics and MTV.* Philadelphia: Temple University Press.

Lewontin, R. C. (1992) *Biology as Ideology: The Doctrine of DNA.* New York: HarperCollins.

Light, Andrew, and Higgs, Eric (1997) "The Politics of Corporate Ecological Restorations: Comparing Global and Local North American Contexts," in *Articulating the Global and the Local: Globalization and Cultural Studies,* eds. Ann Cvetkovich and Douglas Kellner. Boulder, CO: Westview Press.

Livingston, Ira (1997) *Arrow of Chaos.* Minneapolis: University of Minnesota Press.

Lovelock, James E. (1987) *Gaia: A New Look at Life on Earth.* Oxford: Oxford University Press.

Lowenthal, Leo (1961) *Literature, Popular Culture and Society.* Englewood Cliffs, NJ: Prentice-Hall.
Lowrie, Walter (1962) *Kierkegaard.* New York: Harper and Row.
Lukács, Georg (1964) *Realism in Our Time.* New York: Harper and Row.
____________ (1971) *History and Class Consciousness.* Cambridge, MA: MIT Press.
____________ (1980) *The Destruction of Reason.* Atlantic Highlands, NJ: Humanities Press.
Lyon, David (1994) *Postmodernity.* Minneapolis: University of Minnesota Press.
Lyotard, Jean François (1984) *The Postmodern Condition.* Minneapolis: University of Minnesota Press.
____________ (1986–1987) "Rules and Paradoxes and Svelte Paradox," *Cultural Critique* 5: 209–219.
Mandel, Ernest (1975) *Late-Capitalism.* London: New Left Books.
Marcus, Greil (1989) *Lipstick Traces.* Cambridge, MA: Harvard University Press.
Marcuse, Herbert (1955) *Eros and Civilization.* Boston: Beacon Press.
____________ (1964) *One-Dimensional Man.* Boston: Beacon Press.
____________ (1969) *An Essay on Liberation. Boston: Beacon Press.*
____________ (1972) *Counterrevolution and Revolt.* Boston: Beacon Press.
____________ (1978) *The Aesthetic Dimension.* Boston: Beacon Press.
Martin, Biddy (1991) *Women and Modernity.* Ithaca, NY: Cornell University Press.
Marx, Karl (1973) *Grundrisse.* Baltimore: Penguin Books.
Marx, Karl, and Engels, Friedrich (1975) *Collected Works: Vol I.* New York: International Publishers.
____________ (1976) *Collected Works: Vol VI.* New York: International Publishers.
____________ (1978) *The Marx–Engels Reader,* second edition, ed. Robert C. Tucker. New York: Norton.
Mason, Jim (1993) *An Unnatural Order: Uncovering the Roots of Our Domination of Nature and Each Other.* New York: Simon and Schuster.
Matson, Floyd D. (1966) *The Broken Image: Man, Science and Society.* New York: Anchor Books.
Mayr, Ernst (1988) *Toward a New Philosophy of Biology: Observations of an Evolutionist.* Cambridge, MA: Harvard University Press.
McGowan, John (1991) *Postmodernism and Its Critics.* Ithaca, NY: Cornell University Press.
McLuhan, Marshall (1964) *Understanding Media: The Extensions of Man.* New York: Signet Books.
Merchant, Carolyn (1980) *The Death of Nature: Women, Ecology, and the Scientific Revolution.* New York: Harper and Row.
Merquior, J. G. (1985) *Foucault.* Berkeley and Los Angeles: University of California Press.
Miller, Donald, ed. (1986) *The Lewis Mumford Reader.* New York: Pantheon Books.
Miyoshi, Masao, and Harootunian, H. D., eds. (1989) *Postmodernism and Japan.* Durham, NC: Duke University Press.
Monod, Jacques (1974) *Chance and Necessity: An Essay on the Natural Philosophy of Modern Biology.* London: Fontana/Collins.
Moretti, Franco (1996) *Modern Epic: The World-System from Goethe to Garcia Marquez.* London and New York: Verso.

Mouffe, Chantal (1988) "Radical Democracy: Modern or Postmodern?," in *Universal Abandon,* ed. Andrew Ross. Minneapolis: University of Minnesota Press.
Mumford, Lewis (1968) "Architecture as a Home for Man," *Architectural Record* 143 (February): 68–92.
Needham, Joseph (1969) *The Grand Titration: Science and Society in East and West.* Toronto: University of Toronto Press.
Negri, Antonio (1984) *Marx beyond Marx.* South Hadley, MA: Bergin and Garvey.
Newman, Charles (1985) *The Post-Modern Aura: The Act of Fiction in an Age of Inflation.* Evanston, IL: Northwestern University Press.
Nietzsche, Friedrich (1954) *The Portable Nietzsche.* New York: Viking Press.
____________ (1967a) *The Birth of Tragedy.* New York: Random House.
____________ (1967b) *On the Genealogy of Morals.* New York: Random House.
____________ (1968a) *The Will to Power.* New York: Vintage Books.
____________ (1968b) *Twilight of the Idols.* New York: Penguin Books.
____________ (1974) *The Gay Science.* New York: Vintage Books.
____________ (1982) *Daybreak.* Cambridge, Eng.: Cambridge University Press.
____________ (1986) *Human, All Too Human.* Cambridge, Eng.: Cambridge University Press.
Norris, Christopher (1990) *What's Wrong with Postmodernism?* Baltimore: Johns Hopkins University Press.
O'Doherty, Brian (1971) "What Is Postmodernism?," *Art in America* (May–June): 19.
Oelschlaeger, Max (1991) *The Idea of Wilderness: From Prehistory to the Age of Ecology.* New Haven, CT: Yale University Press.
Oliver, Kelly (1995) *Womanizing Nietzsche.* London and New York: Routledge.
Owens, Craig (1983) "The Discourse of Others: Feminists and Postmodernism," in *The Anti-Aesthetic,* ed. Hal Foster. Port Townsend, WA: Bay Press.
____________ (1992) *Beyond Recognition.* Berkeley and Los Angeles: University of California Press.
Parusnikova, Zuzana (1992) "Is a Postmodern Philosophy of Science Possible?," *Studies in History and Philosophy of Science* 23 (1): 21–37.
Passmore, John (1974) *Man's Responsibility for Nature: Ecological Problems and Western Traditions.* New York: Charles Scribner's Sons.
Perkins, Robert L., ed. (1990) *The Corsair Affair.* Macon, GA: Mercer University Press.
Perloff, Marjorie (1989) *Postmodern Genres.* Norman: University of Oklahoma Press.
Peterson, Ivars (1993) *Newton's Clock: Chaos in the Solar System.* New York: W. H. Freeman.
Pfeil, Fred (1990) *Another Tale to Tell.* London and New York: Verso.
Plant, Sadie (1992) *The Most Radical Gesture.* London and New York: Routledge.
Polyani, Karl (1957) *The Great Transformation.* Boston: Beacon Press.
Portoghesi, Paolo (1982) *After Modern Architecture.* New York: Rizzoli.
Poster, Mark (1975) *Existential Marxism in Postwar France.* Princeton: Princeton University Press.
____________ (1990) *The Mode of Information.* Cambridge, Eng., and Chicago: Polity Press and University of Chicago Press.
Poulantzas, Nicos (1973) *Political Power and Social Classes.* London: New Left Books.
Prigogine, Ilya, and Stengers, Isabell (1984) *Order Out of Chaos.* New York: Bantam.
Raban, Jonathan (1974) *Soft City.* London: Pluto Press.

Rabinbach, Anson (1990) *The Human Motor: Energy, Fatigue, and the Origins of Modernity.* New York: Basic Books.
Randall, John Herman (1976) *The Making of the Modern Mind.* New York: Columbia University Press.
Reich, Wilhelm (1972) *Sex-Pol: Essays 1929–1934.* New York: Vintage Books.
Resch, Robert Paul (1992) *Althusser and the Renewal of Marxist Social Theory.* Berkeley and Los Angeles: University of California Press.
Ricoeur, Paul (1970) *Freud and Philosophy.* New Haven, CT: Yale University Press.
____________ (1984) *Time and Narrative: Vol. 1.* Chicago: University of Chicago Press.
Rifkin, Jeremy (1989) *Entropy: Into the Greenhouse World.* New York: Bantam.
____________ (1995) *The End of Work: The Decline of the Global Labor Force and the Dawn of the Post-Market Era.* New York: Tarcher/Putnam.
Ritzer, George (1997) *Postmodern Social Theory.* New York: McGraw-Hill.
Robbe-Grillet, Alain (1965) *For a New Novel.* Evanston, IL: Northwestern University Press.
Rorty, Richard, ed. (1967) *The Linguistic Turn.* Chicago: University of Chicago Press.
____________ (1979) *Philosophy and the Mirror of Nature.* Princeton: Princeton University Press.
____________ (1989) *Contingency, Irony, and Solidarity.* Cambridge, Eng.: Cambridge University Press.
____________ (1991) *Objectivity, Relativism, and Truth.* Cambridge, Eng.: Cambridge University Press.
Rose, Barbara (1967) *American Art since 1900.* New York: Praeger.
Rose, Richard (1988) *The Postmodern President.* Chatham, NJ: Chatham House.
Rosemont, Penelope (1997) *Writings by Women Surrealists.* Austin: University of Texas Press.
Rosenau, Pauline Marie (1992) *Postmodernism and the Social Sciences.* Princeton: Princeton University Press.
Rosenthal, Michael (1992) "What Was Postmodernism?," *Socialist Review* 92(3): 83–106.
Rutherford, Jonathan, ed. (1990) *Identity: Community, Culture, Difference.* London: Lawrence and Wishart.
Sandler, Irving (1970) *The Triumph of American Painting: A History of Abstract Expressionism.* New York: Harper and Row.
____________ (1988) *American Art of the 1960s.* New York: Harper and Row.
Sassen, Saskia (1991) *The Global City.* Princeton: Princeton University Press.
Sassower, Raphael (1995) *Cultural Collisions: Postmodern Technoscience.* London and New York: Routledge.
Schor, Judith (1992) *The Overworked American.* New York: Basic Books.
Schrecker, Ellen (1996) "On the Science Wars," *Lingua Franca* (July/August): 61.
Seidman, Steven (1990) "Against Theory as a Foundationalist Discourse," *Perspectives in Sociological Theory* 13 (2): 131–146.
____________ (1994) *Contested Knowledge: Social Theory in the Postmodern Era.* Oxford: Blackwell.
Seidman, Steven, and Wagner, David G. (1992) *Postmodernism and Social Theory.* Oxford: Blackwell.
Sessions, George (1992) "Radical Environmentalism in the '90s," *Wild Earth* (Fall): 64–67.

Shaiken, Harley (1984) *Work Transformed.* New York: Holt, Rinehart and Winston.

Shapiro, David, and Shapiro, Cecile (1990) *Abstract Expressionism: A Critical Record.* Cambridge, Eng.: Cambridge University Press.

Sheldrake, Rupert (1988) "The Laws of Nature as Habits: A Postmodern Basis for Science," in *The Reenchantment of Science: Postmodern Proposals,* ed. David Ray Griffin. Albany: State University of New York Press.

____________ (1990) *The Rebirth of Nature: The Greening of Science and God.* London: Random Century Group.

Shlain, Leonard (1991) *Art and Physics: Parallel Visions in Space, Time, and Light.* New York: William Morrow.

Singer, Peter (1990) *Animal Liberation.* New York: Avon Books.

Smart, Barry (1992) *Modern Conditions, Postmodern Controversies.* New York and London: Routledge.

____________ (1993) *Postmodernity.* New York and London: Routledge.

Smith, Huston (1982) *Beyond the Post-Modern Mind.* New York: Crossroad.

Snow, C. P. (1964) *The Two Cultures: And a Second Look.* Cambridge, Eng.: Cambridge University Press.

Soja, Edward (1989) *Postmodern Geographies.* London and New York: Verso.

Sokal, Alan (1996a) "Transgressing the Boundaries: Toward a Transformative Hermeneutics of Quantum Gravity," *Social Text* 14 (1, 2): 217–252.

____________ (1996b) "A Physicist Experiments with Cultural Studies," *Lingua Franca* (May): 62–64.

____________ (1996c) "Proposed [but unpublished] article for *The Nation,*" May 6, 1996. Sokal@acf4.NYU.EDU.

____________ (1996d) "Alan Sokal Replies," *Lingua Franca* (July/August): 57.

Sontag, Susan (1967) *Against Interpretation.* New York: Dell Books.

Soule, Michael, and Lease, Gary, eds. (1995) *Reinventing Nature? Responses to Postmodern Deconstruction.* Washington, DC: Island Press.

Spretnak, Charlene (1990) "Ecofeminism: Our Roots and Flowering," in *Reweaving the World: The Emergence of Ecofeminism,* eds. Irene Diamond and Gloria Orenstein. San Francisco: Sierra Club Books.

____________ (1991) *States of Grace: The Recovery of Meaning in the Postmodern Age.* New York: HarperCollins.

Sterling, Bruce, ed. (1986) *Mirrorshades: The Cyberpunk Anthology.* New York: Ace Books.

Swimme, Brian (1988) "The Cosmic Creation Story," in *The Reenchantment of Science: Postmodern Proposals,* ed. David Ray Griffin. Albany: State University of New York Press.

Tafuri, Manfredo (1976) *Architecture and Utopia.* Cambridge, MA: MIT Press.

Taylor, Brandon (1985) *Modernism, Post-Modernism, Realism.* Winchester, Eng.: Winchester School of Art Press.

Thom, Rene (1975) *Structural Stability and Morphogenesis.* Reading, MA: Benjamin/Cummings.

Thompson, Josiah (1973) *Kierkegaard.* New York: Knopf.

Tomkins, Calvin (1965) *The Bride and the Bachelors.* New York: Viking Press.

____________ (1976) *The Scene: Reports on Post-Modern Art.* New York: Viking Press.

Toulmin, Stephen (1982a) "The Construal of Reality: Criticism in Modern and Postmodern Science," in *The Politics of Interpretation,* ed. W. J. T. Mitchell. Chicago: University of Chicago Press.

____________ (1982b) *The Return to Cosmology: Postmodern Science and the Theology of Nature.* Berkeley and Los Angeles: University of California Press.

Trachtenberg, Stanley, ed. (1985) *The Postmodern Moment: A Handbook of Contemporary Innovation in the Arts.* Westport, CT: Greenwood Press.

Turner, Bryan, ed. (1990) *Theories of Modernity and Postmodernity.* London: Sage.

Vaneigem, Raoul (1983) *The Revolution of Everyday Life.* London: Pending Press.

____________ (1985) *The Book of Pleasures.* London: Pending Press.

Varela, Francisco J. (1979) *Principles of Biological Autonomy.* New York and Oxford: Elsevier.

Vattimo, Gianni (1988) *The End of Modernity.* Cambridge, Eng.: Polity Press.

Velikovsky, Immanuel (1950) *Worlds in Collision.* New York: Doubleday.

Venturi, Robert (1966) *Complexity and Contradiction in Architecture.* New York: Museum of Modern Art.

Venturi, Robert; Brown, Denise Scott; and Izenour, Steven (1972) *Learning from Las Vegas.* Cambridge, MA: MIT Press.

Wallis, Brian (1984) *Art after Modernism.* Boston: Godine.

Warren, Karen J. (1993) "The Power and Promise of Ecological Feminism," in *Environmental Philosophy,* ed. Michael Zimmerman et al. Englewood Cliffs, NJ: Prentice-Hall.

Waters, Malcolm (1995) *Globalization.* New York and London: Routledge.

Weatherford, Jack (1988) *Indian Givers: What the Native Americans Gave to the World.* New York: Crown.

Weber, Max (1946) *From Max Weber.* Oxford and New York: Oxford University Press.

____________ (1958) *Economy and Society,* 3 vols. New York: Bedminster Press.

Weinberg, Steven (1996) "Sokal's Hoax," *The New York Review of Books* (August 8): 11–14.

Whitehead, Alfred North (1967) *Science and the Modern World.* New York: Free Press.

Wiener, Norbert (1966) *I Am a Mathematician: The Later Life of a Prodigy.* Cambridge, MA: MIT Press.

Wilden, Anthony (1980) *System and Structure,* second edition. London: Tavistock.

Wilder, Carol, and Weakland, John H. (1981) *Rigor and Imagination: Essays from the Legacy of Gregory Bateson.* New York: Praeger.

Williams, Raymond (1977) *Marxism and Literature.* Oxford: Oxford University Press.

Williamson, Judith (1978) *Decoding Advertisements.* Boston: Marion Boyars.

Wittgenstein, Ludwig (1958) *Philosophical Investigations.* Oxford: Blackwell.

Wolfe, Tom (1981) *From Bauhaus to Our House.* New York: McGraw-Hill.

Wollen, Peter (1993) *Raiding the Icebox: Reflections on Twentieth-Century Culture.* Bloomington: Indiana University Press.

Wollstonecraft, Mary (1975) *A Vindication of the Rights of Woman.* Baltimore: Penguin Books.

Woolf, Virginia (1929) *A Room of One's Own.* London: Hogarth.

Zeeman, E. C. (1976) "Catastrophe Theory," *Scientific American* 234 (May): 65–83.

____________ (1977) *Catastrophe Theory: Selected Papers 1972–1977.* Reading, MA: Addison-Wesley.

Zimmerman, Michael (1994) *Contesting Earth's Future: Radical Ecology and Postmodernity.* Berkeley: University of California Press.

Index

Adorno, Theodor 31, 74, 76, 78–79, 113, 126, 198, 241, 268, 276
Albers, Joseph 167, 169
Althusser, Louis 5, 33
Anderson, Laurie 34, 181, 186–187, 258
Antonio, Robert J. 30, 36, 77
Aquinas, Thomas 18
Arata, Isokazi 188
Aristotle 197–198
Aronowitz, Stanley 227
Artaud, Antonin 277
Augustine 18
Autopoiesis 206–207

Bachelard, Gaston 227
Bacon, Francis 196–200, 234, 269
Bahro, Rudolf 278
Bakunin, Michael 18, 77, 279
Ballard, J. G. 134
Barrett, William 6–8
Barth, John 11, 31–32, 131–132, 180, 253, 256
Barthelme, Donald 132
Barthes, Roland 176, 256
Bataille, George 38, 76
Bateson, Gregory 223, 262
Baudelaire, Charles 126, 189
Baudrillard, Jean x, xiii, 3–4, 6, 11–12, 18, 22, 26, 32, 36, 39–40, 48, 56–57, 60–61, 77, 80, 91, 94–105, 110–119, 122, 124, 155–156, 176, 178, 184, 190, 193, 207, 221, 231, 256–257, 262–264, 268, 271–272
Bauman, Zygmunt xiii, 20–22, 260–261
Beavis and Butt-Head 88
Becker, Paula Modersohn 129
Beckett, Samuel 8
Benedikt, Michael 157–158, 192
Benjamin, Walter 31, 76, 86, 102, 134
Berger, Peter L. 189, 227
Bergman, Peter 77
Bergson, Henri 202
Berkeley, Bishop 235
Berman, Marshall 16, 20, 126, 189
Berman, Morris 197
Bernstein, Richard 25, 31
Bertens, Hans xiii, 191–192, 260
Beuys, Joseph 181
Bhabha, Homi K. 10
Birch, Charles 244, 246
Blake, Peter 141, 192
Blake, William 18, 28, 202, 240
Bloch, Ernst 115, 278
Boggs, Carl xiii, 121
Bohm, David 212–213, 225, 241, 244, 246
Bohr, Niels 212, 214–215, 257
Bookchin, Murray 226
Braderman, Joan 186, 187
Braque, Georges 141, 165, 166
Brecht, Bertolt 93, 128, 132, 136, 186, 277
Breton, André 167
Breuer, Marcel 151
Bridgman, P. W. 215
British Cultural Studies 19
Bronner, Stephen Eric 191, 193
Brown, Denise Scott 151
Bunshaft, Gordon 154
Burger, Peter 190
Burgin, Victor 186
Burke, Edmund 18, 28
Burroughs, William 124
Burtt, E. A. 227
Butler, Judith 10, 192
Byrd, Joseph 172
Byrne, David (and the Talking Heads) 183

Cage, John 124, 136, 170, 172–173, 187, 193, 206, 256
Calinescu, Matei 35, 189, 190
Campbell, Jeremy 209, 257
Camus, Albert 29

Capitalism 13–16, 50–61, 79, 91, 98–100, 104–110, 146–148, 270
Capra, Fritjof 262
Cassatt, Mary 129
Catton, William 265
Causality 68–70
Cézanne, Paul 127, 165–166, 189, 260
Chaos theory x, 68, 216–221, 245–246, 259, 261–262, 264, 266
Chomsky, Noam 210
Christo 181
Clausius, Rudolf 205
Clemente, Francesco 183
Clinton, Bill 111, 122
Cole, Thomas 173
Communism 5, 13
Complexity theory 22, 246, 259, 261
Comte, Auguste 201, 244
Connor, Steven 34–35
Coover, Robert 131
Copernicus, Nicolas 197
Coveney, Peter 259
Crichton, Michael 259
Cronenberg, David 89
Crimp, Douglas 180, 184
Cultural politics 276–280
Cunningham, Merce 124, 136, 172, 206, 256

d'Alembert, Jean Le Rond 199
d'Holbach, Baron Paul 199
Daly, Herman E. 265
Dante 18
Darwin, Charles 202, 206, 225
Dash, Julia 186
Davies, Paul 205
Davis, Douglas 193
Dawkins, Richard 225
de Antonio, Emile 168, 175, 193
de Certeau, Michel 116
de Gaulle, Charles 5
de Kooning, Willem 168–169, 193
de Maistre, Joseph 28
de Maria, Walter 181
Debord, Guy x, 39, 56, 79–96, 98, 100, 103–105, 109–124, 185
Deleuze, Gilles 5–6, 72, 92, 96, 117, 255, 276
Denis, Maurice 165
Derrida, Jacques 6, 22, 33, 72, 75, 96, 157–158, 176, 230–231, 250, 255–258, 268
Descartes, René 36, 134, 196–200, 269
Dewey, John xiv, 6, 25, 227, 257, 260
Diderot, Denis 12
Dilthey, Wilhelm 12, 202
Dine, Jim 175
Dinkaloo, John 152
Dostoyevsky, Fyodor 49, 189
Duchamp, Marcel 136, 167, 170, 172–174, 180, 184, 192
Durkheim, Emile 12, 30

Ebert, Teresa 9, 27, 36, 137
Ecology x, 242, 263–270
Eddington, Sir Arthur 210
Einstein, Albert 127, 211–213, 236, 245, 259
Eisenman, Peter 157, 158
Eliot, T. S. 8, 128, 131
Ellul, Jacques 229
Emerson, Ralph Waldo 28
Engels, Friedrich 15, 20, 34, 77
Enlightenment 6–7, 12, 28, 62, 65, 120, 195, 199, 259, 272–273, 278, 281
Entropy x, 204–211, 265–266
Epstein, Barbara 21–22, 247
Erlich, Paul 245
Existentialism 6–8, 28–29, 42

Faulkner, William 8
Featherstone, Mike 12
Feminism 22, 69, 200, 230, 232, 242, 257, 279
Ferré, Frederick 189, 244, 247, 255
Festa, Tano 173
Feuerbach, Ludwig 79, 91
Feyerabend, Paul 221, 241–242, 246, 251, 255, 257, 264
Fiedler, Leslie 130, 131
Fischl, Erich 184
Fish, Stanley 235–236
Flaubert, Gustave 189
Ford, Joseph 218
Foreman, Dave 230
Foster, Hal 9, 36, 130, 137, 180–181, 190, 193–194
Foucault, Michel xi–xii, 4, 6, 10, 22, 25, 62, 72, 75–76, 92, 96, 113, 117, 143, 176, 202–203, 230–231, 254, 257–258, 268, 272, 274–276
Fourier, Jean Baptiste 203–204

Frampton, Kenneth 139, 155, 157, 163
Frankenthaler, Helen 167
Frankfurt School 19, 39, 62, 72, 76–77, 80, 85, 94, 96, 191, 234, 248–249, 258, 268
Fraser, Nancy 10, 25
Frege, Gottlob 216
Freud, Sigmund 202
Fromm, Erich 53, 56
Fuhrman, Mark 11, 115

Gadamer, Hans-Georg 227
Galileo, 197
Gane, Mike 36
Gates, Bill 16
Gaudi, Antonio 143
Gauguin, Paul 165
Gehry, Frank 158
Georgescu-Roegen, Nicholas 245
Ghirardo, Diane 159
Giacometti, Alberto 8
Gifford, Kathie Lee 91, 121–122
Giroux, Henry 10
Glass, Philip 34, 127
Gleick, James 218, 246
Godard, Jean-Luc 114, 255, 277
Gödel, Kurt 216, 245
Goethe, Johann Wolfgang von 40–41
Gould, Stephen Jay 241
Graff, Gerald 37
Gramsci, Antonio 47, 82, 93, 96, 117, 276
Graves, Michael 139, 152
Greenberg, Clement 165, 168, 169
Griffin, David Ray 223, 225, 241–242, 244, 258, 264, 267
Griffith, D. W. 93
Gris, Juan 166, 192
Gropius, Walter 140, 142–145, 161
Gross, Paul 227–244, 247–250, 258, 269
Guattari, Félix 5, 92, 96, 117, 255, 276, 282

Haacke, Hans 181, 185
Habermas, Jürgen x, 28, 36, 38, 64, 76–77, 127, 132, 224, 273
Hall, Stuart 10
Halley, Jeff 184
Hamamoto, Darrell xiii, 194
Hanson, Duane 178
Haraway, Donna 10, 227, 241, 247
Harding, Sandra 10, 25, 227, 241, 247–248, 258, 264
Hart, Roger 247–248
Harvey, David 3, 5, 25, 34, 56, 191
Hassan, Ihab x, xiii–xiv, 38, 194
Hawkes, John 131
Hawking, Stephen 220–221, 244–246
Hayden, Dolores 163
Hayles, Katherine 227, 247
Haug, W. F. 55
Heartfield, John 185
Hegel, G. W. F. 12, 36, 38, 77, 82, 281
Heidegger, Martin 6–7, 25, 28–29, 38–39, 72–76, 78, 227, 229, 269
Heisenberg, Werner 214–215, 242, 250
Heizer, Michael 181
Hemingway, Ernest 8, 131
Hermeneutics 110–118
Herrnstein, Richard J. 238
Hertz, Heinrich 213
Herzog, Werner 190
Hesse, Mary 227
Highfield, Roger 259
Hilbert, David 216
Hindess, Barry 24, 236
Hirst, Paul 24, 236
Hitchcock, Henry Russell 190
Hobbes, Thomas 199, 200
Hofmann, Hans 167, 168
Holzer, Jenny 180, 187
hooks, bell 10
Horgan, John 3, 224, 262, 246
Horkheimer, Max 74, 76, 78, 198, 241, 268
Howe, Irving 130
Hubbert, M. King 265
Hubble, Edwin 244
Hume, David 12, 69, 134, 201
Hutcheon, Linda 184, 186, 192, 194
Huxley, Aldous 118
Huyssen, Andreas 128, 190, 194

Ice Cube 115
Ice-T 115
Identity politics 9, 33, 273–280
Information theory 207–211
Izenour, Steven 151

Jacobs, Jane 149–150, 160, 163, 191
James, William xiv, 38

Jameson, Fredric xii–xiii, 5, 12, 19, 34–35, 56, 113–114, 124, 134, 153, 169, 255
Jantsch, Erich 206, 211, 219
Jencks, Charles 138, 148, 151, 154, 159, 191–192, 255, 257
Johns, Jasper 164, 170–171, 173–176, 181, 193
Johnson, Philip 124, 136, 148, 151–153, 157, 190
Johnson, Steven 246, 259
Jorn, Asger 93
Joyce, James 8, 127–128, 258–259
Junger, Ernst 75

Kafka, Franz 128, 131
Kahn, Louis 151
Kandinski, Vasily 128
Kant, Immanuel 12, 18, 30, 36, 69, 259
Kauffman, Stuart 207, 219, 221, 225
Kearney, Richard 38
Keller, Evelyn Fox 227
Kennedy, Jackie 177
Kenzo, Tange 188
Kiefer, Anselm 190
Kierkegaard, Søren ix, 6, 28–29, 38–50, 57, 62–63, 74, 77, 80, 91–92, 94, 96, 202
King, Martin Luther 281
King, Rodney 11, 115, 119
King, Ross 158
Klee, Paul 128
Kline, Franz 168–169
Klotz, Heinrich 151, 154–155, 158
Kollwitz, Kathe 128
Koons, Jeff 134, 183–184
Korsch, Karl 82, 94
Kosko, Bart 245
Kostabi, Mark 134
Krauss, Rosalind 180, 181, 184, 194
Kroker, Arthur xiii, 259, 262
Kruger Barbara 181, 186–187
Kuberski, Phillip 259
Kuhn, Thomas xi, 227, 237, 252–254

La Mettrie, Julien Offray de 199–200
Lacan, Jacques 255
Laclau, Ernesto 5, 12, 25, 272–273
Langer, Susanne 227
Laplace, Pierre-Simon 200–201, 219
Latham, William 259
Latour, Bruno 247
Lawson, Thomas 184
Layzer, David 210
Le Corbusier 141–142, 144–145, 149, 152, 155, 159–161, 163, 192
Lefebvre, Henri 81–82, 94, 96, 276
Lenin, V. I. 94, 190
Levidow, Les 249
Levin, Kim 180, 193–194
Levin, Harry 130
Levine, Sherrie 185
Levitt, Norman 227–244, 247–250, 258, 269
Lewontin, R. C. 237–238, 249
Lichtenstein, Roy 175, 181
Limbaugh, Rush 232
Livingston, Ira 259
Locke, John 199
Longo, Robert 183, 186
Loos, Adolph 143
Lorenz, Edward 219
Louis, Morris 174
Lovelock, James 206–207
Luckmann, Thomas 227
Ludic postmodernism 26–27, 36, 127, 184, 257
Lukács, Georg 76–77, 79, 82, 93–94, 277, 282
Luke, Allan 10
Luke, Carmen 10
Lyotard, Jean-François xiii, 3–4, 6, 11–12, 22, 26, 35–36, 92, 96, 156, 230–231, 242, 250, 255, 265, 268, 272, 276

Mach, Ernst 213
Madonna 103
Malcolm X 278–279
Malevich, Kasimir 133
Mandelbrot Benoit 220
Manet, Edwourd 166, 189, 260
Mannheim, Karl 227
Marcel, Gabriel 29
Marcuse, Herbert 56, 76, 79, 94, 114, 247, 268, 277–278
Marx, Karl ix, 6, 12, 15, 18, 20, 30, 34, 38–39, 41, 47, 50–57, 60, 75, 76–77, 79–82, 84–85, 89–92, 95, 98, 100–101, 107, 115, 202, 248, 259, 271, 275, 279

Marxism 3, 5–6, 9–10, 81, 233, 275–276, 278
Matisse, Henri 166
Matson, Floyd 212–214
Maxwell, James Clerk 213, 257
May, Ernst 148
McCollum, Allan 184
McGowan, John 38
McLaren, Peter 10
McLuhan, Marshall 122, 282
Mechanistic paradigm 68, 70–71, 199–202
Milne, E. A. 205
Modern architecture 139–148
Modern art 126–129, 132–133, 165–174
Modern paradigm 18–19, 202
Modern science 196–203
Modernity 29–30, 57–65, 73–77, 269
Mondrian, Piet 168
Monet, Claude 128, 260
Monod, Jacques 246
Monroe, Marilyn 117, 175, 185
Montesquieu, Baron de 201
Moore, Michael 14
Morisot, Berthe 129
Motherwell, Robert 167
Moore, Charles 155
Mouffe, Chantal 5, 12, 272–273
Muldoon, Paul 259
Multiperspectivalism 66–67, 71–72, 281
Mumford, Lewis 138, 148–150, 160
Munch, Edvard 134
Murray, Charles 238

Nabokov, Vladimir 131–132
New social movements 9–10, 274
Newman, Barnett 168–169
Newton, Isaac 18, 97, 200, 218, 236, 240, 245, 248
Nicholson, Linda 10, 25
Niemeyer, Oscar 192
Nietzsche, Friedrich ix, 6–7, 25, 28–30, 36, 38–39, 49, 57–78, 80, 92, 94, 96, 137, 147, 163, 202, 232, 248, 257, 264
Nihilism 64–66
Noland, Kenneth 169, 174

O'Doherty, Brian 180
Oelschlaeger, Max 223, 244, 267
Oldenburg, Claes 175, 178
Olitski, Jules 169, 174
Orwell, George 118
Owens, Craig 150, 184, 185, 194

Paik, Nam June 184
Paradigm shift xi–xii, 253–255
Parusnikova, Zuzana 250
Pavia, Philip 167
Pearson, Karl 213
Peckham, Morse 172
Peirce, Charles 227, 260
Penrose, Roger 225
Peterson, James 201, 218
Pevsner, Nikolaus 191
Pfeil, Fred 11, 34
Picasso, Pablo 127, 133, 141, 166
Pirandello, Luigi 128
Planck, Max 214
Plato 261
Poincaré, Henri 218–219
Polk, Brigid 175
Pollack, Jackson 168, 171, 193
Poons, Larry 169
Portman, John 152
Positivism 200–202, 234, 244, 250
Postcolonial science 242–244
PostFordism 3, 13, 34
Postindustrialism 3, 208
Postmodern architecture 138–164, 205
Postmodern art 133–135, 164– 189, 280
Postmodern literature 130–132
Postmodern paradigm 19, 253–282
Postmodern turn viii–ix, 8, 16, 19, 22, 25, 27, 38–40, 63, 76–77, 96, 130–131
  in arts 124–194
  in politics 270–282
  in science 195–251
Postmodernism of resistance 9, 27, 36–38, 184–185, 190
Postmodernity 20–21, 24, 30–31, 95, 110, 264
Poststructuralism 5–7, 9, 10, 256
Pound, Ezra 8, 74, 126–128, 131
Presley, Elvis 175
Prigogine, Ilya 203, 206, 210–211, 219–220, 225, 241, 244, 246–247, 249–250, 267, 269
Prince, Richard 185
Pulitzer, Joseph 146
Pynchon, Thomas 131, 207, 256, 259

Quantum mechanics 7, 212–216, 255–256, 261, 264

Raban, Jonathan 191
Rainer, Yvonne 186
Rauschenberg, Robert 133, 164, 170–173, 175–176, 181, 189, 193
Reagan, Ronald 103
Reich, Wilhelm 94, 276
Relativity theory 7, 211–213, 257, 259, 261
Repo Man 34
Ricoeur, Paul 227
Rifkin, Jeremy 266, 267
Riggs, Marlon 186
Riley, Bridget 183
Rimbaud, Arthur 74, 126
Rivers, Larry 173
Robbe-Grillet, Alain 256
Roche, Kevin 152
Rorty, Richard 22, 25, 35, 75, 255, 257, 268, 272
Rosen, Ruth 247
Rosler, Martha 185, 186, 187
Ross, Andrew 227, 248
Rothko, Mark 169
Rousseau, Jean-Jacques 279
Ruskin, John 282
Russell, Bertrand 216, 245

Saarinen, Eero 151–152
Sagan, Carl 241
Saint-Simon 190, 201
Salle, David 183, 186
Sartre, Jean-Paul 29
Schaffer, Simon 247
Schnabel, Julien 181, 184, 186–187
Schoenberg, Arnold 127–128
Schopenhaeur, Arthur 63
Schrecker, Ellen 234
Schrödinger, Erwin 249
Science wars 226–244
Seidman, Steven 10–11, 25, 35
Shakespeare, William 261
Shakur, Tupac 115
Shapin, Steven 247
Shannon, Claude 209
Sheldrake, Rupert 226, 244, 267
Shelly, Pierce 129
Sherman, Cindy 185
Shlain, Leonard 260, 282
Spengler, Oswald 75
Spivak, Gayatri 10
Simpson, O. J. 106–108, 111
Sims, Karl 259
Situationist International x, 79–96, 98, 104–105, 115, 117, 120–121, 162–163, 185, 187, 276–277
Smart, Barry xiii, 10, 25
Smith, Huston 32, 195, 213, 244, 264
Smithson, Robert 181
Snow, C. P. 240
Social constructionism 235–240, 249
Sokal, Alan 228, 235–244, 247, 249–251, 258, 269
Sonic Outlaws 194
Sontag, Susan 131, 137, 183
Soule, Michael 282
Spectacle, society of 82–92, 119
Spinoza, Benedict 199
Stella, Frank 169, 174
Stengers, Isabelle 203, 206, 220, 244, 246–247, 249, 267, 269
Stent, Gunther 225
Sterling, Bruce 261
Stoppard, Tom 259
Stravinsky, Igor 128
Structuralism 5, 9
Sullivan, Louis 146

Taaffe, Philip 183
Tadao, Ando 188
Tafuri, Manfredo 147
Talking Heads 34
Taut, Bruno 148
Thermodynamics 68, 203–207
Thomson, William 205
Tomkins, Calvin 172, 193
Toulmin, Stephen 205, 223, 226, 242, 244, 258, 269
Tung, Mao-Tse 175

Ulrich, Roger 250
Utzon, Jorn 152

van der Rohe, Mies 140, 143–145, 148–149, 153–154, 161
van Gogh, Vincent 128, 165

Vaneigem, Raoul 93, 122
Venturi, Robert 122, 124, 128, 149–151, 153, 155–156, 163
Voltaire, 12

Warhol, Andy 101, 112, 124, 133, 164, 173–179, 183, 193
Weber, Max 12, 30, 38–39, 75–77, 80, 99, 144, 147
Webern, Anton 127
Weinberg, Steven 250
West, Cornell 10, 25
Whitehead, Alfred North 202, 216
Whitmore, Hugh 259
Wiener, Norbert 208
Wigley, Mark 157
Williams, Raymond 108
Wittgenstein, Ludwig 6, 254, 260
Wojnarowitz, David 181
Wollstonecraft, Mary 271
Wordsworth, William 28
Wright, Frank Lloyd 139, 143, 144, 152, 157, 161

Young, Iris 10
Young, Robert 249